RESEARCH IN THE SCHIZOPHRENIC DISORDERS

RESEARCH IN THE SCHIZOPHRENIC DISORDERS

THE STANLEY R. DEAN AWARD LECTURES
VOL. II

Edited by

Robert Cancro, M.D.
Department of Psychiatry
New York University Medical Center
New York, New York

Stanley R. Dean, M.D.
The Fund for the Behavioral Sciences
Miami, Florida

MTP PRESS LIMITED
International Medical Publishers

Published in the UK and Europe by
MTP Press Limited
Falcon House
Lancaster, England

Published in the US by
SPECTRUM PUBLICATIONS, INC.
175–20 Wexford Terrace
Jamaica, NY   11432

ISBN-13: 978-94-011-6340-8   e-ISBN-13: 978-94-011-6338-5
DOI: 10.1007/978-94-011-6338-5

Contributors

(current affiliations of contributors to this volume)

Julius Axelrod, Ph.D. Pharmacology Section, National Institute of Mental Health, Laboratory of Clinical Science, Bethesda, Maryland

Robert Cancro, M.D., Med. D.Sc. Department of Psychiatry, New York University Medical Center, New York, New York

Arvid Carlsson, M.D., Ph.D. Department of Pharmacology, University of Goteberg, Sweden

Norman Garmezy, Ph.D. Department of Psychology, University of Minnesota, Minneapolis

Franz J. Kallman, M.D. (deceased)

Seymour S. Kety, M.D. Intramural Program for Basic Research, National Institute of Mental Health, Bethesda, Maryland

John D. Rainer, M.D. Psychiatric Institute, New York, New York

Eliot H. Rodnick, Ph.D. Department of Psychology, University of California, Los Angeles

David Rosenthal, Ph.D. (deceased)

David Shakow, Ph.D. (deceased)

Solomon H. Snyder, M.D. Department of Neuroscience, Johns Hopkins University, Baltimore, Maryland

John K. Wing, M.D. Social Psychiatry Unit, Institute of Psychiatry, London, England

Richard J. Wyatt, M.D. Adult Psychiatry Branch, Department of Health and Human Services, St. Elizabeth's Hospital, Washington, D.C.

Introduction

In the autumn of 1957, Stanley R. Dean proposed the creation of a foundation on schizophrenia to a group of friends. The purpose was to bring the illness of schizophrenia before the public's attention and out of the shameful anonymity in which it was buried. Stanley Dean recognized the major public health importance of this illness and was determined to have an impact on it during his lifetime. By December of that year, the decision was made to create an organization called Research in Schizophrenia Endowment (RISE). Within less than two months articles of incorporation for RISE were issued by the State of Connecticut. By the latter part of March 1958, Dean was successful in getting Congress to provide $1,300,000 for research into schizophrenia. This was the first time that Congress had targeted funds specifically for research on this disorder. Within a year, by April 1959, Dr. Dean was instrumental in getting Congress to increase its appropriation to the NIMH by more than $8,000,000 with the specific instruction that approximately one-half of the total research funds of the NIMH be used for projects relating to schizophrenia.

In the spring of 1960, RISE relinquished its corporate charter and became an integral part of the Research Foundation of the National Association for Mental Health. This merger occurred before RISE had implemented its planned annual award program for outstanding research in schizophrenia. Several members of the Board of Directors of RISE, under the leadership of Dr. James G. Miller and Mrs. Everett Callender, decided to establish the annual lectureship, naming it in honor of Dr. Dean's many contributions. The first award was given in 1962. The award was supported by the Fund for the Behavioral Sciences during the period 1962 to 1968. In 1968, because of the efforts of Raymond W. Waggoner, Sr., the award became jointly sponsored by the American College of Psychiatrists and the Fund for the Behavioral Sciences. The award is both international and interdisciplinary. It is not meant to support research, but rather to acknowledge research contributions made by a scientist who has contributed in a major fashion to research relevant to the schizophrenias. The award consists of a tax-free grant, a commemorative certificate, and an all-expense-paid trip to the meeting of the American College of Psychiatrists at which the award is made.

The people who have been honored constitute a pantheon of schizophrenia researchers, from the clinic to the laboratory, and from the biochemical to the psychosocial. Very little of importance has been done on schizophrenia in the last twenty years that does not involve an awardee.

The list of distinguished recipients is:

1962 – Seymour S. Kety, M.D.	1973 – Joseph Zubin, Ph.D.
1963 – David Shakow, Ph.D.	1974 – Arvid Carlsson, M.D., Ph.D.
1964 – Ralph W. Gerard, M.D., Ph.D.	1975 – Margaret T. Singer, Ph.D.
1965 – Leo Kanner, M.D.	Lyman C. Wynne, M.D., Ph.D.
1966 – Franz J. Kallmann, M.D.	1976 – John Wing, M.D., Ph.D.
1967 – Eliot H. Rodnick, Ph.D.	1977 – Solomon H. Snyder, M.D.
Norman Garmezy, Ph.D.	1978 – Yrjo O. Alanen, M.D.
1968 – Gabriel Langfeldt, M.D.	1979 – William T. Carpenter, Jr., M.D.
1969 – Manfred Bleuler, M.D.	John S. Strauss, M.D.
1970 – David Rosenthal, Ph.D.	John J. Bartko, Ph.D.
1971 – Sarnoff A. Mednick, Ph.D.	1980 – Julius Axelrod, Ph.D.
Fini Schulsinger, M.D.	1981 – Robert Cancro, M.D., Med.D.Sc.
1972 – Theodore Lidz, M.D.	1982 – Richard J. Wyatt, M.D.

It became increasingly apparent to the editors of these volumes that the record of a generation of research activity in the schizophrenias was languishing and was not being brought together for interested students. The lectures of the first ten years had been published as a single volume in 1973,* but the book has long been out of print. It was decided that the first twenty-one years worth of lectures would be compiled into a single source and organized on the basis of their subject matter, rather than on the basis of the year of the award. It was also decided that all of the living award recipients would be given the opportunity to write some additional comments on their award lecture so as to reflect on what has happened in their particular field of interest during the intervening time interval. Obviously, some of the more recent lectures are current and do not benefit from additional comments.

The work is divided into two volumes. Volume 1 focuses on diagnosis and different etiologic theories, with greater emphasis on the role of social and familial factors. Volume 2 explores many of the psychosocial factors which operate in the disorder, including antecedents and consequences, and on the biologic sphere. The volume is organized with the first group of chapters looking at biochemical factors, and the second group at the role of genetic factors. The last chapter attempts to synthesize the information presented in the two volumes.

It is difficult to improve upon Manfred Bleuler's comments on the award. In a personal letter to Stanley Dean several years ago he stated the award "provides a great opportunity to focus on such world-wide progress in research. For nineteen years it has encouraged and invited scientists from different countries and schools of thought to present their theories and results to each other. The nineteen Dean Award Lectures enable us to overview and synthesize the ongoing developments.

---

*S. R. Dean (ed.), *Schizophrenia: The First Ten Dean Award Lectures*. MSS Information Corp., New York, 1973.

I am particularly impressed by the stimulus of the Dean Award for arriving at inter-national cooperation in treating schizophrenias. I congratulate the founders, the sponsors, and recipients of the Award for their valuable influence on progress in psychiatry." This tradition is confirmed by the two award lectures delivered since that letter was written.

Contents

PSYCHOSOCIAL ASPECTS

Psychosocial Deficit in Schizophrenia

David Shakow, Ph.D.

INTRODUCTION

The years since I first began to work with schizophrenia have gone by fast. I am naturally reluctant to count them. I shall try to give at least a glimpse of some of the work I have been involved with during these years, mainly at Worcester State Hospital, but also at the University of Illinois, and most recently at the National Institute of Mental Health. In the face of a problem like schizophrenia, I can only humbly hope that I have contributed at least a few solid bricks to the final structure of understanding.

GENERAL BACKGROUND

Before going on to some of our data and their theoretical implications, I remark briefly upon some of the special difficulties and complexities that we have found in the course of our research on schizophrenia—problems which I believe have plagued other serious investigators in this field. There are so many problems relevant to the conditions for carrying out dependable research in schizophrenia. I cannot refrain, nevertheless, from at least touching upon a few of the problems I have discussed at some length in previous publications [33,36]. I believe that a large part of the many discrepant findings in this large and varied group of disorders stems from failure to take these factors into account. Not that even our own best thought through and carefully executed designs have not ended up in some

Research in the Schizophrenic Disorders: The Stanley R. Dean Award Lectures, vol. 2, edited by R. Cancro and S. R. Dean. Copyright © 1985 by Spectrum Publications.

way inadequate. I have not infrequently been appalled by some of the compromises with experimental design that circumstances have caused me to make.

Both interindividual and intraindividual *variability* are a major source of difficulty in research on schizophrenia. In our psychological studies, groups of schizophrenic patients have quite consistently given coefficients of variation three times that of normal subjects, and individual patients two times that of the normal ones. In studies of physiological functions, these have varied from one and one-half to more than twice that of normal subjects. Of the many sources of variation, I shall only concern myself here with the two important ones of nosology and attitude.

Taken in its broadest sense, *nosology* has many facets. Initially there is the fundamental problem of the diagnosis of schizophrenia and its subtypes. Aware of this problem, we have made every effort, especially in our Worcester studies, to obtain reliable diagnoses of patients. In addition to certain definite exclusion criteria, we had specified standards for the general diagnosis of schizophrenia, as well as for subtype classification. Whenever there was a question about the schizophrenia diagnosis, the patient was not used in the research. If the patient did not clearly meet the criteria established for one of the four subtypes, the categories of mixed, unclassified, or indeterminate were used.

We have a tendency these days to be condescending about subtype classification. In fact, there are even "nihilists" among us who would do away with all diagnostic categories, the less enthusiastic of these calling for at least the casting overboard of the subtypes. If diagnoses are to be made on the relatively careless bases so prevalent in many centers, then I would go along with this point of view. If, however, they are based upon carefully worked out and tested criteria, then they deserve considerable respect. I say this despite my original and continuing "dynamic" bias. For the clear-cut syndromes of behavior, for which such labels are referents, have definite value for research purposes, even carrying consistent dynamic and "style" implications for those interested. I am not claiming that there is not much room for improvement. Perhaps it will come through the factorial techniques which, after some thought, we decided not to use at Worcester. What I do hold is that such perceptive clinicians as Kraepelin and Bleuler saw things that we might also see if we look carefully.

The growing trend for dichotomization in schizophrenia presents us with another aspect of the nosological problem. Although these dichotomies are helpful when used conservatively, analyses based on them too often show a tendency to oversimplify. Absolute constancy and consistency, whether within an individual or within a group of schizophrenics is, of course, illusory. Schizophrenia, particularly in its less chronic phases, is a fairly continuous succession of action and reaction, of regression and restitution—processes which sometime appear at an overt, easily discernible level, but most frequently at a more cryptic level. Should the patient come for study during a reaction phase, one may be led to place him in the opposite part of the dichotomy from that in which he characteristically belongs. For this reason several readings on a patient need to be taken.

These dichotomies and characterizations do have a distinct contribution to make. Although they have led to much discrepancy in the literature, careful examination reveals a surprising amount of overlap and synonymity among the various dichotomies. Still needed, however, are more rigorous criteria for class membership and more dependable methods for defining the criteria.

I should like to make a particular point about the chronic/acute dichotomy. These classes may be roughly separated by such a criterion as period of hospitalization. Even ideally, this measure tends to be approximate and arbitrary because the correlation of the classes of the dichotomy with the criterion is at best mildly positive. To complicate the problem we may be dealing with certain effects on functions which are more attributable to the long period of hospitalization than to the psychosis itself. Certainly many chronic patients show qualities which are not found in acute patients. Whether these are the direct and indirect effects of hospitalization, or whether they are developments of the psychosis that might very well have come about if the patients had not been hospitalized, remains an open question.

Another problem which troubles psychologists perennially is the part played by the *cooperation* or *attitude* of the subject. Almost all psychological tests and experiments require at least the passive participation of the subject. The data from such studies, except those directly investigating functioning at nonoptimal levels (a hazardous procedure, I must point out), carry the implication of having been collected under optimal conditions—external as well as internal. When there is suspicion that nonoptimal conditions are present, justifiable doubt about the validity of the findings arises. The argument may be offered that poverty in cooperativeness is intrinsic to schizophrenia; therefore, any attempt at the separation of its effects is at best academic. This thesis has validity to the extent that poor cooperation *is* intrinsic. The argument, however, runs into difficulty by not making a distinction between the intrinsic effects of attitude and of other temporary or superficial interfering effects. In order to control for this factor in our studies, we consistently used an A to E rating scale which defined various levels of cooperation. The patients used in the studies reported fall mainly into the classes we labeled A and B cooperation, those showing either active interest in the task itself, or active effort because of secondary interest.

At this point a few words seem necessary about the kinds of studies we conducted and the kinds of patients we have generally used.

The studies I shall be concerned with have been directed at answering questions of the *what* and *how* rather than the *why* and *about what* of the psychological functioning of schizophrenics. They are part of an effort to understand the nature of the schizophrenic organism, something about his psychological structure and function. Ego function has been emphasized, particularly single ego functions such as psychomotility and learning.

The patients we generally studied can be described as chronic. They had a mean age of approximately thirty, a mean schooling of nine to ten years, and a

mean hospitalization age of approximately seven years. (Hospitalization age is defined as the time elapsed since first hospitalization for mental disorder.) The major
advantage of working with chronic rather than acute patients is that because of the
relative stability achieved, the intraindividual and indirectly, the interindividual
variability, is reduced.

EXPERIMENTAL FINDINGS

An examination of our actual studies, which ranged from the patellar tendon-
reflex latent time at one extreme to group behavior involving competitive and cooperative activity at the other, revealed four general categories of results. The first
group includes areas in which, from the very beginning, no differences were found
between the schizophrenic and the normal subjects. A second area of results was
one in which differences found between the two groups initially or under ordinary
conditions either disappeared with repetition or were considerably decreased under
certain special conditions. This group of findings I have characterized as showing
"normalizing" trends. By "normalizing," I mean nothing more than that the
originally different results obtained from the schizophrenics came to fall close to
or actually within the range of normal performance. In the third group, differences
were found between the schizophrenic and normal groups which *did* persist.
Actually, a fourth category of "results" could be added here. This would include
the situations in which the patient withdrew and would not cooperate, so that
we were unable to get experimental data.

I shall now go on to present some of the data from each of the first three
groups, based on both published and unpublished studies.

No Differences

Contrary to what seems to be generally true in physiological and biochemical
studies, we have tended to find very few variables in which no differences were
found between groups of schizophrenic and normal subjects. Possibly this is to be
expected since the disorder has been defined in behavior terms.

But even if this is the case, certain cautions must be observed in defining
"differences." We have already seen one in the discussion of "cooperation": the
investigator must be careful in evaluating a response as to whether it is due to a lack
of interest in responding adequately, or to an intrinsic inability to do so. Another
relates to the need for conservatism in evaluating a "response" on the part of a
schizophrenic. Even given satisfactory cooperation, this problem exists in relation
to certain tasks. If one approaches a patient without having sufficient clinical sensitivity, it is quite possible to obtain considerably more pathological, and what may
be termed experimentally "inadequate," responses than one would under "fairer"
conditions. I should like to illustrate this by a relatively simple example from a
study which we did on color-blindness in schizophrenia [25], which I hope will

also indicate a general principle of our way of approaching the study of schizophrenia.

In the course of our routine clinical studies of schizophrenics, in a relatively short time we ran into several patients who could be designated as "red–green" color blind, by Miles' criteria in the Ishihara Test. Since we were intrigued with this finding and its possible genetic implications, we decided to carry out an extensive study with this test on schizophrenics, using other types of patients in the hospital as controls. Employing Miles' criteria (two or more incorrect responses, either color-blind or anomalous, excluding Plates X and XI) we obtained a 13 per cent incidence of color-blindness, as opposed to an eight per cent normal incidence. This meant that we had a significantly greater number of color-blind schizophrenics. We felt uncomfortable, however, about the "anomalous-response" criterion. We recognized that anomalous and doubtful responses were much more likely to occur in a psychiatric population. We experimented with a number of other scoring systems but felt most secure about what we called the Worcester III criteria—two or more definite color-blind responses, excluding Plates X and XI. We felt rather strongly that we could not use the Miles criteria since the Gertrude Steinian law (a response is a response is a response is a response) held only in a limited sense for schizophrenics. In passing, may I mention my prejudice that this is one of the weaknesses in some of the operant conditioning work being carried out with schizophrenics. (And I say this as a very early, if only transient, operant conditioner with schizophrenics in the early 1930s.)

To get back to color-blindness. When we used these stricter criteria we found no difference between the incidence in schizophrenic and normal subjects. I am arguing here for conservatism in making a judgment of pathology in those situations where we are offered a choice. Although most test and experimental situations do not offer us this kind of choice, caution is almost always possible. There are so many differences already; is there any need for increasing them unnecessarily?

Another area where we did not find any differences was in the patellar tendon reflex latent time. Huston [16], when he carefully controlled for height of his subjects, found that there were no differences between schizophrenic and normal control subjects at this simple psychophysical level.

These and other experiments led us to decide that any differences existing between schizophrenic and normal subjects were more likely to be found at more complicated levels than the simple sensory or psychophysical. We therefore turned our experimentation in these directions.

"Normalizing" Trends*

We have characterized as showing "normalizing" trends those findings in which the differences originally found between schizophrenic and normal subjects

---

*A version of this section on normalizing trends was previously presented at the XVth International Congress of Psychology in 1957 [9].

TABLE 1. Constants of the Distribution of Individual Thresholds in Volts of 22 Schizophrenic Patients and 28 Normal Subjects for Two Sessions Three Months Apart

	Patients		Normals	
	Session 1	Session 2	Session 1	Session 2
Min	61	55	65	60
Max	174	156	164	188
Mean	110 ± 5.7	102 ± 5.0	101 ± 4.6	103 ± 4.7
S.D.	26.8 ± 4.1	23.1 ± 3.5	24.1 ± 3.2	24.6 ± 3.3

have tended to disappear or become reduced under certain conditions. The normalizing factors that have struck us most forcefully appear to be of seven kinds. These factors are: (1) repetition, (2) passage of time, (3) cooperation, (4) time for preparation, (5) social influence, (6) stress, and (7) shock. I shall consider the seven factors separately and present one or two studies to exemplify each.

Repetition

In many of our studies—physiological as well as psychological—the first reading taken of a patient's performance in a task was found to be nonrepresentative of subsequent readings. Thus in a study of the threshold for direct current stimulation [15], the results of which are presented in Table 1, during the first experiment the patients had a mean threshold of 110 volts and the normal subjects a mean threshold of 101 volts. In the second session, three months later, the schizophrenic mean was 102 and the normal mean 103. It is thus seen that on repetition, without any essential clinical change in the patients, the patient mean fell essentially to the level of the normal subjects.

Let us take another type of study, this time a learning situation in which a kind of "pursuit meter," the "Prodmeter," was used. In this experiment [19], the subject is required to follow a revolving target with a pointer for ten trials of ten revolutions each. (The turntable stops when the pointer is not on the target.) The subjects were examined on 33 consecutive days, excluding Sundays. The results from Table 2, given in terms of the time taken for the task in seconds, are worthy of note. As indicated in the first column, the mean scores of the schizophrenic subjects for the first day are without exception higher (that is, poorer), and generally much higher, than those of the two normal subjects. On the lowest day, however, the mean scores of the schizophrenic subjects are much closer to those of the normal subjects. An examination of the next two columns, which give the lowest

TABLE 2. Scores of Individual Schizophrenic Patients and Normal Controls on Thirty-Three-Day Prodmeter Experiment

		Co-operation	Mean score first day	Mean score lowest day	Lowest trial score	Day lowest trial score reached
	1	A	37	12	11	22
	2	B	38	13	11	24
	3	B	39	15	12	30
	4	B	34	16	13	17
Patients	5	B	36	16	14	30
	6	C	39	17	13	27
	7	C	64	17	15	33
	8	C	43	18	16	18
	9	C	39	19	16	28
Normals	1	A	24	12	11	26
	2	A	30	13	11	24

trial score and the day on which the lowest trial score was reached, corroborates the general impression that the differences between the two groups tended to be reduced considerably with repetition. In fact, one of the patients actually did better than either of the two control subjects, both of whom were persons of unusually high intelligence and highly skilled in motor tasks. A number of the other patients came close to the scores of the normal subjects. Because of the small number of subjects, the findings can, of course, be considered as only suggestive, and in need of replication. One does gain the impression, however, that given a sufficient amount of practice, schizophrenic subjects can sometimes reach a "physiological limit" not very different from that of normal subjects.

Passage of time

Aside from repetition, with or without practice effects, we have also found what appears to be evidence that mere lapse of time can result in improvement in the performance of schizophrenic subjects. This phenomenon appeared in a pursuit meter learning experiment [18] which called for ten trials of ten revolutions each in following a target with a pointer. This task was repeated two times at intervals of three months. The data for the first two periods are given in Table 3, the scores being in terms of contacts made. (A high score in this case is a good score.) It will

TABLE 3. Pursuit Scores, Trial-by-Trial for 46 Schizophrenic Patients and for 22 Normal Subjects

Period	Trial	Patients		Normals	
		Mean ± S.E.	S.D. ± S.E.	Mean ± S.E.	S.D. ± S.E.
1	1	12.9 ± 1.4	10.0 ± 1.0	30.0 ± 3.1	14.3 ± 2.2
	2	15.3 ± 1.8	12.2 ± 1.2	29.6 ± 2.8	13.3 ± 2.0
	3	15.5 ± 1.6	11.0 ± 1.1	34.0 ± 2.9	13.5 ± 2.0
	4	14.9 ± 1.5	10.5 ± 1.1	34.3 ± 2.8	13.3 ± 2.0
	5	16.1 ± 1.7	12.1 ± 1.2	35.9 ± 2.5	11.7 ± 1.8
	6	17.2 ± 1.8	13.0 ± 1.2	34.1 ± 2.3	10.7 ± 1.6
	7	17.9 ± 2.1	14.5 ± 1.5	40.7 ± 2.7	12.6 ± 1.9
	8	18.7 ± 1.8	13.1 ± 1.3	42.5 ± 2.9	13.6 ± 2.1
	9	19.8 ± 1.9	13.6 ± 1.4	43.0 ± 3.2	15.2 ± 2.3
	10	18.8 ± 1.9	13.6 ± 1.4	43.0 ± 3.2	14.9 ± 2.2
	Mean	16.7 ± 1.6	10.8 ± 1.1	36.7 ± 2.4	11.2 ± 1.7
2	1	24.1 ± 2.2	14.8 ± 1.5	41.0 ± 2.9	13.4 ± 2.0
	2	25.8 ± 2.2	14.8 ± 1.5	45.4 ± 3.4	15.7 ± 2.4
	3	26.3 ± 2.2	15.1 ± 1.6	45.2 ± 3.3	15.4 ± 2.3
	4	25.0 ± 2.1	14.1 ± 1.5	46.4 ± 2.8	13.1 ± 2.0
	5	27.3 ± 2.3	15.5 ± 1.6	48.3 ± 2.7	12.6 ± 1.9
	6	29.6 ± 2.8	19.5 ± 2.0	49.3 ± 3.3	15.6 ± 2.3
	7	28.6 ± 2.7	18.6 ± 1.9	51.2 ± 3.4	15.8 ± 2.4
	8	29.3 ± 2.6	18.1 ± 1.9	52.7 ± 2.8	13.1 ± 2.0
	9	28.8 ± 2.7	18.6 ± 1.9	52.6 ± 2.9	13.7 ± 2.1
	10	31.4 ± 2.6	18.1 ± 1.9	60.0 ± 2.8	13.1 ± 2.0
	Mean	27.6 ± 2.2	15.1 ± 1.6	49.2 ± 2.6	12.1 ± 1.9

be noted that for the patients the mean for the tenth and last trial of Period One was 18.8, and the mean for the first trial of Period Two, following an interval of three months, was 24.1. They thus showed an actual gain of 5.3 points. For the comparable trials the normal subjects, on the contrary, showed a loss of two points—from 43 to 41. The trends of the curves for each of the other periods are in general at the same level. Whether this reminiscence effect is the result of the dropping out of interfering habits and irrelevant ruminations that prevent the patients from being able to show how well they actually learn originally, or whether this effect is due to a kind of "consolidation" resulting from some continuing process, the important point is that the phenomenon appears in the schizophrenic but not in the normal subjects.

· These first two of the seven normalizing factors—repetition and lapse of time—appear to be primarily outside the patient's control. The next series of categories that I wish to consider appear to require his involvement to some degree and are for this reason perhaps different in quality.

Cooperation

In all of our studies in which patients were involved, our general practice was to use a five point rating scale to characterize their participation in the task. For general purposes we labeled this characterization as "cooperation level." The term must be considered one which deals largely with the patient's involvement in the assigned task as determined by his effort and interest. A rating of A indicates active interest and maximum effort—generally the level of response obtained from normal subjects under similar circumstances. A rating of B indicates real effort by the subject but one deriving from sources other than a primary interest in the task itself, probably from friendliness to the experimenter and a wish to be thought well of by him. A C rating indicates a docile, perfunctory, spasmodic effort, one where the patient requires some urging to complete the task. Ratings of D and E represent participation at even lower levels than that rated C. Performances with such ratings have rarely been included in reported data—actually E is never included and D only for very special purposes.

It has turned out in a number of our studies that when the cooperation level of the patient was at approximately that of the normal subject—that is, at the A level—the differences which were otherwise generally found between the total group of schizophrenic and normal subjects were considerably reduced and sometimes actually eliminated. I shall present a few sample studies which bring out this fact.

The data for the first of these studies, one of *steadiness* [17], are shown in Table 4. As you will see from the columns of the means, the performance was markedly different at the various cooperation levels—in fact, the differences were highly significant among the various patient groups. However, when a comparison was made between the performance of the patients rated A in cooperation (118.7) and the performance of the normal subjects (120.0), the difference was not significant—in fact, the scores were almost identical.

Time for preparation

Another experiment which showed a similar trend and involves still another factor is that on tapping [37]. The data are presented in Table 5. It will be noted

TABLE 4. Constants of the Distribution of Weighted Steadiness Scores According to Cooperation Levels for 135 Schizophrenic Subjects and 64 Normal Subjects

	Co-operation	N	Mean	S.D.
Patients	A	26	118.7 ± 11.1	55.3
	B	48	73.2 ± 6.7	45.6
	C & D	61	29.7 ± 4.3	33.9
	Total	135	62.3 ± 4.7	54.3
Normals	A	64	120.0 ± 8.1	65.1

TABLE 5. Constants of the Distribution of Mean Tapping Scores (Ten Trials) for Schizophrenic Patients and Normal Subjects

	N	Mean	S.D.
Patients			
Total	125	19.5 ± .50	5.58
A co-operation	16	25.1	
Paranoid total	24	24.4 ± .94	4.60
Paranoid: A co-operation	8	27.5	
Normals	60	28.9 ± .55	4.24

from the column giving the mean scores that whereas the total schizophrenic group had a mean of 19.5 and the normal group a mean score of 28.9, the patients with an A cooperation rating had a score of 25.1. Although this difference was still significantly different from that of the normal subject group, it was now significantly reduced. Actually, if we take only the paranoid patients (the subtype giving the highest scores) who are rated A in cooperation, then the significant difference is wiped out.

The results obtained from the study of tapping puzzled us when we compared them with the results from a series of reaction-time studies (which we shall have occasion to consider later) that we conducted at the same time. The reaction-time differences remained significant even when cooperation level and subtype of schizophrenia were taken into consideration. Since tapping would, on the face of it, seem to be a form of continuous reaction-time activity actually requiring more persistent effort than isolated reaction-time, we were at a loss to account for the differences in the two sets of results.

For a time the only suggestion that occurred to us was that the distinction lay in a fundamental difference in the nature of the two experimental tasks. In the tapping experiment, the subject was told that he should start tapping at a given signal and to continue tapping (for five seconds) until asked to stop. In several senses, the schizophrenic was able here to set his own pace. He could actually start tapping when ready and was not penalized for any time he took for preparation. In addition, the apparatus was so constructed that the first 750 msec. of the tapping period was not counted. This provided the subject an additional period in which to prepare. In the reaction-time situation the subject was warned that a stimulus was coming. But here, of course, the stimulus called for immediate response, and the subject was penalized for delay. The pace of this task was in no way under the subject's control. It is quite possible that, for the schizophrenic, the opportunity to set his own pace, as opposed to having it determined from the outside, is very important. Obviously a systematic experiment should be designed to deal with this question. For the present, however, we are considering *self-preparation* one of the "normalizing" factors. We shall consider this hypothesis further at a later point in our discussion.

Social pressure

The usual way of obtaining involvement in a task by a patient is through the rapport and influence which the experimenter is able to establish. Another, more rarely used with schizophrenic subjects, is through providing a social situation in which the schizophrenic is a member of a group of peers, where the social situation may press him into involvement. Radlo and I designed a study as yet unpublished [35], which used card sorting for the investigation of schizophrenic learning under conditions of both individual and group competition. The task consists of speedily sorting sixty cards, each marked with a set of five digits, one of which set is either one, two, or three. The cards placed in compartments so labeled. The experiment was carried out with many groups and under a variety of conditions. For our present purposes it is only necessary to consider three groups, closely matched for initial scores. Table 6 shows the data from these groups.

Group one, a group of schizophrenic patients in which no competition was introduced, served as the controls. Note the nature of the learning curve in this group. The asterisks indicate points at which there are significant differences between successive scores.

Group 2A was a patient group in a situation where *interindividual* competition was introduced. After going through trials one to four individually in separate rooms, three patients at a time were brought into an adjoining room for trials five and six and placed under conditions in which competition among them was emphasized. For trials seven to ten, they again went back to their original rooms, where

TABLE 6. Speed in Seconds of Card Sorting by Schizophrenic Subjects Under Different Conditions of Learning

Conditions[a]	Trials											
	1	2	3	4	5	6	7	8	9	10	11	12
1	64.5[b]	60.0[b]	56.5	56.5	53.5	55.0	54.0	54.5				
2A	64.5	63.0	60.0	61.5[b]	43.5[b]	40.0[b]	44.0	48.0	43.5	44.0		
5	63.0	59.5	54.5	55.0	52.0	53.5	52.0	53.5[b]	42.5[b]	38.0	38.5	39.0

[a]1. Learning Control Series (N = 15). 2A. Interindividual Competition Series (N = 6) Trials 5 and 6. 5. Group Competition Series (N = 12) Trials 9, 10, 11, 12.
[b]Significant differences between items.

they worked individually under the previous conditions. It is to be noted that there was a drop in score (a low score, of course, being a good score) for trials five and six when compared with trial four, and a rising, more or less flattened, curve thereafter. The patients appeared generally to retain most of what they had gained during the period of competition. This may be interpreted either that once a patient, for whatever reasons, has achieved a level of performance he is able to maintain it, or that for the schizophrenic patient the competition situation "continues" despite the immediate physical absence of the competitors.

Group five represents patients in a *group competition* situation. In trials one to eight, six patients worked in individual rooms as previously and under similar instructions as in the control situation. (The similarity of the scores of Group five to the control group through these eight trials is quite striking.) In trials nine to 12, two groups of three patients each competed as teams, each team using one set of sorting bins. A comparison of trial nine of Group five with trial eight of Group one (the control group) shows a spurt to a quite different level from the plateau achieved at trials five to eight in both Groups one and five. The differences are significant at this point. In general, we believe these results tend to show the susceptibility of schizophrenics to both competitive and cooperative motivation; that is, they show a greater involvement in tasks and a higher level of performance under such conditions.

Stress

Although this is an area in which we have done a good deal of experimentation, we do not have any satisfactorily completed studies on chronic schizophrenics. We do, however, have a number of studies which are at least suggestive. But again, they call for replication.

Two studies using a targetball stress situation gave results similar to those we have just discussed. The targetball apparatus is designed to study achievement in relation to aspiration level. In addition, however, it is so organized that the experimenter can manipulate the apparatus to make the subject either succeed or fail in relation to a "bogey" score, which represents the level of achievement of the "average" person. Subjects, whether schizophrenic or normal, generally become highly involved in the experiment. In our experiments in which this stress of failure was induced, it was found that such stress tended to improve the scores of the schizophrenic subject and to bring him closer to the normal. The stress appeared to improve the performance of the patients not only in a psychomotor task involving a special kind of pursuit meter [31], but also on a *TAT* test involving thinking. In the latter, the general quality of the stories, the clarity of the thinking, and the consistency with which the ideas were presented, all improved.

Shock

The last of the "normalizing" factors which I shall discuss is "shock." I shall present two experiments which offer evidence along this line.

TABLE 7. Responses of Schizophrenic Patients, Showing the Effect of Metrazol Shock upon a Habit System

Group	Less than 50 percent reversion to habit 1	More than 50 percent reversion to habit 1
Metrazol (N = 21)	7	14
Control (N = 21)	17	4

x^2 = 7.88
$p < .01$
(Yates correction for small n)

The first of these is from a study [27] on the effect of metrazol shock on habit systems. In this experiment, two groups of 21 schizophrenic subjects each were initially trained, as Habit One, to respond with a right-finger movement to a tone of 500 cycles and with a left-finger movement to a tone of 700 cycles. In this first session the subjects were given 100 trials, 50 for each tone. In a second session, 24 hours later, Habit Two was established, both groups being instructed to reverse the direction of response to the tones. This training session consisted of 75 trials. One hour after the second session, the subjects in one group were given metrazol injections. One and one half hours later, both groups were retested to determine which habit was dominant. Ten trials were given, five to each of the tones.

As Table 7 shows, a statistically significant greater number of reversions to the first habit occurred in the group subjected to metrazol shock. Although the habit system here involved is of course of a quite different order from that ordinarily associated with schizophrenic behavior, the value of the experiment lies in indicating that a metrazol shock (and perhaps other shock treatments) does have a differential effect upon older and more recently acquired habits.

A study more closely associated with the complex habit systems of the schizophrenic is that carried out by Schnack, Shakow, and Lively [32]. We administered a battery of tests (the Stanford–Binet, the Kent–Rosanoff Word Association Test, and an aspiration-level test) to 50 male schizophrenic patients before and after treatment with insulin or metrazol. Our comparison of the two sets of scores revealed considerable changes in the direction of improvement of intellectual functions with treatment on most of the measures.

To test the significance of these changes, a further comparison was made with individually matched control patients who had had two successive examinations while under routine hospital care, but who had had neither form of therapy. These results showed that insulin and metrazol could only be held responsible for about one third of the improvement in test scores—two thirds being directly attributable to the ordinary hospital regime and familiarity with the test situation. The data further indicated that those with mental ages below 12 to zero upon first testing seemed to benefit most from metrazol therapy. The patients with originally higher mental ages, however, were apparently disturbed by metrazol. Insulin, on the other hand, though helpful in both groups, seemed more effective with the higher intelligence group, when the changes occurring beyond those shown by control patients under routine care was taken as the criterion. Of the higher intelligence group, those more nearly normal had a better long-term reaction to insulin, but a poorer long-term reaction to metrazol.

Thus if we were to summarize what we have been saying about these normalizing factors we could say that they involve differing degrees of "heroics" which achieve at least "temporary effects"; they (1) allow a period for acquaintance and removal of strangeness; (2) permit personal motivation to enter, either self-induced, or from the outside by providing "stress"; (3) allow the subject to control the situation without his knowing it (if he knows, as we shall see later, he is likely to do his worst work); and (4) use extreme "shock" to bring the person out of his more recently acquired, nonadaptive habits.

Persistent Differences

We have thus far considered two groups of psychological findings: those which showed no differences between normal and schizophrenic subjects, and those in which the differences either disappeared or tended towards the normal level under certain conditions. Let us now examine a third class of findings: those studies in which the differences were found to persist. By "persist" I mean that they endured under conditions generally similar to those which yielded the normalizing trend with the factors we have discussed.

Autonomic Reactivity

Besides motor responses to a stimulus, for example reaction time, which we shall discuss shortly, there is a category of autonomic responses to stimuli in the context of experiments in which the subject is asked to relax and make no overt response. Over the years, we have carried out a series of studies involving galvanic skin response, heart rate, and other such responses affected through repeated stimulations by verbal ready signals, noise, tone, light, and pain. The following are examples from two studies, one carried out a number of years ago at Worcester State Hospital, and another more recently at the National Institutes of Health.

In the first study [7], the aim was to determine the rate of adaptation of schizophrenic and normal subjects to pain—in this case, induced by pressing the skin with an algesimeter. There were ten subjects in each group of schizophrenics and normals.

As you will see from the graph in Figure 1, both groups of subjects started with a mean heart rate of 80. At the end of an hour of repeated stimulation, the

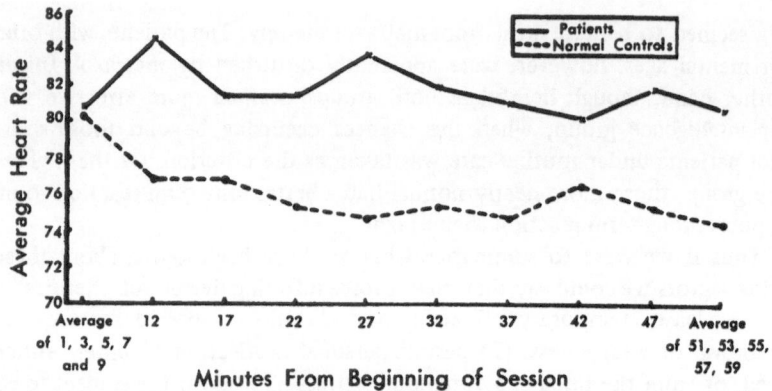

FIGURE 1. Mean heart rate changes of schizophrenic (N = 10) and normal (N = 10) subjects undergoing repeated pain stimulation.

normals gradually came down to a level of 75, whereas the schizophrenic group fluctuated between 80 and 85, giving a reading of 80, the equivalent of their initial mean, in the last trial. Their actual mean level on the ward was 74.

A recent study of *GSR* orienting reactions to visual and auditory stimuli [42], employed a red light and a 300-cycle tone as stimuli. The subjects were 52 chronic schizophrenic and 20 normal subjects, both groups averaging a little over 40 years of age. The stimuli were presented 40 times each, at separate sessions, for a one second duration at one half minute intervals, with appropriate beginning and ending control readings. We might consider the results in two ways: in relation to the level of arousal or activation, and in relation to specific or orienting responses to stimuli.

Base conductance and frequency of nonspecific *GSR*s are two measures of skin resistance that have been found to be related to level of arousal. As is shown in Figure 2, using base log conductance as a criterion of arousal level shows that the schizophrenic subjects had a significantly higher arousal level in relation to the light stimulus. The normal subjects showed a progressive adaptation to both the light and tone stimuli that did not occur in the schizophrenics. The similarity of the curves for the tone stimulus to the one for the heart-rate reaction to pain (see Figure 1) is striking. Let us take the other measure of arousal—the number of nonspecific *GSR*s (drops in skin resistance of 400 ohms or more which did *not* occur within the first three seconds after the stimulus). Here again the patients showed a significantly higher arousal level.

How about the specific or orienting responses? These were defined as drops in resistance of 400 or more ohms beginning within the first three seconds after the onset of the stimulus. As Figure 3 indicates, here again the schizophrenics were

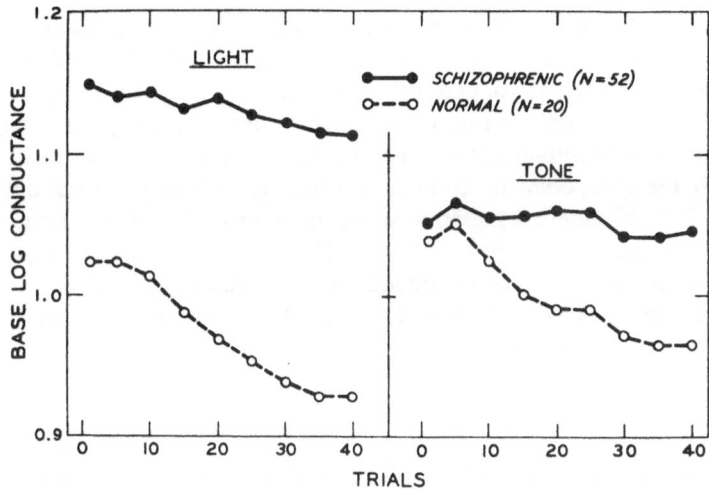

FIGURE 2. Base log conductance of schizophrenic and normal subjects in repeated stimulation by light (Exp. 1) and tone (Exp. 2).

more responsive overall. The striking difference, however, is in the rate of habituation, which was significantly fast for the normal subjects.

An interesting comparison is the ratio of the specific to the nonspecific *GSR* response frequency per unit time. A clear difference existed in the direction of

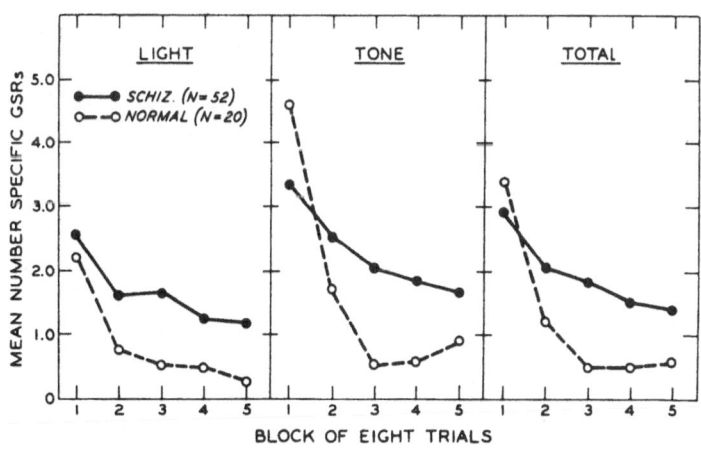

FIGURE 3. Mean number of specific *GSR*s of schizophrenic and normal subjects to repeated light and tone and the total of these.

greater specific to nonspecific responsiveness for the normals. One likely interpretation of this finding is that the influences of internal or self-produced stimulation in relation to the influence of external stimulation is proportionately greater in the schizophrenics. This ratio seems more likely to be related to "preoccupation" rather than to "distractibility," although the latter cannot be ruled out altogether.

Both these experiments seem to indicate that whereas the normal subject shows gradual autonomic adaptation or habituation to the stimuli, schizophrenic subjects do not seem to adapt in this way. The schizophrenic tends to continue to react at the end of the session—frequently for as long as an hour—at approximately the same autonomic level as he had at the outset. In this respect, the schizophrenic remains inordinately unaffected by the preceding succession of stimuli.

Reaction Time

Another area in which we have found persistent differences is reaction time. An early exploratory study of simple visual, simple auditory, and discrimination visual reaction time [20], found that the schizophrenic means in each of these three types of reaction were very much higher than those of the normal subjects, and that there was relatively little change over the three periods of testing, which were three months apart. A similar significant difference was obtained with the minimal reactions, the normal group having a significantly shorter mean minimal reaction time than even the best patients—those rated A in cooperation. Unfortunately we do not have long-term data involving daily practice such as that on pursuit meter learning. Such an experiment would be important in determining whether the schizophrenic can eventually achieve a normal level of performance.

In the first follow-up study on simple auditory reaction time [20], there was, as is shown in Figure 4, significantly slower time for schizophrenic subjects at every one of six preparatory intervals (range 0.5-10 sec.) for both the regular and irregular procedures. In addition, an interesting phenomenon appeared which seemed to be related to what might be considered the maintenance of set. Our schizophrenic subjects seemed to have special difficulty in maintaining a major set, having instead a tendency to react to isolated stimuli, or at best to depend upon minor, less adaptive sets.

We noted in this experiment that whereas normal subjects were able to take advantage of the regular procedure by giving shorter reaction times at all intervals, the schizophrenic subjects had a breakdown point at the two second interval—that is, they gained no advantage from the regular, and of course much simpler, presentation. Beyond two seconds they seemed unable to take advantage of the regularity of the presentation, giving reaction times as long as they did in the irregular series. These results held for all the patients, including those whose cooperation levels were judged to be equal to that of normal subjects.

A third study [29] used simple visual rather than auditory reaction time, and extended the preparatory intervals to 25 seconds. As Figure 5 shows, this study in

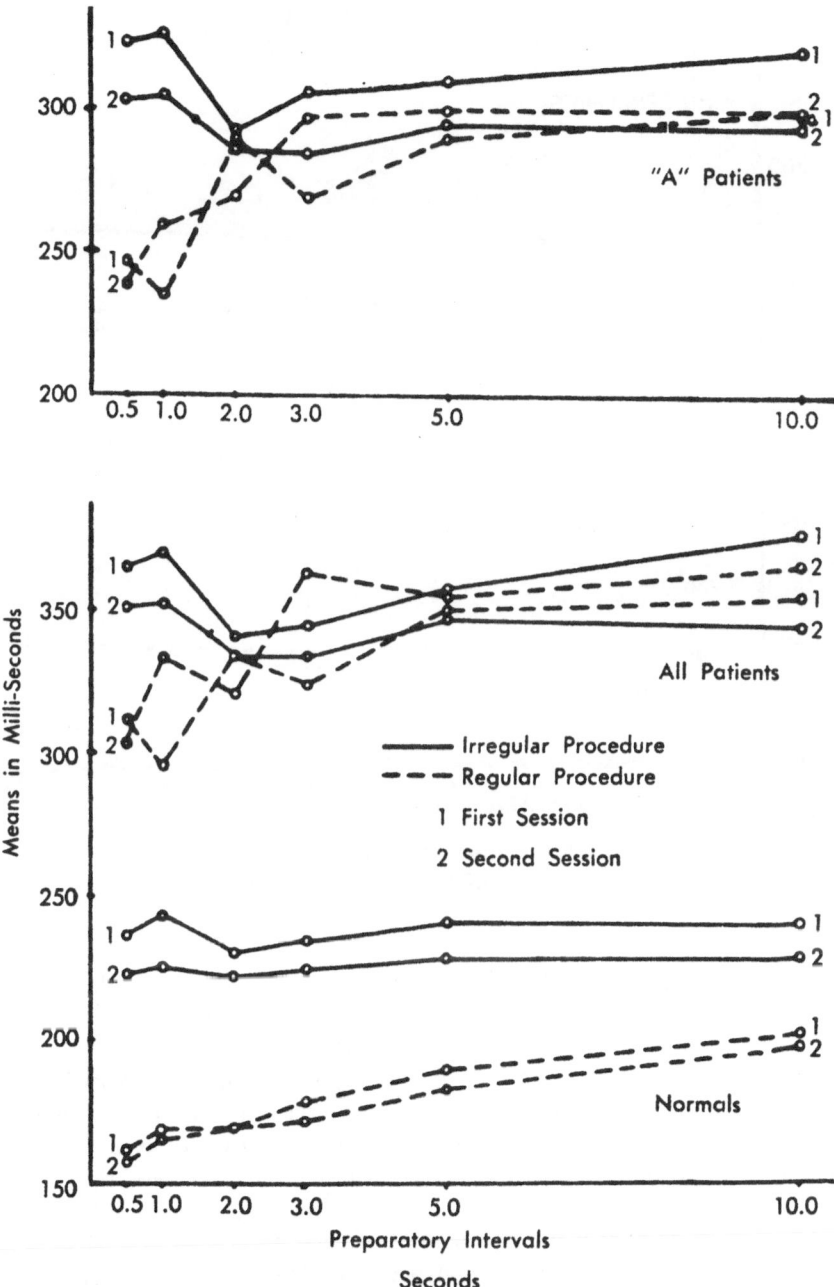

FIGURE 4. Auditory reaction times for total schizophrenics (N = 25), A schizophrenics (N = 16), and normals (N = 18) for six preparatory intervals, regular and irregular procedure.

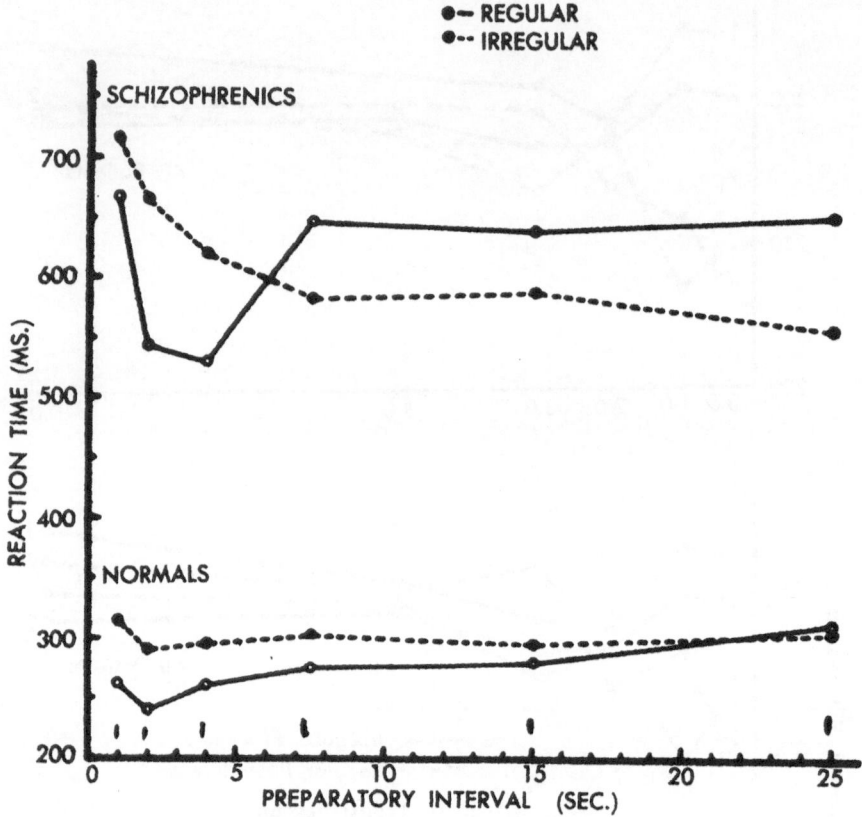

FIGURE 5. Mean visual reaction times of 25 schizophrenic and 10 normal subjects at the various preparatory intervals of the regular and irregular warning procedures.

general corroborates the results of the first, except for a change in the point of juncture of the regular and irregular curves. The curves of the normals came together somewhere between 20 and 25 seconds. We were now able to find a breakdown point in the normals. When the preparatory interval reached somewhat over 20 seconds, they were no longer able to take advantage of the regular procedure. In contrast, this point was reached by the patients at the five to six second level. Reaction times for both the regular and irregular procedures remained significantly longer at every interval for the schizophrenic patients. The major additional contribution of this study was the development of what we called a *set index*, based upon length of reaction time and the relationship between the regular and irregular periods. This index was able to differentiate the schizophrenic and normal subjects without overlap. This is the only instance that we know of in which such a finding on psychological phenomena has been reported in the literature.

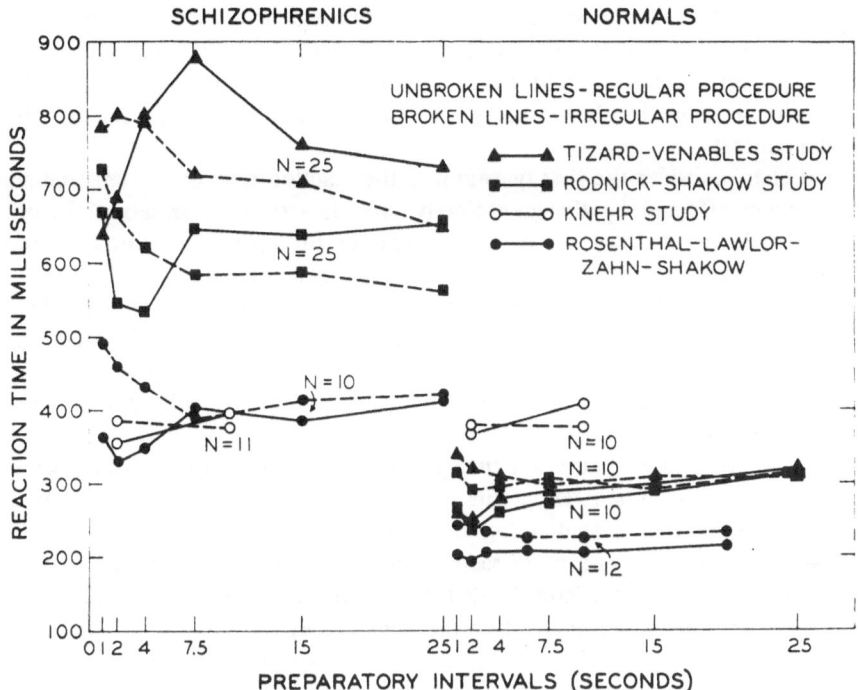

FIGURE 6.   Reaction times of schizophrenic subjects during regular and irregular procedures in four separate studies.

Since this last study, which was done some time ago, numerous additional experiments have corroborated these findings. Huston and Singer [21], Tizard and Venables [39], and several additional studies [30,43-45] from our own laboratory at the National Institutes of Health, have given the same pattern of results. Figure 6 shows most clearly the similarity in findings for regular and irregular procedures between several of these studies and the Rodnick-Shakow data. Of course, differences in reaction time level because of differences in modality (and context) exist and must be partialled out. Only one study [22] has obtained discrepant findings. The conditions and subject samples of this study seem so aberrant compared with those described in the other studies, however, that the results can probably be accounted for on the basis of these differences.

The set hypothesis we had formulated from our early studies posed two general questions: (1) How basic and pervasive was this disturbance in the ability to maintain a major set? and (2) With what other characteristics was the disturbance associated?

We know from our reaction time studies involving regular and irregular procedures that the schizophrenics, even at their own relatively slow speed, are unable to take advantage of the simplification that the regular procedure provides. But how widespread is this disturbance? I shall first deal with this question as it is reflected in performance in relation to two other major aspects of the reaction time situation: the immediately preceding stimulus situation, and the preceding experimental context.

Let us consider the first question: Is the reaction time of the schizophrenic inordinately affected by the immediately preceding stimulus situation? In other words, is the generalized set he must maintain for optimal performance more readily disturbed by what has happened immediately before?

Woodrow [41], in his early studies on preparatory interval, was able to obtain such an effect in normal subjects. In our own first study, on the other hand, we did not obtain this finding clearly in either normal or schizophrenic subjects. The fact that our experiment was not designed to test this hypothesis may account for this, however.

A recent auditory reaction time study in our laboratory at NIH [44] was directed primarily at answering this question. Twelve schizophrenic and twelve normal subjects were tested, using the irregular procedure with six preparatory intervals ranging from one to 25 seconds. Some of the results obtained from this experiment are seen in Figure 7. *PI* refers to the preparatory interval. The *PPI* refers to the preparatory interval of the immediately preceding trial. Thus at the four second point, the *PPI* curve represents the mean of all the reaction times for *all* intervals which were preceded by a preparatory interval of four seconds. This contrast with the *PI* four second point, which represents the mean of the reaction times given specifically to the four second interval.

For both normal and schizophrenic subjects: (1) the *PI* and *PPI* slopes were significantly different from zero; (2) when the *PI* was short and reaction times were longer, whereas when the *PI* was long the reaction times were shorter; and (3) for the *PPI* the opposite effect was true—when *PPI* was short, reaction times were short; when *PPI* was long, the reaction times were long. When we contrasted the two subject groups, (1) we found a significant difference in absolute levels of reaction time; (2) the slopes of the *PI* curves were also significantly different, shown by the marked steepness of the schizophrenic curves as compared with those of the normals; and (3) the schizophrenics seemed to show a greater effect of *PPI*, though (because of group variability) the trend did not quite reach statistical significance. In a replication experiment, however, the difference was found to be significant.

The relationship between *PI* and *PPI* is best seen by looking at the two extremes of the curves—the one second and the 25 second intervals. At the one second interval on the *PI* curve, the schizophrenics gave the longest reaction time, but when the one second interval preceded other intervals (one second point on the *PPI* curve), the shortest reaction times appeared. At the 25 second interval, the effect was just the opposite: the shortest reaction times were given to the 25 second interval on the *PI* curve; but when the 25 second interval preceded other intervals, the reaction times were the longest. Although the same general pattern

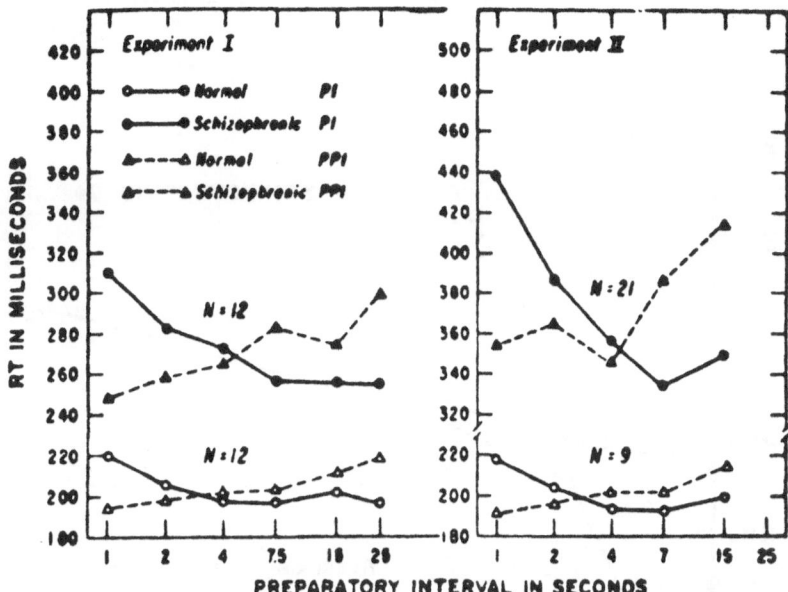

FIGURE 7.   Reaction time as a function of the preparatory interval (*PI*) preceding preparatory interval (*PPI*) for schizophrenic and normal subjects.

was found in the normal subjects, there was a significant difference between the two in the degree of the effect.

The answer to the question we raised earlier, then, seems to be "yes": schizophrenics are inordinately affected by the immediately preceding stimulus situation in this type of task. I might say in passing that the replication study that I just mentioned gave essentially the same results.

Let us now turn from a consideration of the effect on set of a preceding single stimulus situation, to that of the effect of the preceding context—a repeated series of single, similar preceding stimulations. (So named, of course, for the use of the regular procedure.) To test the relationship between set and this broader experimental context, we designed two auditory reaction time experiments [43] using regular prepratory intervals arranged in both ascending and descending series.

As the results from these studies are essentially similar, we need only discuss one of them here. This experiment, which used twelve schizophrenic and nine normal subjects, called for five preparatory intervals ranging from one to 15 seconds. Three sessions were given: a descending series beginning with the longest preparatory interval and ending with the shortest; an ascending series beginning with the shortest interval and working up to the longest; and a repetition of the descending series. Some of the results of this experiment can be seen in Figure 8.

We will limit the discussion to the Ascending and the Descending, two curves

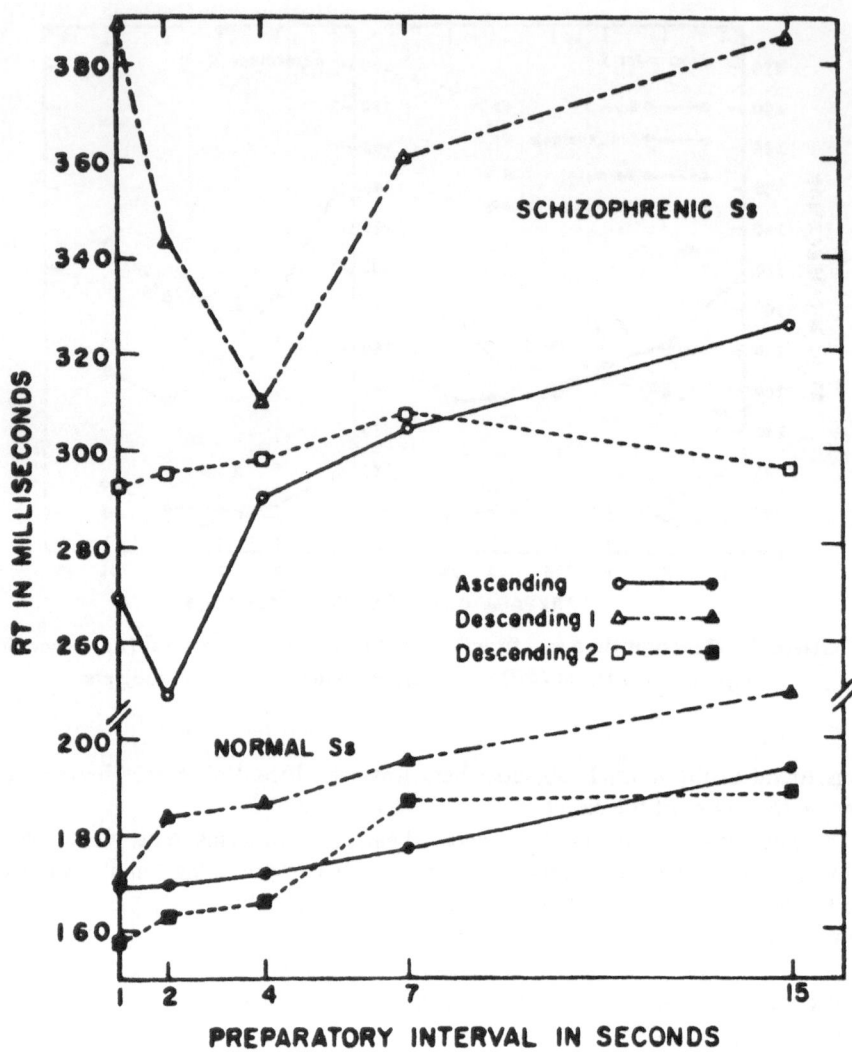

FIGURE 8. Reaction time as a function of the prepratory interval for two
descending orders and one ascending order of presentation.

of both the schizophrenic and normal subjects. It will be seen that the schizophrenic reaction time for the smaller intervals was much shorter in the ascending series
than in the descending series. In contrast, regardless of whether small intervals
appeared first or last in the series, the normal subjects tended to give shorter reaction times on the shorter intervals than on the longer intervals. They did not seem
to be influenced by the previous succession of longer preparatory intervals.

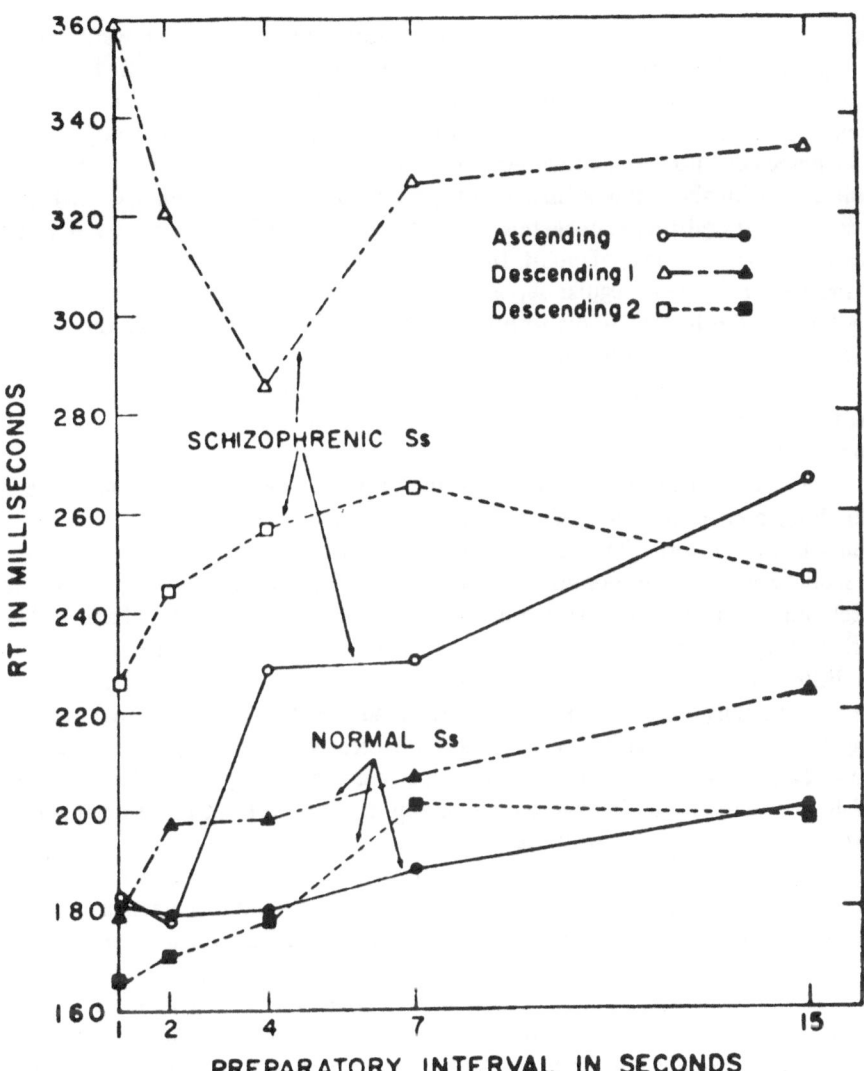

FIGURE 9. Reaction time as a function of the preparatory interval (subgroups of six schizophrenic and seven normal subjects matched on reaction time at two second preparatory interval).

Figure 9 brings out this point in an even more striking fashion. Again let us limit ourselves to the Ascending and Descending two curves. For this analysis, the six fastest reacting schizophrenics were matched with a group of normal subjects (the normals turned out to be the seven slowest subjects) at the level of optimal performance in the ascending series—the two second interval. A mean reaction time of approximately 180 msec. was obtained in both groups. It later appeared that this matching held for the one second interval as well. This is the same trend shown in Figure 8, but it is even more marked for the schizophrenic curves. The major point to be made here, however, is that despite the matching of normal and schizophrenic subjects, the breakdown of schizophrenic performance in the ascending curve began at the next tested preparatory interval—the four second level. Reaction time increased significantly at that point. In a sense, this is a corroboration of the findings of the earlier regular-irregular series studies, which also suggested that this is approximately the length of preparatory interval at which the ability of the schizophrenic to maintain a set breaks down.

We can say, then, that repeated exposure to a condition—the broader experimental context—appears to have a much greater effect on schizophrenic than on normal subjects.

In passing, I might point out that the disproportionately long reaction time with long preparatory intervals is not a function merely of the slow tempo of events in such a series. One might expect that a schizophrenic would more easily be seduced into slow responses by a setting that is somewhat "drawly." In order to check out this factor, we [45] compared particularly the *RT*s of normal and schizophrenic subjects under the "usual" long *PI* condition with reaction times under a condition in which the tempo of events was the same, but where the *intertrial interval* was long and the *PI* short. The results showed that reaction time under the long *ITI*-short *PI* condition was virtually identical with that under an "optimal" short *ITI*-short *PI* condition for both groups, and significantly faster than the short *ITI*-long *PI* conditions for the patient group. So, we do seem to be dealing here with a problem connected with the length of the preparatory interval.

Having gained some notion of the pervasiveness of the phenomenon, we now can go on to consider some of the conditions which appear to be associated with this difficulty in maintaining set, as well as several conditions which do not seem to be so associated.

Despite Knehr's statement to the contrary, reaction time, and we may add set index, do not seem to be highly correlated with intelligence. In our Worcester studies we repeatedly obtained correlation coefficients of about .30 between reaction time and IQ. When cooperation was partialled out, these correlations fell to about zero. In one study at NIH [30], the correlation between set index and Progressive Matrices score was only .26, a nonsignificant correlation. Thus, intelligence seems to be a negligible factor.

Likewise, cooperation which served as a "normalizing" factor with other functions such as speed of tapping and steadiness, cannot account for the difference between schizophrenic and normal reaction time performance. In the Worcester studies, the correlation between cooperation and various kinds of reaction time consistently ran about .50. Despite this fairly high correlation, the patients we rated as *A*, those who presumably had a cooperation level not much different from that of normal controls, still showed a significantly longer reaction time—both simple and discrimination—than the normal subjects. The differential effect of preparatory interval on simple reaction time also held in spite of optimal cooperation. Thus cooperation, although related to level of reaction time *within* the schizophrenic group, does not seem to account for the differences between schizophrenic and normal performances.

In contrast to IQ and cooperation, there are two factors which seem more centrally related to reaction time performance and set.

The first of these may roughly—and broadly—be called "mental health." In the first study in which we had used the set index [29] we had found that the two patients closest to the normal subjects in index were the two least deteriorated in the group. Because this was only a passing finding, we felt the need of a systematic study of set index in relation to mental condition. Therefore, as part of an NIH study of the reaction times and set index scores of a group of 13 schizophrenic patients, we had eight judges, five attendants and nurses, one psychologist, and two psychiatrists who had been in close contact with the patients, rate their mental health. (With the psychiatrists, we used the term "ego intactness" rather than "mental health.") We used a method of paired comparison so that every patient was matched with every other. There was high reliability, the median interrater correlation being .81. The correlation of "mental health" with set index was .89, and with reaction time, .82.

In addition to what we called mental health, or ego intactness, and probably related to it, is another factor which may be called "autonomy." The background of our interest in this function follows.

As has been mentioned, our early Worcester studies included a tapping test that gave surprising results. Tapping had seemed to us to be only a more complicated and repetitive reaction time task. But we found that, in contrast to reaction time differences between normal and schizophrenic subjects which held at significant levels despite excellent cooperation, tapping did not hold up as a differentiating function. The differences tended to disappear with a high level of cooperation, especially with paranoid patients. The most likely hypothesis accounting for this seemed to relate to an intrinsic difference in the two experimental tasks. The reaction time situation was entirely experimenter controlled, performance being measured from the point when the experimenter gave the signal to react. The tapping situation, on the other hand, allowed the subject a certain autonomy—of which, however, he was not aware. Although he was told to start when the ready signal was given, his performance was measured from the time he himself initiated the tapping. His own first tap activated the mechanism for counting taps.

A more recent auditory reaction time study [8] has relevance to this problem. The procedure called for three conditions: an autonomy condition in which the subject had freedom of choice of preparatory interval and freedom to start each trial on his own initiative; a *controlled* condition wherein the subject was told the length of the preparatory interval to come, but the interval was chosen by the

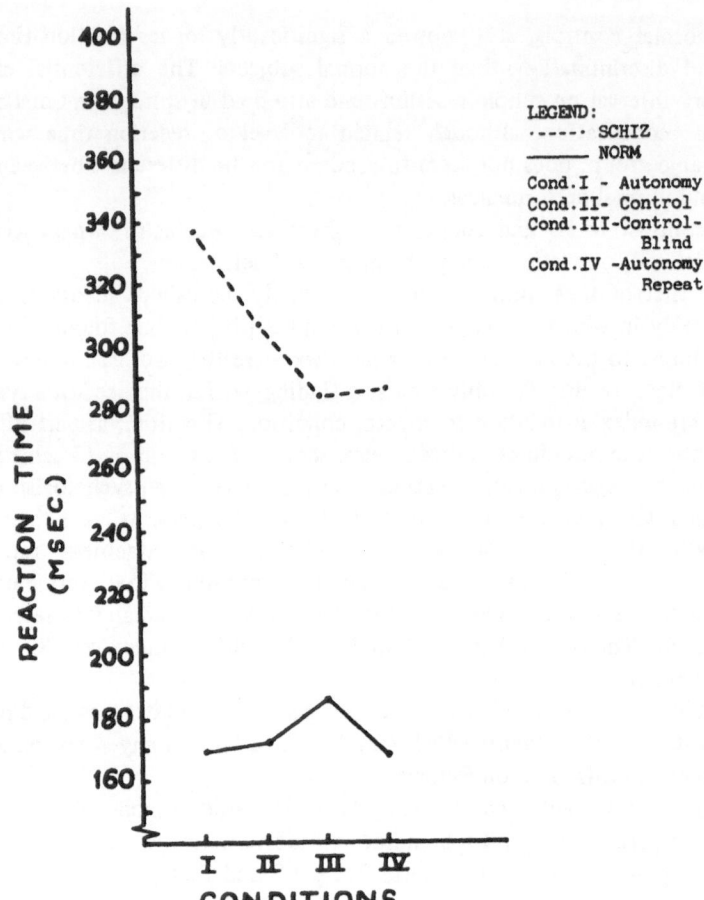

FIGURE 10. Reaction time as a function of experimental conditions I
(autonomy), II (control), III (control-blind), and IV (autonomy)
for normals and schizophrenics.

experimenter, and each trial was initiated by the experimenter; and finally the
control-blind condition which did not provide the subject with freedom to choose
the preparatory interval, information about it, or freedom to initiate the trials.
The last is, of course, the ordinary procedure in reaction time experiments.

Figure 10 gives, in order of presentation, the four conditions used (including
a repeat of the autonomy condition). The upper curve represents schizophrenic,
and the lower, normal subjects. It will be seen that the normal subjects performed
best under autonomy conditions and poorest under the control-blind; the schizophrenics, in contrast did their best work under the control-blind condition. An

index relating the autonomy and control conditions indicated that this difference between schizophrenic and normal subjects was significant at the one per cent level.

After each subject had completed the experiment, he was asked to state his preference for the autonomy or the control condition. The results indicated that preferences were in general consonant with performance in each group: the normal subjects stated a preference for the autonomy condition, whereas the schizophrenic subjects either preferred the control condition or had no preference. This difference in personal preferences of the normals and schizophrenics was significant at better than the one per cent level.

The conclusion that seems to come from these two autonomy experiments—the tapping and the reaction time studies—is that when the schizophrenic is made aware of his responsibility for a situation, he does most poorly. However, when he is unaware that he is in a situation of autonomy, he does surprisingly well.

Kent-Rosanoff Association

Another of our studies [14], parts of which were reported in Shakow [35], dealt with the material from still another area of function, the association test. Table 8 presents data on sixty schizophrenic patients who were given the Kent-Rosanoff Association Test five times at three to four month intervals. The categories of response call for some clarification. "Most common" refers to responses which fall among the hundred most frequently given by the Kent-Rosanoff standardization group, those having the highest frequency in the tables. "Individual responses" are defined as those subsumed under the classificatory criteria given in the Appendix to the *Test Manual* but not actually given as a response by any one of the thousand standardization subjects. "Unusual responses" are those which are neither in the tables nor subsumed under the categories of the Appendix. The "composite index" is a weighted score which takes into account all three measures, the higher index scores being the more aberrant.

It will be noted that there was practically no change in the mean scores of the schizophrenics over the period of 16 to 17 months which elapsed between the first and the last examinations. The intraclass correlation for the patients was .67. Unfortunately we do not have similar extended data on normal subjects. However, we did give two successive examinations two weeks apart to an equally large group of normal subjects. There was a significant drop in the composite index of these subjects, from 10.2 to 8.0. This occurred despite the fact that the intervening period was shorter, and the perhaps more important fact that in successive tests the normal subjects gave 42 per cent of their responses in identical form and the schizophrenics gave only 24 per cent of identical responses. The intraclass correlation for the normal subjects was .80.

Is there a common factor in these two situations—the reaction time and the Kent-Rosanoff Association Test—in which the schizophrenic appears to be particularly vulnerable? Although there may be many more aspects of difference than

TABLE 8. Distribution of Categories of Response on the Association Test of Sixty Schizophrenic Patients for Five Successive Examinations (3–4 Months Average Intervals) and of Sixty Normal Subjects for Two Successive Examinations (2-Week Interval)

	Examination	Most Common		Individual		Unusual		Composite Index	
		R	Mean and S.E.	R	Mean and S.E.	R	Mean and S.E.	R	Mean and S.E.
Schizophrenics	1	0–38	19.5 ± 1.3	0–32	8.0 ± 0.8	0–82	15.8 ± 2.3	1–66	20.7 ± 2.0
	2	2–37	20.4 ± 1.2	1–32	7.7 ± 0.7	0–71	13.2 ± 1.9	0–62	18.6 ± 1.8
	3	0–45	20.8 ± 1.3	0–29	7.5 ± 0.7	0–97	14.7 ± 2.2	1–58	19.2 ± 2.1
	4	0–44	18.7 ± 1.4	0–24	8.1 ± 0.8	0–77	16.9 ± 2.4	2–69	21.1 ± 2.2
	5	0–41	20.0 ± 1.3	0–21	7.4 ± 0.8	0–82	14.6 ± 2.4	1–72	18.9 ± 2.1
Normals	1	3–61	27.8 ± 1.5	0–15	3.7 ± 0.4	0–58	9.7 ± 1.5	3–35	10.2 ± 1.1
	2	1–63	28.9 ± 1.6	0–13	3.0 ± 0.3	0–55	8.2 ± 1.4	3–37	8.0 ± 1.1

commonality, one common aspect does seem to exist: *the need for quick response to a demand from the environment.* Our data have suggested that there is an optimal preparatory interval for response, depending upon the complexity of the task. In normal subjects the optimal period seems to be of considerable range, going down to quite small periods. For the schizophrenic, however, this range is quite narrow, falling at the smaller end of the scale, but not the very small. If the stimulus follows too quickly upon the warning, the schizophrenic subject finds difficulty in choosing among the numerous (in the context of the task, irrelevant) associations which the stimulus arouses. If the stimulus does not come quickly enough after the warning, then there is the opportunity for irrelevant stimuli to obtrude and delay the reaction. Only if the time relationships are just right can one get an optimal solution from the schizophrenic subject. This situation may hold for the schizophrenic when compared with normal subjects, or may hold relatively for the schizophrenic within his own low level of performance.

Aspiration

The findings of a study of aspiration behavior carried out by Radlo and myself [35] may provide some hint of the underlying dynamics behind the schizophrenic's reaction in the last two situations or, for that matter, in many of the variety of experimental situations that I have described. There were two parts to the experiment, both using the same subjects. In one, a pinboard was employed; in the other a tapping apparatus. Since the results were almost identical for both situations, I shall report only some aspects of the pinboard study. The apparatus was a Johnson O'Connor pinboard. The subject was required to place single metal pins as rapidly as possible in a board with a metal face having holes of the proper size for a smooth fit of the pins. After a practice period, he was asked to indicate how many pins he thought he could place in the allotted time, this estimate being his aspiration level. He was then instructed to place as many pins as he could within the time limit. This was his achievement level. The analyses were based on the means of ten aspiration/achievement scores per subject.

Table 9 gives some of the results from this experiment. The patients were classified into three types: Type One consisted of those patients who set an average aspiration level above achievement level of one or more pin placements; Type Two consisted of patients who set an average aspiration level equal to or up to one above their achievement level; Type Three were patients who set an average aspiration level lower than their achievement level. You will note the differences between the normal subjects and the patient subjects, particularly between the normals and the Type Two and Type Three patients, in their ability to tolerate failure. There was a decidedly greater percentage of failures among the normal subjects. The details of the experiment are brought out more clearly in Table 10. Here the actual responses to success and failure are given. In all cases the differences between normal and

TABLE 9. Success and Failure in Attainment of Aspiration Goals (Pinboard) of Schizophrenic and Normal Subjects

Group	N	Percent of successes	Percent of failures
Normals	20	24	76
Patients, Type 1	17	35	65
Patients, Type 2	13	59	41
Patients, Type 3	7	79	21

TABLE 10. Aspiration Responses after Success and Failure in Attainment of Aspiration Goals (Pinboard) of Schizophrenic and Normal Subjects

Group	N	Percent after success			Percent after failure		
		Up	Same	Down	Up	Same	Down
Normals	20	83	17	—	8	76	16
Patients, Type 1	17	76	22	2	6	54	40
Patients, Type 2	13	66	30	4	—	45	55
Patients, Type 3	7	36	55	9	—	40	60

patient subjects and between patient groups were statistically significant. Note particularly the reactions after failure of the patients.

What is the meaning of these data? It may be that the aspiration situation, because of its evaluative judgmental character and its relative clearness in indicating success or failure, approaches some of the dynamically important factors in schizophrenia in a more obvious and direct way. There is a possible relationship here between these studies and the studies of censure and implied censure by Rodnick and Garmezy and their students [28]. We perhaps have evidence here of the attempt of the schizophrenic to play safe, to avoid failure and affect, because these have played such important roles in the development and maintenance of his condition. These are, of course only speculations growing out of our findings. They are also, however, hypotheses which we must try to test experimentally if we are to discover the relationship between genetic developmental factors and the contemporary structure of the personality of schizophrenic subjects.

SUMMARY OF FINDINGS

General

In addition to the situation in which the patient withdraws and does not cooperate, we have considered a sample of experiments showing three types of findings about schizophrenic response in relation to normal response: (1) studies in which no differences from the normal were found from the very beginning—in which the schizophrenic patient responded adequately to the experimental situation; (2) those in which initial differences were found but, under certain conditions, tended to disappear; and finally, (3) those in which initially-found differences continued to persist despite the provision of special conditions.

Let me summarize, through Table 11, some of the generalizations from the experiments I have reported and others I have not taken the time to report. From this sampling of response, we can get a picture of the range of schizophrenic inadequacy.

I have already discussed the problem of set, the slowness of response as reflected in the reaction time situations, and the slowness of adaptation shown in the GSR and heart rate reactions, as presented in the first three items. While schizophrenics, when given enough time, appear in learning situations to be able to

---

TABLE 11. Areas of Schizophrenic Inadequacy

---

1. Longer to establish set; set difficult to maintain long—RT/interval; RT/ discrimination

2. Slowness of response—RT

3. Slowness of adaptation—GSR/simple stimulus; heart rate/pain

4. Slowness of learning—prodmeter and pursuit meter learning; Ferguson and Worcester $2C$ formboard; LI (Stanford-Binet)

5. Unrealistic perception—Rorschach: W-, F-, O-; Tachistoscope: No. error, uncommon error

6. Associative thinking—TAI (Stanford-Binet); TAS (Stanford-Binet)

7. Conceptual thinking—TC (Stanford-Binet); Alpha 3; Wegrocki

8. Weak goal behavior—interruption; substitution; play; $As \gg Ac$

9. "Weak ego"—tautophone: $1 < 3$

10. Loose affect—Rorschach: experience type

11. Individuality of response—Kent-Rosanoff; Rorschach: O, P

12. Variability—RT; prodmeter; tapping; etc.

---

achieve a "physiological limit" close to that of normal subjects, there is still a slowness in the learning process. Item four lists tasks in which we have found evidence for this: prodmeter and pursuit learning experiments; an extended series of repeated administration of two complicated form boards (one of the Ferguson series and one developed at Worcester State Hospital—the Worcester 2C); the immediate learning items of the Stanford-Binet. Schizophrenics as a group also showed (item five) a higher degree of unrealistic perception as was indicated by the larger percentage of minus responses in the "whole" (W), "form" (F), and "original" (O) categories of the Rorschach, as well as in the number of errors and the number of uncommon errors in a tachistoscopic experiment [1]. In associative thinking items on the Stanford-Binet (item six), whether immediate (TAI) or sustained (TAS), they also fall down as they do (item seven) on conceptual thinking (TC). Conceptual thinking difficulty was also revealed by results on Test Three of the Army Alpha, and in a special study [40] involving generalizing ability. Weaker goal behavior (item eight) was found in a Lewinian experiment on interruption [26] and in an experiment on play. Radlo and I [35] obtained similar results in the study on aspiration level. We have interpreted as "weak ego" response (item nine) the significant trend in schizophrenics toward giving third-person rather than first-person responses on the tautophone, and we have considered the higher-experience type index on the Rorschach (item 10) among the schizophrenics as evidence of "loose affect." "Individuality of response" (item 11) is quite characteristic of the schizophrenic, as we have already seen in the Kent-Rosanoff Association Test. It was also revealed by the greater number of "original" (O) and smaller number of "popular" (P) responses in the Rorschach. We have already commented on "variability" (item 12), a quality so highly characteristic of the schizophrenic group and so much more marked than in the normal.

Subtypes

I have already mentioned the considerable contribution of subtype to the striking interindividual variation, as well as the intraindividual variation—at least in the psychological realm—found among schizophrenic patients. As I previously pointed out, it was our practice at Worcester, on the basis of careful diagnoses and rediagnoses according to carefully outlined criteria, to use the standard subtype classification, but with a liberal use of the additional categories: mixed, unclassified, and indeterminate. The latter, which is a category we developed, involved a history of clear membership in one of the subtypes but with the symptoms now abeyant.

To give some inkling of the range of such variation in chronic patients, it may be helpful to depict a few aspects of the different personality patterns of what we came to consider the two major subtypes of schizophrenia—the paranoid and the hebephrenic. In 58 measurements which we made on groups of normal, paranoid,

TABLE 12. Paranoid-Hebephrenic Comparison

Paranoid	Hebephrenic
1. Intellectually preserved	Intellectually disturbed
2. Rigid	Loose
3. Persistent	Shifting
4. Limits environment	Broadens environment[a]
5. Responsive to personal meaning	Irresponsive to personal meaning
6. Accurate and cautious	Inaccurate and venturesome

[a] Either by being at the mercy of environment, or by establishing superficial contact with it.

and hebephrenic subjects, we found the paranoid to be nearer the normal in 31 instances and the hebephrenic nearer the normal in only seven instances. These findings related to measurements in which the normal appeared at one end of the distribution. In 20 instances in six quite varied experiments, however, the paranoid and hebephrenic subject scores fell on *either* side of the normals. Thus, these two groups seem consistently to deal with situations in distinctive ways, giving quite different "styles" of response.

Table 12 gives a highlighted qualitative profile of the two groups. The paranoid is quite rigid in his response, is relatively preserved intellectually, limits his environment with accuracy and caution, and has sufficient "pride" to protect his personality against the inroads of the environment. The latter is seen particularly in his mirror behavior, his play constructions, and his behavior in tachistoscopic experiments and in an aspiration study. His play constructions also show a sensitivity to personal reference. The hebephrenic subject, on the other hand, who is quite disturbed intellectually (actually he is at an intellectual level somewhat below that of the general paretic), is inaccurate and venturesome, and as seen in the same mirror and play situations, seems to be at the mercy of the environment, constantly being buffeted about by it. In play constructions he appears unresponsive to personal meanings. He consistently takes the "easy way" out, whether it be in preferring to do something that he has already done successfully, lowering his aspiration level after failure, or shifting from one situation to another without any plan. Even this much abbreviated account should lead to caution against any tendency to disregard the subtype classifications as they have evolved over time.

THEORETICAL CONSIDERATIONS

I thought at first that I would limit myself to the presentation of data. However, I found that I could not leave it at that—that I could not neglect theory entirely. I have elsewhere considered schizophrenia in relation to some theoretical problems [33] and more recently in the context of a theory of segmental set [36]. I would like to elaborate somewhat on these earlier discussions and present a few additional and complementary aspects.

Let us consider some of the implications for theory of the more specific findings we have described. It is in this connection that some aspects of the Coghillian theory of integrated-individuated action and Cannon's hypotheses concerning homeostasis are relevant.

Coghill studied the salamander *Ambylstoma* with particular interest in the development of the nervous system as related to behavior. The relevant theoretical possibilities lie in what he reported of the antagonism between the processes of integration of the whole and individuation of the parts. Let us use Coghill's own words of some thirty years ago from his Presidential Address to the American Association of Anatomists (1933, p. 136): "The mechanism of total integration tends to maintain absolute unity and solidarity of the behavior pattern. The development of localized mechanisms tends to disrupt unity and solidarity and to produce independent partial patterns of behavior. In the interest of the welfare of the organism as a whole, partial patterns must not attain complete independence of action; they must be held under control by the mechanism of total integration. Parts become integrated with each other because they are integral factors of a primarily integrated whole, and they remain integrated, and behavior is normal, so long as this wholeness is maintained. But the wholeness may be lost through a decline of the mechanism of total integration or through the hypertrophy of mechanisms of partial patterns." Coghill offers a suggestion which appears to fit much of what we have been describing for schizophrenia.

Likewise, Cannon's theory, which emphasizes the tendency of the organism to maintain stability at the vegetative level, even at the expense of economy of effort, also has importance in the present context. Cannon said (5, pp. 302-303): ". . . I have called attention to the fact that insofar as our internal environment is kept constant we are freed from the limitations imposed by both internal and external agencies or conditions that could be disturbing. The pertinent question has been asked by Barcroft, freedom for what? It is chiefly freedom for the activity of the higher levels of the nervous system and the muscles which they govern. By means of the cerebral cortex we have all our intelligent relations to the world about us. . . . The alternative to this freedom would be either submission to the checks and hindrances which external cold or internal heat or disturbance of any other constants of the fluid matrix would impose upon us; or, on the other hand, such conscious attention to storage of materials and to altering the rate of bodily processes, in order to preserve constancy, that time for other affairs would be lacking. . . . The full development and ample expression of the living organism are impossible in those circumstances. They are made possible by such automatic regulation of the routine necessities that the functions of the brain which subserve intelligence and imagination, insight and manual skill, are set free for the use of these higher services."

In the light of the findings I have described, it seems reasonable to add to Cannon's description of the wisdom of the body, a correlative description of the "wisdom of the mind." For Cannon has told only part of the story. The same trend towards freeing the higher centers appears to hold within the cerebrospinal system as well.* In the course of life, cerebrospinal activity, at first focal, gradually becomes peripheral, and finally becomes automatized. At the time of automatization, the central control necessary for the particular function becomes minimal and is thus released for higher activities. Only in times of emergency does it become focal again. In both the interofective and exterofective systems, this process is a result of experience and practice. Automatization comes earlier to the interofective system, apparently being established during the early years of childhood; in the exterofective system the process continues in varying degrees throughout life. Somewhere Whitehead has said that civilization advances by extending the number of important operations which we can perform without thinking about them. He had reference presumably to this process of automatization.

On the psychological side[†] we can see certain overlappings and analogies with the physiological phenomena considered by Cannon. At our present state of knowledge, however, satisfactory correlation of these aspects is, of course, difficult. But modern neurophysiology, with its extensive uncovering of neural areas apparently having crucial relationships to emotion and attention, its acceptance of a dynamic, constantly active complex of interconnecting systems, its emphasis on the high degree of interrelationship between the phylogenetically old and the neocortex, and the evidence it has uncovered for important internal (largely inhibitory) control mechanisms which exist throughout the system, is opening new vistas for correlation of the phenomena described above [34].

What, then, may one say about the general principles which seem to be behind the behavior of the schizophrenic as we have seen him? Without becoming specific about the actual physiological or psychological structures, I shall attempt to summarize these principles in line with my own thinking and with what seems to me the spirit of both Coghill's and Cannon's theories.

The main point is that in schizophrenia one sees a distinct weakening of the control center that serves the integrating and organizing function and provides for the establishment of what I have called "generalized" or "major sets."

---

*In Dempsey's [9] discussion of homeostasis the implications of this principle for the exterofective system are also considered. The broader implications of homeostasis are discussed at some length in various sections of the volume edited by Grinker [11].

[†]A detailed consideration of the physiological homeostatic difficulties of schizophrenic patients has been presented by Hoskins [12]. In the main, these data derive from essentially the same patients I have been considering.

Accompanying this weakening is a tendency for the individuated patterns Coghill has described—I have referred to them as "segmented" patterns—to come to the fore and become inordinately important. This may occur in both the interofective and exterofective systems of both the psychological and physiological realms.

The process is, of course, not a simple, straight-line change. There is much "to-ing and fro-ing." In fact, this is one of the striking qualities of schizophrenia. But in general, the trend is expressed through the following stages which I shall describe briefly. The behavior shown is, of course, not exclusively limited to the behavior predominantly associated with each particular stage. Much overlap occurs and it is only these prominent features that characterize a stage. There is first a tendency to split off into individuated patterns, followed by a reactive strengthening effort on the part of the central control mechanism to control these split-off systems. This is, in most cases, unsuccessful. Then follows a diminution of the direction of energies to the outside environment, and finally the establishment of equilibrium through the dominance of the segmental patterns.

Associated with these stages are a variety of both experimental and clinical manifestations, some of which we have already discussed: The marked variability of response that we have found reflected in lower correlation coefficients across psychological functions; the inability to maintain set that subsumes the variety of phenomena we have considered particularly under reaction time; the slowing of response time as well as of adaptation and learning time; the difficulties with sustained and conceptual thinking; the weak goal behavior; the unrealistic nature of perception; and the individuality of response—all these appear to be evidences of this tendency towards segmentalization, or sometimes of the attempt to overcome it.

I have not hitherto discussed the more clearly defined experimental physiological findings. It might therefore be desirable to present just a few to round out the picture. Among the findings relating to physiological function there are some which have always impressed me. I refer to a set of significant differences in correlations relating to blood pressure and to body temperature. In carefully controlled studies by our group at Worcester under Hoskins we found the following correlations between individual systolic and diastolic blood pressures [13].

Normal

Basal (N = 323)	.43
Nonbasal (N = 1398)	.45

Schizophrenic

Basal (N = 180)	.62
Basal (N = 100)	.62

In body temperature [23] we found correlations between the oral and rectal readings of individuals to be:

	Normal	*Schizophrenic*
Mean of individual correlations	.56	.73
Correlations of mean ratings	.41	.91

Thus, we see in these significantly higher correlation coefficients in schizophrenics, what seems to be a kind of "robotization"—a "hydrostatic" type of relationship in blood pressure and a "thermostatic" type of relationship in temperature. Both would appear to be reflections of a process of segmentalization that takes these functions out from under the normal adaptive and modulating control. These processes become "independent" and less amenable to the central control needed for most effective adaptation. Our own experiments on *GSR* and heart rate, together with the "physiological withdrawal" documented by Angyal, Freeman, and Hoskins [3], and the evidences for *high* central nervous system and autonomic nervous system *spontaneous* activity ("spontaneous" is here defined as the response called for by irrelevant stimuli) documented by Malmo, Shagass, and Smith [24], are further examples of this segmentalized reactivity. There is, on the other hand, evidence for *low* central nervous system and autonomic *directed* activity—"directed" being defined as the adaptive activity called for by a relevant stimulus. This is seen, for instance, in diminished nystagmic response to caloric and rotatory stimulation, in lessened "sway" response to rotation [2,4,10], and in other phenomena, many of which are described by Hoskins [12].

On the clinical symptomatic side we see parallel segmental phenomena which can best be described in clinical terms. Segmental cravings, that are ordinarily not satisfied while total integrated control is effective, are now satisfied. There results a perverted use of the already automatically matured devices of the organism to satisfy these needs. Clinical symptoms take the form of preoccupation with ordinarily unconscious bodily processes—with the mechanics of processes rather than with their ends—sensory disturbances represented by hallucinations, peculiar thought patterns, and delusions. This immense variety of schizophrenic symptoms can in one sense be viewed as different expressions of only partial integration, or individuation, or breakdown of major sets—in other words, of segmentalization. The defensive goal-seeking of the schizophrenic ranges from almost total disintegration to highly organized but localized patterns of behavior. Only very rarely is total integration in the Coghillian sense present. There is an increased awareness of, and preoccupation with, the ordinarily disregarded details of existence—the details which normal people spontaneously forget—train themselves, or are trained, rigorously to disregard. These, rather than the biologically adaptive functional aspects of the situation, appear to take on a primary role. It is only when a patient develops a persisting aversion to food because the cafeteria menu lists a common item which we read as "soup" but which he can only see in its excretory significance as "so-u-p," that we begin to realize how very many of the thousands of details of daily existence get by ordinary normals!

If we were to try to epitomize the schizophrenic person's system in the most simple language, we might say that he has two major difficulties; first, he reacts to old situations as if they were new ones (he fails to habituate), and to new situations as if they were recently past ones (he perseverates); and second, he overresponds when the stimulus is relatively small, and he does not respond enough when the stimulus is great. With regard to reactivity, the chronic schizophrenic is certainly not a "seething caldron." He resembles instead the "simmering pot" on the back of the stove which perpetually simmers at a low level. But it is a pot that does not ever provide one with a tasty *pot-au-feu*.

There is little doubt that the schizophrenic's is an inefficient, unmodulated system, full of "noise," and of indeterminate figure-ground relationships. What a confusing world must be the schizophrenic's when such basic modes of relating to the world are so seriously disturbed!

Nevertheless, we do at times see evidences for recoveries from some of these pathological characteristics of the psychosis. We have seen them, for instance, in the group of our own experimental studies that we have labeled "normalizing." Whether they appear spontaneously or as the result of "heroic" measures, we see them clinically in the not-too-rare occurrence of the "off-on" phenomenon—those instances in which the schizophrenic patient seems to have varying periods of relative clarity or normality. The central question for therapeutics is, then: What can we do to make these "on" periods persist? This, I am afraid, is a question that will remain unanswered for at least a few years more.

REFERENCES

1. Angyal, A. F. Speed and pattern of perception in schizophrenic and normal persons. *Charac. Pers. 11*:108–127, 1942.
2. Angyal, A. and Blackman, N. Vestibular reactivity in schizophrenia. *Arch. Neurol. Psych. 44*:611–620, 1940.
3. Angyal, A., Freeman, H., and Hoskins, R. G. Physiologic aspects of schizophrenic withdrawal. *Arch. Neurol. Psych. 44*:621–626, 1940.
4. Angyal, A. and Sherman, M. A. Postural reactions to vestibular stimulation in schizophrenic and normal subjects. *Am. J. Psych. 98*:857–862, 1942.
5. Cannon, W. B. *The wisdom of the body* (Rev. ed.). New York, Norton, 1939.
6. Coghill, G. E. The neuro-embryologic study of behavior: Principles, perspective and aim. *Science 78*:131–138, 1933.
7. Cohen, L. H. and Patterson, M. Effect of pain on the heart, rate of normal and schizophrenic individuals. *J. Gen. Psychol. 17*:273–289, 1937.

8. Cromwell, R. L., Rosenthal, D., Shakow, D., and Zahn, T. P. Reaction time locus of control, choice behavior, and descriptions of parental behavior in schizophrenic and normal subjects. *J. Pers. 29*:363–379, 1961.
9. Dempsey, E. W. Homeostasis. In: *Handbook of experimental psychology*, edited by S. S. Stevens. New York: John Wiley & Sons, 1951.
10. Freeman, H. and Rodnick, E. H. Effect of rotation on postural steadiness in normal and in schizophrenic subjects. *Arch. Neurol. Psych. 48*:47–53, 1942.
11. Grinker, R. R. (Ed.). *Toward a unified theory of human behavior*. New York: Basic Books, 1960.
12. Hoskins, R. G. *The biology of schizophrenia*. New York: Norton, 1946.
13. Hoskins, R. G. and Jellinek, E. M. The schizophrenic personality with special regard to psychologic and organic concomitants. *Proc. Assoc. Research Nerv. Ment. Disease 14*:211–233, 1933.
14. Huebner, D. M. Effects of repetition on the association test in schizophrenic and normal subjects. Unpublished Master's thesis, Johns Hopkins University, 1938.
15. Huston, P. E. Sensory threshold to direct current stimulation in schizophrenic and in normal subjects. *Arch. Neurol. Psych. 31*:590–596, 1934.
16. Huston, P. E. The reflex time of the patellar tendon reflex in normal and schizophrenic subjects. *J. Gen. Psychol. 13*:3–41, 1935.
17. Huston, P. E. and Shakow, D. Studies of motor function in schizophrenia. III. Steadiness. *J. Gen. Psychol. 34*:119–126, 1946.
18. Huston, P. E. and Shakow, D. Learning in schizophrenia. I. Pursuit learning. *J. Pers. 17*:52–74, 1948.
19. Huston, P. E. and Shakow, D. Learning capacity in schizophrenia; with special reference to the concept of deterioration. *Am. J. Psych. 105*:881–888, 1949.
20. Huston, P. E., Shakow, D., and Riggs, L. A. Studies of motor function in schizophrenia. II. Reaction time. *J. Gen. Psychol. 16*:39–82, 1937.
21. Huston, P. E. and Singer, M. M. Effect of sodium amytal and amphetamine sulfate on mental set in schizophrenia. *Arch. Neurol. Psych. 53*:365–369, 1945.
22. Knehr, C. A. Schizophrenic reaction time responses to variable preparatory intervals. *Am. J. Psych. 110*:585–588, 1954.
23. Linder, F. E. and Carmichael, H. T. A biometric study of the relation between oral and rectal temperatures in normal and schizophrenic subjects. *Hum. Biol. 7*:24–46, 1935.
24. Malmo, R. B., Shagass, C., and Smith, A. A. Responsiveness in chronic schizophrenia. *J. Pers. 19*:359–375, 1951.
25. Millard, M. S. and Shakow, D. A note on color-blindness in some psychotic groups. *J. Soc. Psychol. 6*:252–256, 1935.
26. Rickers–Ovsiankina, M. Studies on the personality structure of schizophrenic individuals. II. Reaction to interrupted tasks. *J. Gen. Psychol. 16*:179–196, 1937.
27. Rodnick, E. H. The effect of metrazol shock upon habit systems. *J. Abnorm. Soc. Psychol. 37*:560–565, 1942.

28. Rodnick, E. H. and Garmezy, N. An experimental approach to the study of motivation in schizophrenia. In: *Nebraska symposium on motivation*, edited by M. R. Jones. Lincoln, NE: University of Nebraska Press, 1957.

29. Rodnick, E. H. and Shakow, D. Set in the schizophrenic as measured by a composite reaction time index. *Am. J. Psych. 97*:214–225, 1940.

30. Rosenthal, D., Lawlor, W. G., Zahn, T. P., and Shakow, D. The relationship of some aspects of mental set to degree of schizophrenic disorganization. *J. Pers. 28*:26–38, 1960.

31. Sands, S. L. and Rodnick, E. H. Concept and experimental design in the study of stress and personality. *Am. J. Psych. 106*:673–679, 1950.

32. Schnack, G. F., Shakow, D., and Lively, M. L. Studies in insulin and metrazol therapy. II. Differential effects on some psychological functions. *J. Pers. 14*:125–149, 1945.

33. Shakow, D. The nature of deterioration in schizophrenic conditions. *Nerv. Ment. Dis. Monogr.* No. 70, 1946.

34. Shakow, D. How phylogenetically older parts of the brain relate to behavior. Paper read at the American Association for the Advancement of Science, Washington, D.C., December 1958.

35. Shakow, D. Normalisierungstendenzen bei chronisch Schizophrenen: Konsequenzen für die Theorie der Schizophrenie. *Schweiz. Z. Psychol. Answend. 17*:285–299, 1958.

36. Shakow, D. Segmental set: A theory of psychological deficit in schizophrenia. *Arch. Gen. Psych. 6*:1–17, 1962.

37. Shakow, D. and Huston, P. E. Studies of motor function in schizophrenia. I. Speed of tapping. *J. Gen. Psychol. 15*:63–106, 1936.

38. Shakow, D. and Rosenzweig, S. Play technique in schizophrenia and other psychoses. II. An experimental study of schizophrenic constructions with play materials. *Am. J. Orthopsychiat. 7*:36–47, 1937.

39. Tizard, J. and Venables, P. H. Reaction time responses by schizophrenics, mental defectives, and normal adults. *Am. J. Psych. 112*:803–807, 1956.

40. Wegrocki, H. J. Generalizing ability in schizophrenia: An inquiry into the disorders of problem thinking in schizophrenia. *Arch Psychol.* No. 254, 1940.

41. Woodrow, H. The measurement of attention. *Psychol. Monogr. 17*, No. 5 (Whole No. 76), 1914.

42. Zahn, T. P., Rosenthal, D., and Lawlor, W. G. GSR orienting reactions to visual and auditory stimuli in chronic schizophrenic and normal subjects. Paper read at Society for Psychophysiological Research, Denver, October 1962.

43. Zahn, T. P., Rosenthal, D., and Shakow, D. Reaction time in schizophrenic and normal subjects in relation to the sequence of series of regular preparatory intervals. *J. Abnorm. Soc. Psychol. 63*:161–168, 1961.

44. Zahn, T. P., Rosenthal, D., and Shakow, D. Reaction time in schizophrenic and normal subjects as a function of preparatory and intertrial intervals. *J. Nerv. Ment. Dis. 133*:283–287, 1961.

45. Zahn, T. P., Rosenthal, D., and Shakow, D. Effects of irregular preparatory intervals on reaction time in schizophrenia. *J. Abnorm. Soc. Psychol. 67*:44–52, 1963.

AUTHOR'S COMMENTS

On this occasion, almost 20 years after my talk, there has accumulated such an abundance of high grade research that it leads me to add some further words on the implications of segmental set theory for therapy and on its relation to other aspects of theory. These developments in schizophrenia research in the last several years have been in biochemical-genetic studies, neurophysiological studies, a variety of psychological studies, beginning ones in behavioral medicine, children-at-risk studies, and family studies. All have contributed in greater or lesser degree to progress in the clarification of issues, but some in a sense have helped to complicate them. However, one feature we must insist on. In the end the *individual* patient is the test. He is the one that is oftimes forgotten in the attempt to be "scientific" about our results.

The question of therapy in schizophrenia above all raises several issues which are of profound importance. On the one hand, there are the claims of the *echt* and extreme "organicists" who maintain that once the physical components of schizophrenia are discovered (genetic or otherwise, a claim which is, for them, only a matter of time, considering the amount of research devoted to it) the problems will be solved with the administration of a drug or some other organic recourse. At the other extreme, there are those psychologists and sociologists who cannot perceive what part the physical aspects may play in an organism which has gone through so many long years of schizophrenic habit formation by pathological factors in the person and in the environment. For them, the essential personality change could only come about through a surrogate reliving of the person's psychological development in an environment which, on the one hand, breaks down the already acquired pathological habits and on the other, helps to build up normal ones. This psychosociocultural approach, they claim, is the only effective device against schizophrenia.

On the whole, it must be said that both of these extreme views seem somewhat simplistic. We are dealing with a psychophysical or physiopsychological organism. In line with Eccles [3] this universe may be said to be made up of three worlds: World One, the world of the brain; World Two, the world of consciousness; and World Three, the world that uniquely relates to man. Psychopathy in all its various forms which are peculiar to it, is one of the characteristics from this last world which relates to man. The schizophrenic psychosis above all offer subtleties that the extreme "organicists" and the extreme psychologists-sociologists are at fault to presume their solutions as definitive. The schizophrenic needs to be treated as a total person, as an individual in his own right—both physiological and psychological—with all that it implies.

Although it is true that at the extremes of the distribution, one or the other of these views may hold. In most instances there is a good deal to be said for the combination of these extremes. For me, the question becomes, how then can segmental set in schizophrenia be overcome: *once psychopharmacological or other physical intervention restores the capacity for integrated functioning, generalized set may not appear automatically but must be learned and encouraged in a variety of ways.*

To overcome segmental set one must deal with two kinds of anxieties and needs: *secondary anxiety* that has accumulated through the longstanding preoccupation with segmental ways of behaving—the person has hoarded up a vast accumulation of inadequate reactions and anxieties which have to be dealt with and *basic anxiety*, which is related to the underlying inability to meet current archaic needs. *Archaic needs* are represented by habit formations of an old kind, which require the breakdown of the segmental acts (which, in turn, must be overcome). The new *underlying needs* are constructive needs and habits which take the place of the old needs and habits and lead to the development of generalized sets.

Stated in another way, the extensive disintegration observed in the schizophrenic, with its inevitable accompaniment of what I call segmental sets, does not necessarily imply utter pessimism. His strange behaviors carry their own kind of literal or symbolic meaning. Schizophrenic behavior, usually split-off acts of the moment, whether in the form of frights, flights, or aggressions, often indicate a reaction to underlying anxiety and deep frustration towards what the schizophrenic perceives as a lack of sympathy and understanding on the part of the environment. These feelings grow out of his underlying unsatisfied need for *love*, the never-met need for primitive security that became established during the very earliest of infantile relations, perhaps because of genetic inadequacies, or those connected with feeding and mothering. Schizophrenics, even when chronologically adult appear to have a peculiar and exaggerated need for complete acceptance which has to be demonstrated to them in "infantile" ways, to meet the basic infantile needs which they believe they have never experienced, as is evidenced by the cases of Sechehaye [8] and Hayward [4] and Milner [6]. In addition, there are the Schiff [7] and Honig [5] cases. In each of these, it is demonstrated that profoundly anxious and frightened persons can only achieve security at a slow rate— often maddeningly slow—through the gradual development of the feelings of safety that come when understanding and love are provided, even to the extent of surrogate breastfeeding by persons in their environment whom they consider sufficiently strong to provide them [7].

In the miasma of schizophrenic disturbance, there are islands of integrity which seem to be untouched. One may, with effort, find underlying capacities for normality, both physiological and psychological. We have seen them, for instance, in our "normalizing" studies [9] which provided evidence that schizophrenics possess functions which can reach a normal capacity level and that there is therefore something to work with, even though the levels achieved are limited to the functions studied and are not transferable. In the clinical situation the "off-on" phenomenon is not uncommon where the schizophrenic patient has periods of relative clarity or normality. These may be the result of psychotherapy, or even of "heroic" shock or drug measures. And, finally there are those rarer instances when recovery, or even a "cure," appears to have taken place, a "cure" which may actually make the schizophrenic more mature than he was before his illness. The ego of the patient appears to have been rescued and reorganized. How has this come about? How has the organism after such marked disintegration become reintegrated?

Many have a faith in the power and magic of drugs, analogous to the giving of L-dopa to some patients with Parkinsonism. I have frequently wondered if drugs alone [10] aside from their often great advantage in breaking through the formidable barriers of schizophrenia, are sufficient, even at their best, unless they are also accompanied by the underlying sympathy and understanding (biochemists and physiologists especially note!) that the building of any human relationship calls for. Only in this way, through the encouragement gained by a lack of overwhelming anxiety, can generalized set occur.

Others have a faith in the power and magic of a deep psychotherapy, one that recognizes the need for going through the complex process of rebuilding the person through the successive steps of normal development: imitation, introjection, identification, and differentiation, in both their literal and symbolic aspects, carried out in surroundings of love and strength. It is only in this reconstructive way that the tendency toward segmental set can gradually be overcome, and the ultimate conceptual strengths of generalized set established. (What is being reestablished is Allport's [1] generalized set to "perceive accurately and respond appropriately" in its fullest meaning—in other words, to conceptualize properly.) The process is desperately long but the rewards are great.

What we can say more specifically in the earlier phases of the treatment, especially when the patient is much distraught? Drugs may be used to quiet him and reduce his anxieties, particularly of the secondary kind. But this should be accompanied by psychotherapy as a means of dealing with the primary anxiety, breaking down old interpersonal ways of behaving and developing new ones. It has been stated clearly by Dyrud and Holzman [2]: "Drugs will reduce thought disorganization, quiet an unruly and excited patient, or mobilize a withdrawn patient. It remains for psychosocial interventions to teach and to train, to reassure and to raise self-confidence, and to help with skills for living that some patients may never have learned or may have learned badly."

What is learned is interpersonal skills which are dependent upon generalized sets: thought disorder is reduced, social capabilities are reactivated, gradually the lack of self-confidence is erased, and the skills for living which for so long have been in abeyance are built up again. The underlying anxiety has been reduced and permits generalized set to gain prominence. But it must again be emphasized that it is a long, drawn-out process because the schizophrenic disturbance is profound.

REFERENCES

1. Allport, F. H. *Theories of perception and the concept of structure.* New York: John Wiley & Sons, 1955.
2. Dyrud, J. E. and Holzman, P. S. The psychotherapy of schizophrenia: Does it work? *Am. J. Psych. 130*:670–673, 1973.
3. Eccles, J. C. *The understanding of the brain*, 2nd Ed. New York: McGraw-Hill, 1977.
4. Hayward, M. L. and Taylor, J. E. A schizophrenic patient describes the action of intensive psychotherapy. *Psych. Quarterly 30*:211–248, 1956.
5. Honig, A. *The awakening nightmare.* Rockaway, NJ: American Faculty Press, 1972.
6. Milner, M. *The hands of the living God.* New York: International Universities Press, 1969.
7. Schiff, J. L. and Day, B. *All my children.* Philadelphia: M. Evans & Co., 1970.
8. Sechehaye, M. *Autobiography of a schizophrenic girl.* New York: Signet Books, New American Library, 1970.
9. Shakow, D. Normalisierungstendenzen bei chronisch schizophrenen: Konsequenzen für die theorie der schizophrenie. ("Normalization" trends in chronic schizophrenic patients: Some implications for schizophrenia theory.) *Schweiz. Z. Psychol. Anwend. 17*, 285–299, 1958.
10. Wyatt, R. J. Biochemistry and schizophrenia. Part IV: The neuroleptics—their mechanism of action: A review of the biochemical literature. *Psychopharmacol. Bull. 12*:167–242, 1976.

Antecedents and Continuities in Schizophreniform Behavior

Eliot H. Rodnick, Ph.D.

INTRODUCTION

Several lines of research in recent years consistently and empirically support the view that the conditions underlying schizophrenic behavior clearly implicate interpersonal relations within the family [14,15]. Even though the identification of factors etiologically significant for schizophrenia, as they may be ultimately delineated, probably will involve the interplay of many conditions not necessarily independent of one another, it is very likely that the familial environment will be found to be of prime importance in eliciting schizophrenic behavior. Yet such a statement is too nonspecific to have much use for mapping the factors contributing to the etiology of schizophrenia. We are still at the stage of identifying and elaborating the conditions that appear to have some reasonable probability of contributing to its development. Systematically exploring some of the variables and the conditions surrounding them in any particular investigation need not necessarily imply that more than a limited portion of the variance is being investigated. At this stage of development of our knowledge about schizophrenia, wise strategy still favors the systematic parametric exploration, in some depth, of quite disparate domains of variables, for which there is both empirical evidence and theory to indicate their probable involvement in the underlying mechanisms. The investigator, in following up a particular strategy of empirical inquiry, may choose to avoid

Research in the Schizophrenic Disorders: The Stanley R. Dean Award Lectures, vol. 2, edited by R. Cancro and S. R. Dean. Copyright © 1985 by Spectrum Publications.

conceptual overload by temporarily ignoring other variables for which evidence exists that they also may be important, but which would greatly complicate the research if they were prematurely brought into a research plan which is poorly designed for those variables. In other words, a research strategy which prematurely seeks to explain too much, may end up accounting for too little.

This paper is concerned with describing aspects of a research strategy which seeks to identify some factors which may be related to the development of schizophrenia. From the standpoint of empirical evidence, it is at present perhaps premature and presumptuous to discuss the actual significance of such factors for the etiology of schizophrenia. We are merely seeking to identify some consistent regularities in the interpersonal environment of schizophrenics which appear to co-vary with the incidence of the condition and to its outcome. Whether they are causally related will depend on the evidence from other studies which are specifically designed for that purpose. If factors can be identified in the interpersonal, familial environment of the premorbid phase of adult schizophrenics, which are consistently found to relate to the occurrence of schizophrenic behavior, we can then seek to delineate the attributes of those adolescents who are exposed to comparable intrafamilial environments. Is there similarity in the behavior of both groups? Does the behavior of such adolescents or young adults appear to be prodromal for schizophrenia? Such a strategy, if workable, could serve as a step toward identifying attributes of adolescents who may be high risk for schizophrenia as they move into adulthood. There are problems, however, with this strategy. There are too many difficulties involved in gaining access to the privacy of the intrafamilial environments of families, particularly those which may resemble the familial attributes reported for adult schizophrenics.

An alternative is to investigate aspects of the intrafamilial environment of adolescents who are already showing sufficient difficulty in adjustment to warrant referral to a clinic for help. We can then investigate the relationship between the patterns of the interpersonal environment of this sample of families and the type of disturbance the adolescent shows. Are there consistent similarities or systematic differences in familial relations for the various patterns of adolescent disturbance? For those who show behavioral attributes which bear some similarity to those of schizophrenics, how comparable are the intrafamilial environments? Do these adolescents constitute a group which is high risk—or vulnerable—for schizophrenia later on as they move into adulthood? What if they were to be followed over a period of years? One advantage of investigating a relatively random sample of adolescents seen in a clinic is that we have controls already at hand who can be studied as intensively as those who may be showing schizophreniform behavior. Our controls can be contrasting intrafamilial environments, or contrasting adolescent behavior.

Our group at UCLA took as a starting point the evidence from a variety of sources [4,12,17], which indicates that the intrafamilial environment of schizophrenics is probably discriminably different from that of other psychopathological conditions, and that the specific attributes of this intrafamilial environment are probably antecedent to the development of schizophrenic behavior. The basic model we were elaborating was that if this intrafamilial environment is a precondition for the occurrence of at least some forms of schizophrenia, it should be found to occur in adolescents with a high risk for schizophreniform behavior. Not all children with such intrafamilial environment need necessarily develop schizophrenia, since vulnerability for schizophrenia is a function of many attributes ranging from genetic liability, biochemical and physiological dysfunctions, early experiences, through any pattern of circumstances which may shape the development of social competence. We were merely seeking to establish at an empirical level whether those adolescents with behavior that bears some similarity to that of the prodromal schizophrenic have characteristic intrafamilial environments comparable to those reported for schizophrenic patients.

The research described here is the work of a group at UCLA associated with Michael Goldstein and myself. For the early phases of this work we were joined by Drs. Lewis Judd, Edward Gould, and Armand Alkire. The group at present includes Drs. Jerome Evans, Sigrid McPherson, Kathryn West and several graduate students.

RESEARCH DESIGN

We decided at first to study adolescents between the ages of 13 and 18 in interaction with their parents under standardized and controlled conditions which would assure reasonable comparability across subjects. The adolescents were a random sample of clients referred to the Psychology Department Clinic. Most were self-referrals by their parents because of difficulties in school, behavioral maladjustment within the family, or because of pressure from the juvenile court and probation authorities. The only restriction on the sample was that the family be intact and that the adolescent and both parents come together to the clinic for at least five to six sessions for an assessment period, after which recommendations for referral or treatment would be discussed with them.

We were aware of the risks of bias in this sample and its probable nonrepresentativeness of disturbed adolescents in the community. We decided nevertheless to accept the bias, with its restrictions on the subject sample. The requirement of intactness of family, and the cooperation of the parents and adolescent involved in coming together for a series of sessions, probably results in selecting a sample which favors more passive adolescents and more involved parents. We hoped to compensate for the bias, by comparing one segment of the sample with another, using as the independent variable the pattern of behavioral disturbance or particular patterns of parent-child interaction. At first our intent was to use four-person families: parents, target child and a comparison sibling, but postponed doing so until the procedures were developed, pretested, and found to be sufficiently productive to warrant the additional complexities introduced by adding the sibling as a within-family control for the target child in the particular set of techniques we were employing. The first 50 families were hence three person families, but we are now expanding the design to permit us to study four person and in selected cases even five person families.

We decided to explore the interaction within the family by having the members interact with one another in role-playing their actual spontaneous responses to problems which both child and parent had identified as significant issues for either or both of them. In addition, they were important enough to be closely related to the reasons for their having come to the clinic for assistance.

The heart of the procedure, which covers Sessions two through five, involves an identical interview given to each person separately, parents as well as child, covering some eight areas of child–parent interaction which might provide a basis of conflict between them. The objective was to elicit idiosyncratic conflicts which were characteristic of the child and his parents, intense enough to evoke defensive coping behavior, and yet under sufficient control to keep them in the situation with little regulation by the investigators. The conditions for eliciting the material were standardized, and in a form which permitted objective assessment of the attributes of the interaction. The areas of interaction tapped by the interview were achievement, sociability, responsibility, communication with parents, response to frustration, autonomy, sex and dating, and overall family tone. During each area of the interview, the interviewer probes for conflicts specific to each family. Once a specific problem has been identified for the area, the parent or child, as the case may be, is instructed to role-play a specific instance of this problem with the intent of attempting to influence the other family member. That is, he addresses the other person, as if he were present, and as the problem existed at that particular moment. These cue statements are recorded and serve as the stimuli to which the other family member is subsequently asked to respond. We also obtain from the subject the actual words he expects the other person to use in response to his own remarks. A comparison with the actual remarks which the responder makes later provides a measure of how well parent and child know one another, as well as of compliance and yielding under conflict.

These recorded cue statements and expectancies of each directed toward the other provide the materials for the following session. At that time, the edited cue statements are played to the appropriate member of the family. As he hears the problem over the speaker, in the actual voice of the member of his family directing remarks to him, his task is to respond as if the situation were an actual one, with the other person present. His responses are recorded, both before and after he hears what the other person expected him to say.

This array of cue statements, responses, and expectancies provides the basis for a family interaction in a free modification of the revealed difference technique in which the various members of the family are brought together in dyads and a triad to hear cue statements and responses relevant to the dyad. They are instructed to discuss for a maximum of five minutes the simulated interaction they had just heard, and in a discussion with each other to indicate why each said what he did and how he feels about it. They are then to initiate a discussion of how to resolve the problem issues in a mutually satisfactory manner. Since we elicit and record a number of problem situations involving each member of the family, we are able to use different problems in each dyad and in the triad which are relevant to the particular family members concerned. Since each tape segment contains material that one of the members of the dyad has not heard before, this confrontation interaction has a considerable amount of spontaneity and novelty. The confrontations are videotaped, and provide the material for the next session, when the members of the family view the videotaped interactions of the session. At that time they rate one another in segments of the dyads and triads as they appear on the videotape of the previous confrontation session. The rating instrument used is one developed by our colleagues, Kaswan and Love [10] as an extensive modification of the Osgood semantic differential.

The conditions for making the ratings also serve to introduce a guided discussion period which follows when they discuss the behavior of each person on the video screen in objective terms by avoiding the use of personal pronouns. Each, in turn, must talk about what he observes about himself (but as "That child," "That woman," etc.), then about each of the others. Finally, each discusses what he would like changed in himself, and in each of the others.

In essence, then, the procedure is designed to bring into the laboratory samples of personalized conflictive interaction among members of the family, under the same conditions of eliciting the conflict and with a suitable balancing of roles of initiator and responder for each member of the family. We hoped that this design would permit a comparison of mother and father in the role they played in relating to the child in this sample interaction. We also hoped that we would be able to assess the comparative strengths of the child's relationship to each parent. One aspect of the design appeared very important to us. We wished to minimize as much as possible the appearance that we were allies of the parents against the adolescent—from his viewpoint perhaps allies of the aggressor. We therefore took special pains to ensure that parents were subjected to the same interview as the adolescent, and that he had as much of an opportunity as his parents to be the initiator of cue statements. On the basis of the degree of cooperation we have enjoyed from our adolescents, we believe that we were reasonably successful.

ATTRIBUTES OF THE DISTURBED ADOLESCENTS AND THEIR PARENTS

Since the objective was to explore the specificity of the interaction between parents and adolescents for various patterns of adolescent pathology—particularly of those who might show schizophreniform behavior—the design was dependent upon some method for characterizing the adolescent sample in terms of subgroups.

In order to avoid preconceived biases regarding a suitable nosology for characterizing adolescent psychopathology, which might force too heavy a reliance on a formal mental status examination which could bias the research conditions, we decided to be *crassly* empirical. Each parent was asked at the time of admission to write down briefly and succinctly in a paragraph, his perception of the child's problem which brought him and the adolescent to the clinic. We found that these statements could be sorted into four groups with satisfactorily high reliability. For relatively few of the cases was there sufficient ambiguity in making the sort to force arbitrary group assignments.

Some of the emerging findings for the first 24 to 30 families have been reported in a series of papers on social power and influence, galvanic skin responses in the confrontation, analysis of TATs, and family ditantiation and closeness in the triadic interaction [1,6,7,13]. We have found significant discrimination for a variety of characteristics among these four symptom groups of adolescents comprising our research sample. Based on an overall impression from the parents' description of the child's problem, the four groups were:

Group I—Aggressive, antisocial, characterized by poorly controlled, impulsive and acting out behavior toward authority in the community.

Group II—Active family conflict, characterized by a defiant, disrespectful stance toward the parents.

Group III—Passive-negative adolescents, characterized by negativism, sullenness and indirect forms of hostility toward parents or school authority.

Group IV—Withdrawn, passive, isolated adolescents, characterized by marked social isolation, general uncommunicativeness, and excessive dependence on one or both parents.

This last group comes closest to the poor premorbid schizophrenic, as identified by the Phillips Scale, even though most of the cases were clearly free of overt psychotic signs at the time they were seen. Since the clinic is a nonmedically oriented clinic, clients with florid psychotic or somatic symptoms are unlikely to be referred to or accepted by the clinic. In other respects the cases are fairly typical of the usual run-of-the-mill community clinic. Since it is housed on a university campus some distance from low-income areas, there tends to be a bias toward a middle class clientele, despite efforts to reduce the bias.

James Armstrong, Dr. Kathryn West and I have recently analyzed in considerable detail statements of the problem descriptions for the first 50 families. It is clear that the discrete phrases comprising the statements can be readily classified, with high reliability into the four sets of child attributes, which at first had been classified on an overall general impression. When the phrases were sorted out for the descriptive information conveyed, these short paragraphs could be classified into child attributes (as reported by the parent) and parent attributes as each parent may have commented on his role in the child's problem. We were prepared for the likelihood that a fine-combed analysis of the statements would even more sharply distinguish among the four groups of children. The attributes as reported by the parents of each of the four groups could be used as a checklist, and then integrated to arrive at an overall description of each of the adolescent groups. We were not prepared, however, for the amount of information these brief problem statements conveyed about the parents. We found that a comparable checklist of parent attributes could be developed which was derived solely from the terse problem-description paragraph which appeared to differentiate the adolescents into the four groups. It is particularly relevant that these statements of the parents regarding themselves, and their relation to the child's problem are entirely spontaneous and gratuitous. They were merely asked to write a brief description of the child's problem which brought them to the clinic.

Some of the flavor of the correspondence between the child attributes and the parental attributes for each of the four groups is indicated by the categories into which the phrases could be sorted for both child and parent. The summary of the problem description statements for each of the four adolescent groups with an actual example follows.

Group I

Child Attributes	*Parental Attributes*
Core quality is one of externalizing and acting out; this may be expressed in any of six ways:	Core quality is one of distantiation of relationship between parent and child; this is expressed in three areas:
Ic1 Opposition to and/or rejection of authority in general, including parents.	Ip1 Parents lack the role-appropriate techniques for controlling the child; they have no sense of parenting.
Ic2 Rejection of the avoidance of parents, specifically.	Ip2 Parents report a total, rather than partial, absence of meaningful communications and/or interactions among the members of the family; tension in the home in general.
Ic3 Antisocial (illegal) behavior; inability to follow rules of social conduct.	
Ic4 General truculence and hostility; initiated aggression; impulsive and/or unmodifiable behavior, with lack of concern for consequences; self-concept of the "born loser."	Ip3 Parents externalize responsibility for problems; correction is expected to come from an external agency; the suggestion that causality may lie within the family is not the idea of the parent,

Group I (Continued)

Child Attributes	*Parental Attributes*
Ic5 Inability to accept responsibility.	Ip3 (Continued) but is suggested to him by an external agent; if the parent calls for help it is to solve his problem rather than to abet his serving as a change agent for the child.
Ic6 Drug taking as part of an anti-social syndrome.	

Example of Group I

Family 07 (Father-Son)

"Serious indications of my son, . . . , who, while having excellent potentials, seems to be rebelling or has little respect for that which we consider proper—that is—respect for law, order, proper authority, school, church, parents, etc. Feels rules and regulations do not apply to him."

"I would like to have his thinking and attitude readjusted so that he can become a worthwhile citizen and reach his capabilities, rather than see him get progressively worse which we have been told will occur unless corrections are made."

Group II

Child Attributes	*Parental Attributes*
Core quality is one of rebelliousness and opposition to parental controls or regulation (and perhaps to school), with active struggle for independence, and concern for the dissonance created by the struggle; this is expressed in four ways:	Core quality is the expression of affiliational ties between parent and child; psychological mindedness; empathy; concern for child; sense of personal responsibility; this is expressed in seven areas:
IIc1 Overt acts of rebelliousness and opposition to parents; turbulence in home between parents and child.	IIp1 Recognition that child has personal problem to solve.
IIc2 Requests and/or demands that parent give (or permit) more freedom.	IIp2 Expression of desire to help child solve his problem for his own sake.
	IIp3 Recognition that as parent he is causal or contributory to the

Group II (Continued)

Child Attributes	*Parental Attributes*
IIc3 Indications of child's concern about family discord.	IIp3 (Continued) child's problem; self-critical as parent.
IIc4 Variability in the child's expression of love or warmth toward the parent.	IIp4 Expression of desire to receive help personally in order to be able to help child.
	IIp5 Recognition of parental responsibility for tutelage and direction, and to control the child with understanding.
	IIp6 Expression of concern over difficulty in communicating with child.
	IIp7 Expression of desire for increased closeness as a family.

Example of Group II

Family 04 (Mother-Son)

". . . is not happy. He is insecure and unhappy at home and in school. He feels that his parents are very overbearing and tend to treat him like a baby. He doesn't exhibit maturity enough for independence but feels he should be completely free of parents. We want to find out what we are doing wrong as parents. We also feel . . . has deep psychological problems which we hope can be uncovered. We feel . . . needs a *great* deal of help and that we as parents may also need a great deal."

Group III

Child Attributes	*Parental Attributes*
Core feature is oppositional behavior in school and/or home:	Core feature is one of trying to control the child, to manipulate him, entitled by right of parenthood:

Group III (Continued)

Child Attributes	Parental Attributes
IIIc1 Oppositional tendencies, including non-cooperation and manipulative behavior, outside the home, primarily in school.	IIIp1 Parental effort to control child is through manipulation or material rewards and the effort fails; parental role is viewed in legalistic manner rather than with empathic qualities.
IIIc2 Oppositional tendencies, including non-cooperation and manipulative behavior in home; may include overt anger or temper display.	IIIp2 Outside agents (school, clinic, drugs) seen as responsible for either causing child's difficulty or correcting it; parents are not seen as responsible.
IIIc3 Antisocial behavior occurs only if entirely a function of drug taking.	IIIp3 Parents feel unappreciated.

Example of Group III

Family 11 (Father-Son)

"I feel that our son should be doing better in school subjects; that he should be able to stay away from marijuana peddlers at school; and to keep out of trouble from his instructors at school—I can't seem to improve his study habits; work habits; my efforts to control him just don't seem to work."

Group IV

Child Attributes	Parental Attributes
The core features are withdrawal and passivity:	Core feature is parallelism between parents and between parent and child, rather than direct interaction between parent and child; expressed along main dimensions:
IVc1 Withdrawal is expressed by asociality and isolation, especially from peers and from members of the opposite sex; withdrawal may manifest itself within the home by child isolating himself from family.	IVp1 Parents view their role as agent for solving child's problem but not themselves as part of the problem; they seek assistance in finding the reason or explanation of the child's problem; allusions to other parent impersonal (that
IVc2 Secondary to withdrawal, child seems preoccupied with and/or	

Group IV (Continued)

Child Attributes	*Parental Attributes*
IVc2 (Continued) devaluing of self; he is reported as insecure or unhappy, within the context of being passive and dependent.	IVp1 (Continued) is, his mother, rather than my wife and I).
IVc3 Secondary to withdrawal, child behaves eccentrically and/or expresses concern over loss of control (insanity).	IVp2 Parent's attitude toward child is one of infantilizing him; they describe or judge him to be grossly immature.
IVc4 Passivity is expressed by non-involvement or severely limited involvement, by low level of energy, or by distractibility and lack of persistence.	IVp3 Parent alludes to own history in reference to child.
IVc5 Passivity extends to participating in school, but through refusal to attend school rather than active rebellious behavior about school.	
IVc6 Child is source of unhappiness and "pain" experienced by other family members; parents report that their efforts are rejected or unappreciated.	

Example of Group IV

Family 06 (Father-Daughter)

"Our daughter is withdrawn and unable to actively relate to her peers socially. She has worked hard to overcome this but seems only to reap rebuffs and heartache. She has no real friends, social life or close relationships at all with the opposite sex. Our desire is to locate the roots of her problem to free her and allow her to relax into a more nearly normal life."

There are consistent differences among the allusions to parental behavior—the underparenting and distantiation of Group I; for Group II the *expressed parental concerns* for closeness and affective involvement with the child in their day to day relationships, while being troubled by what they consider to be premature efforts at independence striving of the child; the stress on control and manipulation of Group III; and the special quality of blandness of affectivity in the interaction of parent-child in Group IV, combined with a criticalness towards the immaturity of the child.

In order to simplify the discussion of the issues, this discussion will concentrate primarily on Groups II and IV. Comparing parental attributes of these two groups, we find some striking differences: when the parents of Group II refer to themselves, they tend to stress the existence of and need for affiliation ties between themselves and their child. There is some recognition of their parental responsibility, and that they may be causal or contributory to the child's problem. In contrast, implicit in the statements of Group IV parents is a parallelism rather than direct interaction with the child. They do not recognize themselves as part of the problem, but rather see themselves as the agent to solve the child's problem by calling attention to some explanation of the child's behavior, or some intellectualized causal factor. Though both are concerned with the child's immaturity, this concern is expressed quite differently. In Group II, the parents feel the child is not ready to be granted the independence he is demanding; in Group IV, they are often critical of the child for not being more mature, yet they continue to infantilize him and to keep him dependent.

These differences among the parents of the four groups which are reflected even in these inadvertent miniphrases which slip into the description of the dominant attributes of their children, come through consistently in various interactions in the sessions themselves. Three quite separate analyses which have been reported elsewhere corroborate this finding of differences between Group II and Group IV families.

In order to keep the contrasts among the various patterns of adolescent disturbance in focus for the purposes of this paper, the comparison will be restricted primarily to Group II and IV families although consistent differences have been found among the four groups. For both groups the presenting problem tends to center in the home situation, rather than as rebelliousness or hostile behavior in school or with the juvenile authorities.

In a report on the TAT stories of the parents and child of the first 24 families [6] we commented:

"Parent of the active family conflict (Group II) . . . adolescents appear willing to see a family unit, but they see it full of *both* negative and positive involvements. This suggests that a high degree of emotional involvement is perceived among the family members, but one which may be ambivalent and conflictual . . .

"The withdrawn children (Group IV) were willing to perceive family relationships on the cards but indicated that the familial involvements they saw were almost invariably harmful and detrimental. These two characteristics did set them apart from the other groups of adolescents. Their parents similarly saw the involvement between the characters on the cards as being universally negative. The perceptions by the fathers and mothers of the withdrawn adolescents were quite different even though each parent projected relationships in whch one family member was excluded. In their stories, the fathers often left out the child, while the mothers often eliminated the fathers. This struggle on the level of a projective test to see unilateral alliances within the family is certainly compatible with clinical studies on families of schizophrenics. . . . This suggests that familial conflicts over alliances may stimulate the child to use withdrawal as a means of coping with this type of family conflict" (p. 363).

In another report [1] on the analysis of social influence and control methods used in the cue statements and responses in the role playing phase of the research, we commented:

"Parents of adolescents whose problems are manifest primarily in the home setting (Groups II and IV) avoided power-assertive techniques, and relied primarily on the indirectness of informational power. . . . Active family conflict adolescents do not discriminate between parents and expect *neither* parent to yield to him. . . . The most striking discrepancy is in the withdrawn group in which the adolescents expect their fathers to be quite weak but that their mothers will be quite resistant to compromise. . . . The maternal social influence style in the latter group involves pervasive restrictive questions, a more covert style of control which in turn is less apt to lead to direct confrontation by the child. . . . The pattern for the withdrawn group is particularly interesting by virtue of the resemblance to the subtle maternal domination and paternal passivity noted so frequently in studies of (poor premorbid) schizophrenics. . . . In the active family conflict group, the child is confronted with subtle maternal power in the form of restrictive questions similar to those received by the withdrawn child. However, unlike the latter, the father seems to ally with the mother and both parents fail to yield. It is this non-yielding by the father which distinguishes this group from that of the withdrawn child. . . (We commented at that time that) it is interesting to note that both of the within-home groups (active family conflict and withdrawn groups) are characterized by maternal use of restrictive questions which in essence forces the adolescent to justify his behavior. It raises the question of whether this style may not be a condition for increasing guilt and ultimate internalization of conflict" (pp. 39-41).

(This heavy use of questions has been reported by others for the families of schizophrenics [11].)

Finally, McPherson [13], in her analysis for the first 28 families of the "intents" toward each other expressed in the statements of the triadic interactions during the confrontation session, found that:

Group II: Father: Controlling and hostile toward child but not toward mother

           Mother: Critical and hostile toward child, but combined with little distantiation and disengagement from child

           Child: High controlling behavior toward both father and mother

Group IV: Father: Weak and yielding, with absence of controlling and hostile behavior toward mother

                     Controlling and covertly depreciating toward child

           Mother: Controlling, but not hostile toward child; controlling and covertly depreciating toward father and child

           Child: Low controlling and hostile behavior toward both father and mother

                     High yielding to father

                     Much distantiation from mother

She also found that in Group II families both parents consistently accepted and shared the parental role, even though the child may have tended to communicate more commonly with one parent. In Group IV, however, the mothers covertly depreciated their husbands by speaking to them with questioning and domineering intents at a rate far exceeding that found with mothers in the other problem groups.

The Group I and III families differed both from Group II and IV, as well as from each other. The emphasis of parental control and acceptance of parental responsibility which was most prevalent in Group II families seemed to go along with the adolescent's coping style of confining his acting out behavior to the home. For the Group IV families, the most striking characteristic was the covert depreciation of fathers by their wives in the triadic interaction and the persistent domination of both toward the child.

RELATION BETWEEN GROUP IV ADOLESCENTS AND
POOR PREMORBID SCHIZOPHRENICS

The Group IV adolescents, as indicated earlier, are comparable in pattern of social adjustment to the poor premorbid schizophrenics as defined by the Phillips Scale. This limited sampling of illustrative findings indicates that the pattern of parental–child interaction may possess some similarity to that reported in the Duke studies on premorbidity in schizophrenia, on which Garmezy and Rodnick and their associates [4,16] had reported some years ago. Especially noteworthy are the findings of maternal domination for the poor premorbid patient and paternal domination for the good premorbid schizophrenics [3] and the yielding and conformity acquiescence reported by Clarke [2] for poor premorbid schizophrenics.

What is missing so far in this report is any evidence that the Group IV adolescents are prodromal, high risk, or highly vulnerable for schizophrenia. The selection of the adolescents tended to preclude actively psychotic adolescents, although two showed borderline psychotic, or at least possible prodromal evidence of a schizophrenic break. The difficulties involved in assessing borderline, incipient or prodromal signs of schizophrenia in adolescents with adequate validity and objectivity are severe. We decided therefore to avoid this issue at the time the adolescents and their families were seen in the clinic. In fact, we wanted to avoid studying a family interaction containing an active psychotic child, since the interaction could be dominated by the psychosis rather than being indicative of basic interpersonal dynamics. We chose to rely instead on the follow-up of the adolescents several years later when the occurrence of pathology could be evaluated more objectively in terms of the usual incidence criteria—namely through behavioral disturbance sufficiently severe for some treatment agency to be seen, and by indications of actual psychotic behavior as reported by members of the family, or as elicited in an evaluative and diagnostic interview with the young adult. We also obtained an MMPI at the time of this interview.

Our research plan includes the regular follow-up of the adolescents seen on the project, which now includes the first 25 adolescents five years after they were originally seen in the clinic. Three and possibly four may have had a psychotic-like break, or currently show signs of borderline psychotic behavior depending on the criteria used. Considering that these patients have not yet entered the period of highest risk, the base rate seems higher than would be expected on a random basis. This is not the place to comment on this finding, beyond indicating that the adolescents of greatest vulnerability are probably in Group IV, followed by Group II. Until we have more cases and the data analyzed more thoroughly, this should probably be given no more credence than hearsay evidence.

Another source of data which bridges the findings for our disturbed adolescents and those of adult schizophrenics is provided by a second thrust of our research program. We are following up acute schizophrenics seen in a nearby mental health center for short periods of hospitalization. We chose a small county, geographically isolated from Los Angeles, and with few population centers, which was covered by a central mental health center. The follow-up includes periodic interviews with them and with members of their families in order to identify possible relationships between posthospitalization social adjustment of the schizo-phrenic and premorbid behavior. In addition, we are exploring factors in the post-hospitalization familial environment which may be related to the continuing clinical condition, and the role which phenothiazine therapy may play in this process. This follow-up investigation of the posthospitalization phase of acute schizophrenia affords the opportunity to compare the social adjustment of diagnosable schizo-phrenics with the follow-up of the disturbed adolescents seen in the family project.

The findings of this research, some of which have been reported [5,8], are interesting and germane to this discussion. We found further strong support for the utility of the Phillips premorbid scale for controlling for the heterogeneity of the response of schizophrenics to phenothiazine therapy. For example, in a controlled double-blind drug-placebo study of acute schizophrenics, the good premorbids remained in the hospital for a mean of 63 days in contrast to a mean of 100 days for the poor premorbid patients. This represents about a one third shorter hos-pitalization for the good premorbid patients. This merely confirms a number of other reports of the predictive significance of the Phillips Scale for hospital remissions. A more surprising finding is that at the end of the first 28 days of hospitalization (the first point at which these research patients could be considered for discharge) of those patients considered by the ward staff to be ready for dis-charge from the hospital, 61 per cent of the good premorbid but only 14 per cent of the poor premorbid patients were in the placebo group (p < .05). This finding raises the question whether phenothiazines might be an indicated therapy primarily for the poor premorbid patients. Despite the fact that there appear to be no readily discriminable differences in pattern or amount of schizophrenic symptoms between the two groups at the time of admission, by the time of discharge from the hospital, the ward team (which ignored premorbid data) recommended phenothiazine medication for only 26 per cent of the good premorbid patients in contrast to the recommendation of drug therapy for 88 per cent of the poor premorbid group (p < .01). This provides firm evidence that the good premorbid patients not only clear up rapidly from the more obvious psychiatric symptoms which brought them to the hospital initially, but that the two groups are probably at different levels of remission at the time of discharge. The staff considers the good premorbid patients as less in need of phenothiazine medication, the poor premorbid patients are being sent home while still in need of active therapy.

These data serve to highlight the importance of considering the characteristics of social competence in the premorbid history of schizophrenics as significantly related to the course of the schizophrenic disorder, the responsivity to pheno-thiazine medication, and the character of the posthospital social adjustment. Any attribute of the pre-psychotic adjustment such as level of social competence prior to the overt display of schizophrenic symptoms, which carries this much information regarding remission potential, duration of hospitalization, and even responsivity to phenothiazine medication merits intensive scrutiny and systematic research inquiry.

We have decided to concentrate on the chronological dimension. Hospitalized schizophrenics are being followed through the period of posthospital adjustment in order to look at close hand at the relation between social competence and rehospitalization. In the other direction we are following a group of adolescents who may be high risk for schizophreniform as well as other varieties of maladjustment to see whether there is any consistent relation between characteristics of the intrafamilial environment and *both* particular patterns of disturbed adolescent behavior and attributes of adult schizophrenic paychosis.

A more direct linkage between the disturbed adolescent and the adult schizophrenic is provided by some initial data of interviews of parents of our posthospitalized schizophrenics regarding the premorbid adjustment of their schizophrenic child *during the adolescent period*, long before the occurrence of overt psychotic behavior. The interview schedule we are using provides a direct comparison with data being obtained with the disturbed adolescent group.

One of the weaknesses of the simplified Phillips Scale is that the scale lends itself readily to assessing the premorbid heterosexual and peer relations of males, but is insensitive for women, since marriage and heterosexual relationships are not adequate discriminators of premorbid level for women. Susan Sturzenberger of our group is finding that peer relationships during adolescence, however, are a good marker variable for women when treated as comparable to the heterosexual and peer relations attributes used for male patients. Using these criteria the same relation is found for both men and women between premorbidity, prognosis and phenothiazine medication. No differences were found in behavior in the adolescent period between men and women.

The following significant differences were found by Sturzenberger between 18 good and 16 poor premorbid schizophrenics at the time they were adolescents.

Q14.   As a teenager what jobs did your child hold?

Good	Poor	P
2.56	1.67	.01

Q70.   How many close friends did your child have?

Good	Poor	P
2.28 (many)	3.00 (few)	.05

Q76.   Was he a leader?

Good	Poor	P
3.00 (occasionally)	4.31 (rarely)	.01

Q57. Arguing with parents

Good	Poor	P
3.25	2.25	.10

Q5. Lack of school motivation (rebellious)

Good	Poor	P
3.39 (unconcerned)	2.56 (concerned)	.05

Thus, the poor premorbid schizophrenics at the time they were adolescents, look suspiciously like Group IV problem adolescents, while the good premorbid patients have strong resemblances to Group II.

Significant differences are found between the good and poor premorbid groups in social withdrawal, lack of ambition, being a follower, and fear of leaving home. These are all characteristics of our Group IV adolescents. The other differences, such as holding jobs, argumentativeness with parents, very poor motivation in school, are higher for the good premorbid patients, and tend to characterize the Group II adolescents.

These findings, if they continue to be confirmed and further corroborated by other data, may provide a basis for identifying in adolescence potential schizophrenics who even at that time show the distinctive pattern of premorbid adjustment, which serves as such an important marker variable in research with adult schizophrenics. The findings which are emerging from our work with the families of problem adolescents provide evidence that the premorbidity differences may in turn be related to consistent differences in the interpersonal environment within the home of the adolescent. In time we may begin to identify the data needed to describe the mechanisms involved in integrating factors in gene pool, early experiences, and psychosocial transmission of parental, and hence familial styles of thinking and coping, which increase the vulnerability for inadequate social competence and schizophrenic modes of adaptive functioning. It is already clear that these disturbed adolescents, particularly those of Group II and IV should be studied systematically, to see whether they show other response attributes of schizophrenia, such as the response to stimulus overload, autonomic arousal patterns, the development of segmental sets, the presence of incipient schizophreniform cognitive styles, etc., as part of a systematic inquiry into prodromal indicators of schizophrenia.

REFERENCES

1. Alkire, A. A., Goldstein, M. J., Rodnick, E. H., and Judd, L. L. Social influence and counterinfluence within families of four types of disturbed adolescents. *J. Ab. Psychol.* 77:32–41, 1971.

2. Clarke, A. R. Conformity behavior of schizophrenic subjects with maternal figures. *J. Ab. Soc. Psychol.* 68:45-53, 1964.
3. Farina, A. Patterns of role dominance and conflict in parents of schizophrenic subjects. *J. Ab. Soc. Psychol.* 61:31-38, 1960.
4. Garmezy, N. and Rodnick, E. H. Premorbid adjustment and performance in schizophrenia: Implications for interpreting heterogeneity in schizophrenia. *J. Nerv. Men. Dis.* 129:450-466, 1959.
5. Goldstein, M. J. Premorbid adjustment, paranoid status, and patterns of response to phenothiazine in acute schizophrenia. *Schizo. Bull.* 3:34-37, 1970.
6. Goldstein, M. J., Gould, E., Alkire, A., Rodnick, E. H., and Judd, L. L. Interpersonal themes in the thematic apperception test stories of families of disturbed adolescents. *J. Nerv. Men. Dis.* 150:354-365, 1970.
7. Goldstein, M. J., Judd, L. L., Rodnick, E. H., Alkire, A., and Gould, E. A method for studying social influence and coping patterns within families of disturbed adolescents. *J. Nerv. Men. Dis.* 147:233-251, 1968.
8. Goldstein, M. J., Judd, L. L., Rodnick, E. H., and LaPolla, A. Psychophysiological and behavioral effects of phenothiazine administration in acute schizophrenics as a function of premorbid states. *J. Psych. Res.* 6:271-287, 1969.
9. Goldstein, M. J., Rodnick, E. H., Judd, L. L., and Gould, E. Galvanic skin reactivity among family groups containing disturbed adolescents. *J. Ab. Psychol.* 75:57-67, 1970.
10. Kaswan, J. W., Love, L. R., and Rodnick, E. H. Information feedback as a method of clinical intervention and consultation. In: *Current topics in clinical and community psychology*, Vol. 3, edited by C. Spielberger. New York: Academic Press, 1971.
11. Lennard, H. L., Beaulieu, M. R., and Embrey, N. G. Interaction in families with schizophrenic child. *Arch. Gen. Psych.* 12:166-183, 1965.
12. Lidz, T., Fleck, S., and Cornelison, A. *Schizophrenia and the family*. New York: International Universities Press, 1965.
13. McPherson, S. Communication of intents among parents and their disturbed adolescent child. *J. Ab. Psychol.* 76:98-105, 1970.
14. Mishler, E. G. and Waxler, N. *Family progress and schizophrenia*. New York: Science House, 1968.
15. Rodnick, E. H. The psychopathology of development: Investigating the etiology of schizophrenia. *Am. J. Orthopsych.* 38:784-798, 1968.
16. Rodnick, E. H. and Garmezy, N. An experimental approach to the study of motivation in schizophrenia. In: *Nebraska Symposium on Motivation*, edited by M. R. Jones. Lincoln, NE: University of Nebraska Press, 1957.
17. Wynne, L. C. and Singer M. T. Thought disorder and family relations of schizophrenics: I. A research strategy. *Arch. Gen. Psych.* 9:191-198, 1963.

Competence and Adaptation in Adult Schizophrenic Patients and Children at Risk

Norman Garmezy, Ph.D.

INTRODUCTION

Two decades have passed since the inception of the Duke studies in schizophrenia. The passage of time provides a vantage point from which to evaluate that ten year period of research, and to record, in the light of subsequent investigations, the extent to which the findings of that project can be viewed as stable or unreliable.

Two articles, descriptive of the program, appeared toward the close of the 1950s [26,53] and contained descriptions of a set of experiments that broadly encompassed six areas of research relevant to schizophrenia:

1. The first, and undoubtedly the most significant, transcended any individual experiment and came to serve as the basic framework for subject selection. This was the effort to test the experimental consequences of differentiating patients along a continuum of good to poor premorbid competence;

2. Within that context of patient selection, emphasis was placed upon the creation of experimental-laboratory analogues that made use of constructs that were conceptually significant in psychodynamic formulations of the development of schizophrenia;

Research in the Schizophrenic Disorders: The Stanley R. Dean Award Lectures, vol. 2, edited by R. Cancro and S. R. Dean. Copyright © 1985 by Spectrum Publications.

3. Specifically, the stimulus focus became the depiction of early family relationships that emphasized the power and punishment aspects of the mother-child and, to a lesser extent, the father–child relationship. Later in the program, ways were sought to test variations in such patient-parent relationships at a function of the premorbid interpersonal competence of the patient;

4. Interest in the reinforcing properties of punishment led to studies designed more broadly to examine the effects of censure *per se* on the level of performance of schizophrenic patients in comparison with normal control subjects;

5. The study of such effects suggested an ubiquitous pattern of avoidance-responding that appeared to characterize patients exposed to censure stress;

6. Finally, when the so-called "tranquilizers" began to assume therapeutic ascendancy in the mid and late 1950s, preliminary assessments were begun on the influence of the phenothiazines on the patient's avoidance-response systems under conditions of stress.

THE DUKE STUDIES: A RETROSPECTIVE APPRAISAL

What is the current status of these six areas of research? To what extent have the early studies proven contributory and to what extent have they been found wanting? In the section that follows, a brief (and hopefully) unbiased summary will be presented. In a number of instances, such as those related to the role of premorbid competence and the effects of punishment, the summary statements are based upon determinate reviews. In other cases, however, the judgment reflects opinion that is rooted in less searching reviews of the literature.

Schizophrenic Performance and the Continuum of Premorbid Competence

The Phillips Scale of Premorbid Adjustment in Schizophrenia [51] early became the instrument by which we sought to distinguish behavioral differences between "poor" and "good" premorbid schizophrenic patients. This dimension of early interpersonal competence we recognized to be isomorphic with earlier distinctions that had been espoused in the clinical study of schizophrenia: dementia-praecox-schizophrenia; process-reactive; typical-atypical; chronic-episodic; evolutionary-reactive; true-schizophreniform. Although cognizance of this basic dichotomy was clearly evident in clinical appraisals of schizophrenia, the distinction had not been systematically applied to experimental data prior to the Duke program. In the course of that research program, the Phillips Scale clearly demonstrated its power for reducing the highly variable group data that was inevitably provided by acutely disturbed schizophrenic patients.

Surveying the current investigations of schizophrenia, one must conclude that of the six areas studied by the Duke group, the demonstration of the power of this poor-good premorbid dichotomy for reducing data variance has been the most significant and lasting contribution. Today the use of the distinction is widely accepted and the Phillips Scale has become a stable aspect of design considerations in the experimental study of schizophrenia. The power of the variable is attested to by evaluations that have appeared recently in a number of review articles and books [23,24,35-37,52].

Experimental Analogues of Parent-Child Relationships in Schizophrenia

Experimental analogues in the Duke studies took two different directions: one group of studies required patients and controls to cope with stimuli designed to serve as symbolic cues for deviant parent-child relationships. In a second group, censuring signals ("Wrong") were used to provide information to the schizophrenic patient of his inadequate performance on tasks involving learning, perception, discrimination and judgment.

An aside seems desirable here. Our use of such content-laden analogues did not constitute espousal of a psychogenic theory of schizophrenia. Others, however, came to perceive us as advocates of a psychological model of etiology. In truth, our research orientation was far simpler and infinitely less presumptuous. We asked merely what would be the behavioral consequences for the schizophrenic patient were we to bring successfully into the laboratory symbolizations of the type of early family stressors that had been suggested in the formulations of a number of distinguished clinical investigators, including Fromm-Reichman, Lidz, Bateson, Jackson and others. We never failed to appreciate the possibility that the patient's performance with regard to these symbolic cues could reflect his disorder and not necessarily the presence of a significant pre-illness antecedent.

Looking back to these earlier studies, one perceives their simplicity and not in an entirely favorable light. To study the complexities that inhere in family relationships demands a reasonable approximation to such complexity in the laboratory. Were this not to be achieved, it is unlikely that the experimental domain could provide a meaningful test of the role played by the family in the development of the patient's psychopathology. Reviewing the adequacy of the stimulus materials used to depict maternal scolding in the studies of Dunn [14] or Harris [33] or the family problems that demanded resolution by parents of schizophrenic patients in the research conducted by Farina [16] and Farina and Dunham [17], it is evident that the themes are universal; the stimulus contents simple. A line drawing suffices to illustrate scolding; a briefly expressed problem understates a theme of parent-child conflict. The contrast with the more elaborate methods used by Rodnick, Goldstein, Judd and their colleagues to study the interpersonal, familial environment of disturbed adolescents is striking. (See Rodnick's paper in this volume.) "Real" parents now become the object of scrutiny, and the cue representations of disturbing events within the family are made intensely personal and immediate by the use of contents derived directly from clinical interviews with parent and child. The description by Goldstein et al. [28] of the decision to focus on themes that are idiosyncratic in content, albeit universal in context, bears repetition:

> . . . each of the three family members (mother, father, child) separately experiences a structured, standardized interview which covers seven general areas of adolescent behavior, selected for the high likelihood of conflict between the adolescent and his parents. The interview is designed to pinpoint a *specific* problem between parent and child in each of the following areas: *achievement, sociability, responsibility, communication (with parents), response to frustration, autonomy, and sex and dating.*
>
> During each area of the interview, the interviewer probes for conflicts, specific to each family. Originally, it was hoped to use a common set of standard problems for each area to elicit emotionally meaningful interactions between parents and children. However, pilot data with this type of problem were very disappointing, and it was found that idiosyncratic expressions of general problems were more effective in stimulating family interaction. In many families much of the emotional charge of a broad problem area becomes focused around a specific conflict situation and this conflict is chosen as the symbolic battleground for a more general principle (p. 236).

It is these situations that subsequently become the significant cue stimuli for subjects used in the Rodnick–Goldstein–Judd project. And, if one cannot help but contrast the greater richness and meaningfulness of such situations to the earlier ones used in the Duke studies, it is appropriate to observe that a procedure, not unlike one of successive approximations, produced the shift to this heightened experimental reality. Thus the line of succession begins with line drawings of mother–child interactions [14,33], moves to taped recordings of hypothetical parent–child interactions that demand resolution [16], proceeds to more dramatic tapes of pseudo-mothers describing child-rearing practices [12] which, in turn, leads to the introduction of real mother directly into the experimental situations [60] and, finally, uses actual clinical content and the family members involved in those family exchanges as the basis for the selection of cue stimuli and subject matter [28]. Science often moves with measured tread and this sequence of a growing sophistication in method reflects such a trend.

Note, too, the shift from retrospective inquiry to concurrent examination of significant familial events that are here-and-now concerns. To have patient and parent retrace the past on the basis of a faith of their retrospective recall is to insure conclusions that are fraught with potential error [32,67]. Thus, a decision to focus on the child predisposed to disorder but not yet actively caught up in consequences resulting from patienthood that irrevocably distort the past seems the wiser course of inquiry. Consideration of this position has been treated in an earlier paper [23].

But what of the stability of the findings derived from these earlier experimental analogues? I would render several judgments. First, the contribution would appear to derive less from the specific findings and more from the *demonstration* of the power of the laboratory analogue in research in schizophrenia. In the 1957 Nebraska Symposium volume, Marshall R. Jones, editor of the series, ventured to predict with reference to the Duke project that "we will see a whole new crop of Ph.D. dissertations in this area." It is indeed the case that the laboratory study of schizophrenia has become a significant domain of psychology and a fruitful content area for dissertation research in psychopathology.

Second, on the substantive side, the empirical data with regard to familial patterning in schizophrenia is more equivocal. Sensitivity to maternal cues on the part of poor premorbid patients finds both refutation [11] and support [60]. With regard to the issue of method, it may be the more sophisticated experimental treatment of the latter study that may account for the positive findings. Stoller used the patient's mother as the direct conveyor of criticism of poor performance. Using a procedure not too dissimilar to one used earlier [19], the patient was first presented information as to the accuracy of his perceptual judgments through the use of "Right" and "Wrong" signals. In a third condition, however, the mother was in the room with the subject; upon a coded signal from the experimenter, which could not be seen by the patient, she informed the patient whether he was wrong or right on a particular trial. In addition, the mother predicted aloud whether she thought her son would be correct or incorrect on the following trial. Under this condition, poor premorbid patients, unlike their nonschizophrenic siblings or good premorbid counterparts, showed a significant increase in stimulus generalization, confirming a differential reactivity to mother's criticism. Furthermore, unlike the good premorbid's mother, the poor premorbid patient's mother significantly overestimated the performance of her psychotic son who, in turn, produced more errors following her predictions, whether these expressed her belief in his success or his failure.

As for the presumed differential sensitivity of good premorbids to paternal cues and poor premorbids to maternal cues (despite Goodman's supportive findings [30]), this has been refuted with sufficient frequency [49] to suggest the unreliability of that specific formulation. Farina, whose early research, in part, provided the basis for the hypothesis, has reported a failure to replicate his earlier findings [18], although some of the prior data on dominance and conflict relationships in schizophrenic families have been reaffirmed.

Censure and Performance in Schizophrenia

There are two questions associated with the effect of censure on task performance of schizophrenic patients. How powerful are the effects of punishment?

Does punishment invariably induce performance deficits as suggested in the 1957 paper? The answer to the first question is "Powerful, indeed!" The answer to the second is a resounding negative.

With regard to the second question, it may be helpful to position prominently our original formulation which, in the light of more than 150 studies, has proved to be grievously incorrect. The following quotation contained the critical initial statement:

> We shall try to demonstrate that schizophrenic patients can and do respond adaptively in tasks of considerable complexity and difficulty provided that these tasks have been made sufficiently interesting to insure the cooperation of the patient. This adaptability, however, is a tenuous one which can be disturbed by the introduction of minimal censure into an experiment. Under these conditions the deficit behaviors which have been described as characteristic of the schizophrenic patient appear [53].

There appears to be no reason for retreat from the first sentence of the above paragraph. Complex experiments demonstrably can involve schizophrenic patients, provided experimenters approach the construction of their studies with innovativeness and creativity. The problem lies with the latter part of the paragraph; the hypothesized censure-deficit formulation proved to be incorrect, primarily because we failed to analyze the response demands of our experiments in relation to the ubiquitous and powerful avoidance-response systems that the schizophrenic patients exhibited following the censure experiences.

In a major paper on psychological deficit in schizophrenia, Buss and Lang [8] appropriately observed:

> The thesis that punishment invariably disrupts performance is clearly not tenable. Both a negative evaluation and specific verbal or physical punishment for errors can lead to significant improvement in performance rather than further deficit (p. 11).

The authors, however, then proceed to assign, as the basis for such improvement, the factor of the greater amount of information that is provided by punishment.

> The fact that this improvement occurs in both the presence and absence of socially or personally punitive conditions suggests that the significant factor is information about inadequate responses rather than the interpersonal context.
>
> Knowledge of results is important to any task in which improvement is expected with practice, and in general the normal subject

recognizes the correctness or wrongness of a response as soon as it occurs, and no assistance is required. However, schizophrenics seem to be less able to instruct themselves and less able to maintain and usefully alter a response set. Furthermore, studies of incidental learning reveal that relative to normals, schizophrenics fail to observe objects or relationships towards which their attention has not been specifically directed. Thus, informational cues introduced by the experimenter have greater importance for the psychotic subject in certain tasks (pp. 11-12).

Since it is clear from existent data that deficits or increments in performance can follow in the wake of censure, the more fundamental question to be asked is what mechanism, if any, operates to explain these divergences. An explanation based upon greater informational cues cannot suffice and is clearly refuted by other studies in the literature. (Contrast, for example, Stoller's three conditions involving information when correct, information for extreme error and information provided by mother on *every* trial. The poorer performance of patients under the last named condition would suggest that the number of informational cues alone cannot predict enhancement of a patient's performance.)

Elsewhere, I have detailed in a review of the literature of punishment [20] a mechanism that may help to explain the findings of a diverse number of studies. Unfortunately the review has not received wide circulation and, as a result, investigators continue to ascribe to both Rodnick and to me continued advocacy of the 1957 censure-deficit formulation.

Within the experiments conducted by the Duke researchers, as well as by others, there were recurrent and marked indications of the pervasive nature of patients' avoidance tendencies—responses that were as evident in the laboratory as they were in clinical practice. But the analysis of the congruence or noncongruence of this type of response pattern in relation to the response demands of the experimental task was not attempted until there was clear evidence of the shortcomings of the earlier formulation [2,9,41]. The suggested interaction of these two variables—the patient's disposition to avoid and the demands of the task—can be set forth by quoting from the 1966 review:

In the light of these clinical hypotheses, it seems reasonable to conclude that the use of censure in laboratory tasks should heighten the evocation of withdrawal behavior in schizophrenic patients. Thus, responses elicited in experimental tasks that are followed by some signal suggesting criticism or disparagement should undergo a marked reduction in response strength. Although we can predict a modification of responsiveness on the basis of censure, we cannot predict the adequacy of the patient's performance on a task without knowledge of the response alternatives that are available to him. Alternative choices may be one or many, structured or unstructured.

In terms of available alternatives, we can define those situations that are most and least evocative of performance deficits.

1. The likelihood of behavior decrements will be heightened if the experimental procedures provide simply for a general devaluation of the patient's performance at some point in the experimental sequence; such decrements will be accentuated if the task permits a multiplicity of response alternatives. Since the criticism is nonspecific with regard to the inappropriateness of the subject's previous responses, the schizophrenic patient can be expected to become either less responsive or more grossly inappropriate in his behavior as the experiment proceeds. Compounding the effect of punishment will be the oft-noted difficulty of the patient in maintaining a mental set and his propensity for discounting task-oriented activity.

2. In many studies, the experimental procedures that have been employed have involved a punishing signal that is administered immediately following an experimenter-designated, incorrect response. As we have seen, such a signal, in some studies, takes the form of a noxious (pain) stimulus that is terminated only when S makes the appropriate response. This probably constitutes the *optimal* condition for behavior improvement since the stimulus literally *forces* the patient to respond correctly—a situation that doesn't readily permit irrelevant response sets to be retained.

However, in other studies, censure (rather than a physically noxious stimulus) has been used in the form of a "Wrong" signal following an error in responding; but the termination of such psychologically noxious events is *not made dependent* upon the subject's choice of the correct alternative. In such paradigms the effectiveness of the censure is best revealed on the trial that *follows* the one on which the patient's response has been punished. Given the simplest situation in which there are two highly structured response alternatives (such as those requiring simple motor movements), one of which is followed by a punishing signal, we can predict that the schizophrenic patient will show more frequent responding with the alternative response that avoids the censure.

If this is a highly probable consequence, the prediction of deficit or incremental behavior in schizophrenia becomes a rather straightforward one: *if the response alternative that permits avoidance of the censuring signal is a task-congruent response, behavior facilitation should follow; if the response alternative is noncongruent, performance deficits should ensue* [20] .

The remainder of the earlier paper was devoted to a consideration of the power of the formulation in terms of a reanalysis of a number of empirical studies, seeking to test its efficacy by applying the hypothesis to findings of published studies.

Finally, what about the broader issue of the presumed motivational power of punishment relative to reward that the Duke research group asserted was a characteristic of the schizophrenic patient? Countless studies of the differential effects of these two classes of reinforcers on the behavior of schizophrenic patients have appeared in the psychological literature. In 1968 in a seminar on Experimental Psychopathology at Minnesota, one of the student participants, Raymond Knight,* set out to analyze statistically 91 studies dealing with the effects of punishment on the performance of schizophrenic patients [40]. Unfortunately, his definitive review is still unpublished, although Knight intends to revise the monograph for publication. His review encompassed the majority of studies that had appeared through 1967—studies that he found to be "riddled with sampling problems, lack of proper controls, inadequate statistics and the confounding of variables."

Knight's procedural review was as follows: First he coded and punched on Unisort cards the reinforcement contingencies, subject samples, types of tasks, situational variables and results for each of the studies. His defined reinforcement conditions included: 1) *positive reinforcement* in which a response or a response class was followed by a reward; 2) *punishment* in which responses were followed by an aversive stimulus; 3) *negative reinforcement* in which a correct response terminated an aversive stimulus; and 4) *mixed reinforcement* involving a reward that followed a correct response and a punishment that was presented after an incorrect response.

Results were coded into four categories: 1) *positive* (+), when the subject's performance was improved by the reinforcer; 2) *negative* (-), when it was impaired; 3) *zero* (0), when it had no statistically significant effect ($p > .05$); and 4) *questionable* (?), when the reinforcement effect could not be analyzed because of an absence of a control group or as a result of inadequate data analysis. Coded studies were then sorted into different combinations to calculate the frequency of occurrence of the various results. X^2 analyses were used to test whether different combinations of independent variables produced significantly different distributions of results across studies.

With Doctor Knight's permission I quote at length from the concluding section of his monograph, for it provides a thorough appraisal of a very complicated and not always "rewarding" literature:

---

*Dr. Knight is now Assistant Professor of Psychology, Brandeis University.

Through the review and analysis some tentative . . . relationships were revealed and several suggestions for future research emerged. In general, schizophrenics were found to be more reactive to punishment than to reward, remaining unchanged in their performance in fewer instances when punished than when rewarded. Also, negative reinforcement was found to be the most successful reinforcer for schizophrenics, producing far more positive and fewer negative results than any other reinforcer.

An examination of sampling problems led to the suggestion that the variety of control groups employed be increased, and that the various samples of subjects be stratified across certain dimensions to reduce their heterogeneity and eliminate error due to the response of opposing subgroups cancelling each other out. It was also hypothesized that the good-poor premorbid distinction, which seems to be a promising mode for stratification, has elicited contradictory results because it has often been confounded with the acute-chronic dimension. The separation of the effects of these two means of dichotomizing samples may clear up some of the confusion extant in the literature.

Schizophrenics seemed to be more able to take advantage of the motivation increments afforded by both reward and punishment when the task was simpler than when the task was more complicated. More difficult tasks produced more differential responding among groups of schizophrenics, with the less intact groups being more disrupted by punishment than the more intact groups. Finally, a statistical analysis of the effects of the four most prevalent tasks (paired-associate learning, concept formation, reaction time, and rote simple tasks) suggested that punishment might cause more differential results among various tasks than reward.

Although the studies dealing with the effects of social variables on the performance of schizophrenics had somewhat perplexing and contradictory results, some tentative generalizations worthy of investigation became evident. It was found that:

1. Schizophrenics, especially acutes, tended to have more difficulty with tasks involving social rather than neutral stimuli;
2. Schizophrenics were more sensitive to stimuli representing censure than normals, but the hypothesis that poor and good premorbid schizophrenics are differentially sensitive to maternal and paternal censure was not consistently supported;
3. The experimenter was an aversive stimulus for the schizophrenic subject, even in reward conditions;
4. Strong social praise produced more negative results than weak social praise, suggesting along with the previous finding that a warmer,

more positive, attitude of an experimenter aroused interpersonal anxiety in schizophrenics;

5. Social reward occasioned fewer positive and more negative results in schizophrenics than in normals, and in schizophrenics it produced more zero and fewer positive results than social punishment;

6. Schizophrenics tended to be more reactive to social censure than normals, having fewer zero results and more negative and slightly more positive results under this condition;

7. Nonsocial punishment was a less effective motivation (had more negative and fewer zero outcomes) than nonsocial reward;

8. The social–nonsocial findings seemed to be more the result of the presence or absence of the experimenter rather than due to the verbal element in the censure.

Knight also reported that contingency and information feedback were important in determining the quality of performance of schizophrenics. Thus, both contingent punishment and reward led to more positive results and fewer negative ones than did noncontingent reinforcement, but only the differences for punishment proved to be statistically significant.

For normals, contingent and noncontingent conditions produced equivalent results. Information feedback was less powerful than the contingency of the reinforcement. For the schizophrenic patients, no information produced more negative and fewer positive results (a difference which approached significance) in comparison with test conditions in which information was provided about performance. By contrast, normal subjects were less dependent on feedback regarding their performance.

Dr. Knight concludes that no single experiment can determine the effect on the schizophrenic patient's performance of the many complex variables suggested by his review. He believes that only through a program of carefully delineated experiments will it be possible to arrive at definitive generalizations of the effect of punishment on performance. Knight's review thus provides a glimpse into the complexity of the problem and raises anew the question of the ultimate viability of any encompassing formulation of the relationship of punishment to performance, including the more extensive 1966 statement quoted earlier.

THE INFLUENCE OF PHENOTHIAZINES ON AVOIDANCE BEHAVIOR

The advent of the phenothiazines as a major therapeutic weapon against schizophrenia occurred at the midpoint of the Duke studies. Undoubtedly, some portion of later failures to replicate earlier findings may be related to the fact that it is now difficult to find drug-free patients who are not long-term chronic, "burned-out" schizophrenics. This critical lack of control in experiments clearly freighted with motivational variables can be expected to be a telling one. To suggest that the effect of the drug variable is attenuated because good and poor premorbid patients do not differ in the dosage level received [39] or that the "main effects of all phenothiazine drugs are essentially the same" [39] or that "all drugs belonging to the phenothiazine family are about equally effective in alleviating psychotic symptoms" [11] is to fail to ask the critical and far more penetrating question: Do the effects of such drugs differentially affect response systems relevant to performance in good and poor premorbid patients? Earlier studies gave evidence that experimental effects were attenuated in the drug schizophrenic groups in comparison with nondrug patients [1,6,15] in ways meaningful to an interpretation of differential deficit functioning in goods and poors. More recently, Goldstein et al. [27,29] have reported a comparison of a fixed dosage of phenothiazine and a matched placebo on a variety of performance measures for a sample of 38 acute, newly admitted male schizophrenics divided into good (N = 22) and poor (N = 16) premorbid subgroups. The drug and placebo conditions were run for a period of seven to 21 days, following a "drying out" placebo period extending seven days after admission.

Marked differences were found between the two groups of patients under the drug and placebo regimens. For the poor premorbid schizophrenic patients three weeks of phenothiazine medication produced a marked reduction in behaviors that reflect disorganization—a finding that did not hold for the placebo group. Drug poors showed a lowered level of general arousal, reduced psychophysiogical reactivity to stimuli, more comprehension of ongoing events, "*less avoidance behavior to potentially anxiety-arousing words*, fewer remote associations and lower self-rated anxiety" (p. 283). Good premorbid patients on extended phenothiazine medication showed seemingly paradoxical effects that included "greater psychophysiological reactivity, *more avoidance behavior to verbal anxiety stimuli*, less of a decrease in distant associations and more autistic associations" than the good premorbid placebo group.

The authors conclude that the phenothiazines affect good and poor premorbid schizophrenics differently:

> In the goods, the general reduction in arousal is associated with greater responsivity to specific changes in the environment and avoidance behavior to threat. In the poors, the same reduction in arousal is associated with lessened responsivity to the environment and vigilance for threat. If we consider that good and poor premorbid patients were probably differentially responsive to their environments prior to their psychotic episodes, it may be that phenothiazine medication returns each type of patient to his premorbid style of coping, permitting relatively more appropriate behavior to reappear (p. 284).

Such data are congruent with several of the later studies in the Duke program in which goods and poors had been separated into drug and nondrug cases in an effort to compare avoidance patterns under censure within each subgroup. Avoidance proved to be most characteristic of drug-free premorbid patients. However, the confounding effects of acuteness are present in these early studies, since such patients were more likely to receive drug therapy [20].

It is obvious that there is a need for carefully controlled investigations to specify the action of ataractic drugs on specific response parameters in good and poor premorbid schizophrenic patients. Until such data are forthcoming, interpretations of differences among studies comparing patient groups on behaviors that clearly implicate motivational systems should only be advanced with marked caution.

In the section that follows, I turn from adult schizophrenia to my current research interest in studying the adaptation of children who are *at risk* for the disorder. The conceptual tie between the earlier Duke program and the research on vulnerability now underway at the University of Minnesota and the University of Rochester is to be found in the unifying theme of *competence*. In the Duke program competence was the explicit criterion for determining a patient's status as a "good premorbid" or a "poor premorbid" patient. In the studies of risk the criterion for competence is broadened but its function is the same—to delineate the adaptive potential of risk children not as a measure of their prognosis for recovery from disorder but rather as an index of their resistance to its emergence.

COMPETENCE AND THE STUDY OF CHILDREN AT RISK

The study of premorbid *competence* has not only helped to resolve the riddle of recovery from schizophrenia, but has been instrumental in delineating qualities of the "healthy personality." Freud's twin criteria of *lieben und arbeiten* couples with an aphorism attributed to Whitehorn by Grinker [31]:

Concerning mental health and the absence of a concept of positive mental health, I would like to give you a kind of cookbook definition developed by John Whitehorn, although it is not a complete statement. What I frequently tell my students is that people who are mentally healthy usually work well, play well, love well and expect well. I think that "expect well" is the most important in that there is an anticipation that the future will have something of value to them (p. 23).

To work, play and love "well" foretells recovery from mental disorder. The designation of *good premorbid* and *poor premorbid* schizophrenic patient which became the central independent variable in the Duke studies, summarized the attributes of low and high scores on the Phillips Scale of Premorbid Adjustment in Schizophrenia [51]. That this differentiation appears to be isomorphic with the process-reactive dimension simply enhances the power of the experimental correlates that accompany the dichotomy [21]. The Phillips Scale emphasizes the form and quality of the patient's patterns of love and play during the period preceding the onset of his disorder: adjustment in the areas of sex and interpersonal relations, social aspects of sexuality during adolescence and adulthood, and his history of personal relations from childhood onward. This emphasis primarily upon sex and friendship patterns can predict recovery from schizophrenia because the process-type schizophrenic, typically, is characterized by a tragically low ceiling of premorbid adaptation. However, to broaden the base of prediction to include recovery from more diverse forms of psychopathology, or to apply the concept of competence to both normal and other psychiatric populations requires that the criteria for effective functioning be extended to include level of intellectual development and cognitive functioning, achieved occupational and educational status, regularity of employment, participation in community organizations and constructive use of leisure time. In the volume, *Human Adaptation and Its Failures* [62], Phillips discusses this extension of the criteria that define the nature of competence:

> In our view adaptation implies two divergent yet complementary forms of response to the human environment. The first is to accept and respond effectively to those societal expectations that confront each person according to his sex and age. Included here, for example, are entering school and mastering its subject matter, the forming of friendships and, later, dating, courtship, and marriage. In this first sense, adaptation implies a conformity to society's expectations for behavior. In another sense, however, adaptation means more than a simple acceptance of societal norms. It implies a flexibility and effectiveness in meeting novel and potentially disruptive conditions and of imposing one's own direction on the course of events. In this sense, adaptation implies that the person makes use of opportunities to fulfill internally established goals, values, and aspirations. These may include any of a universe of activities as, for example, the choice of a mate, the construction of a house, or the assumption of leadership in an organization. The essential quality common to all such activities is the element of decision-making, of taking the initiative in the determination of what one's future shall include (p. 2).

This broadened set of criteria, when joined by measures of "normal development" that incorporated such factors as "accountability, obligation and reciprocity" in personal relationships forms a nexus with Whitehorn's definition with its emphasis on effectiveness in work, love and play. If the criterion "to expect well" has not been mapped by Phillips, it is likely because it is essentially derivative; "to expect well" speaks to issues of self-esteem and a positive self-concept and these are founded on skills which are reflections of economic, social and sexual adequacy.

Although it is not possible within the limits of this brief paper to review the literature relevant to the self-concept, the derivative nature of Whitehorn's fourth criterion can be seen in studies of the antecedents of self-esteem. Coopersmith [13], on the basis of his studies of middle class preadolescents, locates the antecedents for esteem in parental acceptance, opportunities for individual expression and successful academic performance. Wenar [64] parallels these observations with a commentary in which the growth of self-esteem in adolescents is made dependent upon how "love-worthy" the child is to his parents, the extent to which he is "respect-worthy" to his peers, and the "grade-worthy" quality of his performance in school. To "expect well," thus, is to anticipate the regard of parents and peers, and to be able to approach problems with the expectation that success and rewards will be forthcoming—qualities that can only be stabilized by a history of prior achievement. White [66], who has contributed so substantially to an analysis of effectance motivation, expresses the relationship of competence to self-esteem in this manner:

> It is important . . . to make allowance for the child's action upon his environment, of the extent to which this action is apt to be successful, and consequently of the confidence he builds up that he can influence his surroundings in desired ways. I use the term *sense of competence* to describe this, and I think that one's sense of competence is an exceedingly important aspect of self-esteem (p. 201).

In his paper on the experience of efficacy in schizophrenia, White writes of the schizophrenic patient's major liability—the manner in which he is victimized by ineffective actions, and his lack of initiative and persistence in problem solving—a liability that extends back to a period that long antedates the disorder:

> . . . weak action on the environment has very great generality in schizophrenic behavior. Poor direction of attention and action, poor mastery of cognitive experience, weak assertiveness in interpersonal relations, low feelings of efficacy and competence, a restricted sense of agency in leading one's life—all these crop out in almost every aspect of the schizophrenic disorder.
>
> I should like now to entertain the hypothesis that this ineffectiveness in action is central not only in the picture of the schizophrenic's ultimately disordered behavior but also throughout his whole course of development—that from the start, it is the future schizophrenic's major liability. It characterizes his behavior from an early point in life, and it leads to a precarious development in all the spheres I have discussed, including interpersonal competence and self-esteem (p. 202).

It is this last paragraph that forms the bridge between the disturbed adult caught up in a network of schizophrenic pathology and the study of children who are at risk for that malignant disorder.

Children-at-Risk

The study of risk has assumed the proportions of a movement, but limitations of space do not allow me to chronicle that growth. Suffice to say that the constraints placed by retrospective data upon theorizing about the antecedents of schizophrenia have led a number of investigators here and abroad to study the developmental course of those who may be predisposed to the disorder. One critical problem is the identification of such vulnerable persons. Predisposition and the heightened probability of later disorder have begun to be defined in terms that largely reflect the etiological models that dominate the study of schizophrenia: biogenesis, family disorganization, socio-cultural deviance, and early neglect and deprivation, for example, birth and pregnancy defect, poor prenatal care, nutritional deficiency, etc. [24].

Studies of children-at-risk typically are longitudinal in design—a method that inevitably imposes logistical and conceptual problems of great complexity. Obviously, these studies share the common goal of attempting to predict disorder by relating deviant outcomes in the adult to earlier patterns of behavior observed in childhood. But two factors militate against the exclusive reliance on such long-term ultimate outcomes in prediction. First, the Ns in these studies tend to be small, as is the proportion of anticipated adult deviance. Viewed from the standpoint of empirical probabilities, investigators who study vulnerability in children born of schizophrenic parentage can anticipate that the morbidity risk for their subjects will approximate 10 per cent [57], but this figure presumably would have to be projected through a risk period extending to age 45. Children who are truly at high risk, faced with the unfortunate circumstance of having been born to parents both of whom have been schizophrenic, show a fourfold increase in morbidity risk that rises to a figure approximately 35 per cent [57].

Were one to broaden the concept of risk to include other forms of maladaptation, the proportions of assumed deviance in children clearly would rise. But in either case, variability in outcome must be anticipated and planned for in the design of risk experiments. This projection of a heightened variability in outcomes is not inherently disadvantageous. It is as important to study risk children who escape disorder as it is to evaluate those who fall victim to it. But the context for doing so may be better served by more circumscribed short-term, longitudinal-prospective studies which are designed to measure the efficiency with which specific behavioral and biological variables differentiate not only high-risk from low-risk children but also separate within the high-risk group those children who show patterns of incompetence and maladaptation from others who appear to be proceeding toward adulthood along paths of mastery and adaptation.

The study of competence versus incompetence in childhood emphasizes the utility of intermediate outcomes, and must be distinguished from research oriented to predictions of an ultimate outcome of disorder versus normality in adulthood. But, if studies of intermediate outcomes are to prove meaningful to risk researchers, one must be able to demonstrate, at some point, that there exists a continuity to competence that extends from childhood into adulthood. Phillips [52] has stated the issue in this form:

> The key to the prediction of future effectiveness in society lies in ask-ing: "How well has this person met, and how well does he now meet, the expectations implicitly set by society for individuals of his age and sex group?" What we need to learn is the person's relative potential for coping with the tasks set by society, compared to others of his age and sex status. Expected patterns of behavior change with the person's age. Presumably, relative potential for meeting these expectations remains far more stable. Thus, to the extent that an individuals's relative stand-ing in adaptive potential remains constant, we should, in principle, be able to predict his effectiveness in adaptation to society. The prag-matic question in need of resolution is the extent to which relative adaptive potential does, in fact, remain constant. Only to this extent are we in a position to predict the person's future (p. 3).

We may as well ready ourselves for the prediction errors that will follow. MacFarlane's [43,44] observations made in the course of the famed longitudinal study of children conducted by the staff of the Institute of Human Development of the University of California, Berkeley, suggest this likelihood. Her subjects are now moving into their 40s. It is evident that many competent children have indeed become competent adults, but others characterized by early academic failure and patterns of social isolation and withdrawal have become outstanding adults, while still others who clearly gave evidence of mastery during the years of childhood are now unhappy adults puzzled by their manifest failure to sustain the pattern of their earlier achievements.

Nevertheless, it is probabilities that we deal with—and the probabilities stand in favor of the continuity of competence. I would, therefore, suggest that a tenable first stage in risk research is to explore the correlation between the behaviors of children presumed to be vulnerable to psychopathology against a criterion of their qualities of competence.

A FOUR-STAGE STRATEGY FOR RISK RESEARCH

Reflecting on the many methodological problems in risk research, I believe it appropriate to think in terms of a four-stage strategy. Basically we are engaged in a search for specific behavioral and biological differentiators that will separate risk from nonrisk children. But the differentiating power of any parameter in the study of predisposition must, as I have indicated, meet an even more stringent criterion of discriminating the competent from the incompetent children who are at risk. It may not be possible to achieve this goal over the entire age range of childhood since adaptive insufficiency within the high risk group will grow more apparent as children reach toward the more stressful years of middle childhood and adolescence. However, this search for predictor variables assumes that external criteria of success or failure in adaptation are available to the investigator.

Stage One. If these criteria are not available, then the first priority in risk research must be to provide age-related indices of competence. This is necessary to cope with a disorder characterized by low incidence rates and a lengthy time period before the appearance of illness. Were individuals to become the victims of schizophrenia with a rapid and inexplicable urgency in which there was an absence of precursor signs, investigators could lean more heavily on the ultimate criterion of presence or absence of disorder. Fortunately, schizophrenia typically does not follow such an unpredictable course, and the more malignant form of the disorder shows clear prodromal indicators. This fact, based upon an extensive empirical literature, justifies the use of intermediate outcomes as a strategically viable procedure for inferring successful or unsuccessful adaptation in later life, although one's predictions, as has been suggested, will be subject to considerable error.

There arises the question as to which aspects of competence/incompetence are most appropriate to study. Listings of the qualities of the healthy personality have never been in short supply in our literature, but a more stringent screening is necessary. I believe that the most effective screen is to study the qualities of stress-resistant children—the "invulnerables" of our society—on the assumption that these children approximate more closely the competent child-at-risk. This method, to be described in the section to follow, has been the procedure used by our Minnesota group.

Stage Two. Once developmental measures of competence–incompetence have been identified, the selection of response parameters that can successfully differentiate between and within risk and control groups becomes the focus of the second stage of research. Such Stage Two studies, I assume, will include variables that have proved to have significant predictive power in the study of adult schizophrenia [22-24]. Thus studies of psychophysiological responsiveness, attention deployment, and cognitive efficiency are examples of the type of variables that would meet this definitional criterion for Stage Two variables. But other variables may be suggested by developmental studies of children on the assumption that developmental lags may forecast maladaptation. A further requirement must be added to all Stage Two studies; a Stage One (competence) variable must be incorporated in order to test the power of a Stage Two differentiator to segregate subjects who differ in levels of achieved competence.

Stage Three. The third stage in risk research would then involve the selection of several powerful differentiators for use in short-term prospective studies comparing adaptation in children at risk with appropriate control groups. Elsewhere, I have suggested that the convergence technique espoused by Bell [4] and successfully utilized by Schaie and Strother [58] in a study of age changes in cognitive behavior be applied in these circumscribed longitudinal investigations. This method involves the use of cross-sectional groups of different ages studied over a sufficient period of time to allow each age group to brook into the next, hopefully to permit inferences to be made about the developmental nature of the variables under study.

Stage Four. The final stage in risk research would involve intervention efforts cast into a nontraditional format. For the critical element in such Stage Four studies would be the use of experimental–clinical techniques designed initially to modify performance on Stage Two factors with later evaluations designed to test the effects of such behavioral changes on Stage One competence factors. Example: if attention deployment is faulty in the risk child, can operant procedures be used to enhance attentional focusing? Assuming that such intervention proves successful and that the behavior can be maintained and transferred beyond the laboratory, does it result in positive shifts in competence in the child?

I do not suggest that these four stages be viewed as a rigid sequence of priorities. A flexible project, hopefully, would move comfortably among these various stages to test new hypotheses and to introduce and elaborate other variables on the basis of new data inputs.

STUDIES ILLUSTRATING THE FOUR STAGE SEQUENCE

Stage One: Competence Indices

The investigator of risk can often turn, with gain, to the work of other researchers who have studied age-related indices of competence. Some illustrative examples: Bruner's ethological-toned research on the development of various types of skills in infancy; Wenar's [63] studies of competence at age one—high "intensity" of behavior, "self-sufficiency," social curiosity and an orientation to complexity; Burton White's [65] elaborate program for studying the pattern of human competence relevant to the first six years of life.

This last named project is particularly significant for students of risk. White's investigation, ethological and anthropological in orientation and intensity, and determinately empirical in context, illustrates the difficult task of developing competence criteria for any given age period:

> Initially, we selected as broad an array of types of preschool children as we could. Our original sample consisted of some 400 three-, four-, and five-year-old children living in Eastern Massachusetts. We reached the children through 17 preschool institutions (kindergarten and nursery schools). These children varied in at least the following dimensions: 1) residence—from rural to surburban and urban, 2) SES (socio-economic status)—lower-lower to lower-upper class, 3) ethnicity—Irish, Italian, Jewish, English, Portuguese, Chinese and several other types. On the basis of extensive independent observations by 15 staff members and the teachers of these children, and also on the basis of their performance on objective tests such as the Wechsler and tests of motor and sensory capacities we isolated 51 children. Half were judged to be very high on overall competence, able to cope consistently in superior fashion with anything they met. The other half were judged to be free from gross pathology but generally of very low competence. We then proceeded to observe these children each week for a period of eight months. We gathered some 1,100 protocols on the typical moment-to-moment activities of these children, mostly in the institutions, but also in their homes. At the end of the observation period we selected the 13 most talented and 13 least talented children. Through intensive discussions by our staff of 20 people, we compiled a list of abilities that seemed to distinguish the two groups (p. 74).

Social and nonsocial abilities dominate the list, whereas sensory-motor capacities proved to be ineffective discriminators. Relevant social abilities reflective of competence include the following: attention-seeking from adults in socially acceptable ways; the ability to express affection and hostility to adults and to use them effectively as resources; the ability to behave similarly with peers, to compete with them and to assume positions both of leadership and followership; to show pride in accomplishment; and to express a desire to grow up, as evidenced, in part, by involvement in "adult-role play behavior."

Nonsocial abilities include linguistic and intellectual competence, attentional ability, and "executive abilities" of the type involved in planning multistep activities and effectively using environmental resources.

The Minnesota Studies of Competence

At Minnesota, we have focused on competence indices that may be applicable both to the middle years of childhood and early adolescence. Our approach initially was to search extensively the literature of the healthy personality to catalogue as many indices of adaptive functioning as could be found. Such listings typically lack an empirical base and readily reveal their armchair origins. We, therefore, used two requirements in selecting our basic criteria: 1) frequency of occurrence in the literature (which meant simply that many had occupied the same armchair); and 2) satisfactory application of a more stringent and meaningful standard. To meet the latter, we generated another literature review—one which emphasized empirical studies of competence in children who could also be considered to be at risk. Using sociocultural disadvantage as the risk component (a choice made necessary by the lack of a substantive literature available on *able* children exposed to genetic risk, early neglect or familial disorganization), Nuechterlein [50] reviewed a largely un-published literature contained in the studies and documents to be found in ERIC (Educational Resources Information Center). Nuechterlein's summa thesis, The Competence Disadvantaged Child: A Review of Research, provided us with competence indicators in selected groups of children exposed to poverty. We used these indicators as a basis for narrowing the wide-ranging list of attributes assigned to the healthy personality, reasoning that children who had mastered environments weighted with stress and strain were an appropriate criterion group against which to study children vulnerable to psychopathology.

Applying the findings of Nuechterlein's review of the ERIC literature, we selected six indices of competence for use in our risk project, believing these to be the most relevant for adaptation during the period of middle childhood and early adolescence.

The Six Competence Criteria are:

1. Effectiveness in work, play and love; satisfactory educational and occupational progress; peer regard and friendships; behavior;
2. Healthy expectancies and the belief that "good outcomes" will follow from the imposition of effort and initiative; and orientation to success rather than the anticipation of failure in performing tasks; a realistic level of aspiration unbeclouded by unrealistically high or low goal-setting behavior;
3. Self-esteem, feelings of personal worthiness, a proper evaluative set toward self and a sense of "fate control," that is, the belief that one can control

events in one's environment rather than being a passive victim of them (an internal as opposed to external locus of control);

4. Self-discipline, as revealed by the ability to delay gratification and to maintain a future-orientedness;

5. Control and regulation of impulsive drives; the ability to adopt a reflective as opposed to an impulsive style in coping with problem situations;

6. The ability to think abstractly; to approach new situations flexibly and to be able to attempt alternate solutions to a problem.

To reduce these criteria to ratable proportions, our research group has devised a graphic Competence Rating Scale for Teachers.* The scale was designed for teachers for several reasons: teachers provide a common base for observations of target and control children; the schools have been very cooperative in allowing us access to records and classroom (with informal consent from parents); the children we observe share a common arena for observation of their competence qualities. Two items from the scale may provide an indication of its format.

A. *Relationship of achievement to ability level*: The extent to which the child's performance is consonant with his ability level. The level of performance can be high, moderate or low; rate only its consistency with ability level.

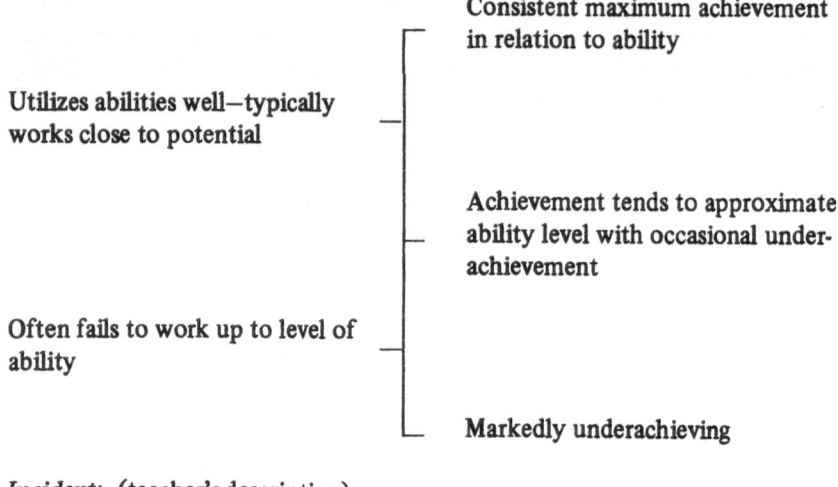

Incident: (teacher's description)

*Members of Project Competence who have been instrumental in the development of the Competence Rating Scale for Teachers include Beverly Kaemmer, Roger Lazoff, Lee Marcus, Keith Nuechterlein, Paul Sanders, and Susan Sanger.

B. *Control of aggression with peers*: Two factors are involved; the intensity of the child's anger and the appropriateness of it in terms of external events.

A bully; often cruel; threatens others without provocation

Overly aggressive; at times may fight or be verbally abusive with little provocation

Generally even-tempered; on occasion may overact with aggression but regains control quickly

Usually well-controlled; can be provoked but rarely gets angry without cause

Rarely antagonistic; anger, when shown, is justified.

Incident:

Stage One Study: Social and Academic Competence of Children Vulnerable to Schizophrenia and Other Behavior Pathologies* [56]

We place a high valuation on the first criterion of competence:

Effectiveness in work, play and love; satisfactory educational and occupational progress; peer regard and friendships.

It is rooted in the literature of prognosis in schizophrenia, socialization in childhood and studies of competence amidst poverty.

Contained within this first criterion is the important constellation described earlier—economic (or academic), social, and sexual competence. This factor became the initial study of Project Competence[†] in which Rolf [55] sought to measure the academic and social competence of six groups of elementary school children:

---

* Dr. Rolf is currently Assistant Professor of Psychology, University of Vermont.

[†] The project title has undergone a series of changes designed largely to underplay our concern with the child at risk. Originally titled *Vulnerable Children*, the research program is now identified as *Project Competence*. The change in title is not simply opportunistic, since we have begun to formulate studies of "invulnerable children" concurrent with ongoing studies of children at risk [25].

Four of these groups shared in common a potential vulnerability to divergent forms of psychopathology. One consisted of children whose mothers had had a previous history of hospitalization for schizophrenia (N = 31); another had mothers whose prior history included hospitalization for depression (N = 26); a third group was comprised of children who were being seen in psychiatric clinics for externalizing (acting-out) symptoms (N = 36), while a fourth group of clinic children presented internalizing (phobic, withdrawn) symptoms (N = 27). As far as could be determined by an appraisal of case histories, these clinic children did not have mentally ill parents. These four groups of target children, when located, were attending 37 schools and were in 113 different classrooms of a city public school system. Within each classroom, Rolf read all the cumulative records of the pupils to select two control children who in the opinion of principal, teacher or social worker were behaving competently. These two children and the target child comprised a triad consisting of a child at risk, a competent child matched in terms of his or her comparability to the target child on such factors as age, sex, grade, social class, intactness of the home, school achievement, and IQ (when available), and a competent, random control child of the same sex and grade but unmatched on the other demographic variables. Thus, the total number of subjects that comprised Rolf's triads was 360 (targets N = 120; matched controls, N = 120; random controls, N = 120).

Measures of social competency were derived from the ratings of peers and teachers. Sociometric ratings built around Bower's [5] "Class Play" were based on the judgments of some 3400 children who participated in the project. (Every child in each of the 113 classrooms successfully joined in the study.) The typecast assignments given to the members of the triad by their peers was the basis for determining peer approval or disapproval. Teachers completed rating scales designed to reflect positive academic behavior, emotional stability and maturity, social agreeableness and positive social extroversion. Academic competence was also analyzed by a thorough-going appraisal of the children's cumulative records and included an application of Watt et al. [62] method for quantifying dimensional attributes derived from qualitative descriptions of the children written by teachers over successive years. This last analysis is still in progress and cannot be reported in detail here. To date, the variables of yearly grade reports, health status, standardized achievement and intelligence test scores, grade reports, notations of absences and tardiness, socioeconomic status of the family and the frequency of address changes per academic year have been abstracted and readied for computer analysis. Overall grade averages for the most recent academic years do indicate that the sons of schizophrenic mothers appear to be significantly less achieving than their respective controls. Rolf further plans to subject the teachers' descriptive comments to a computerized analysis that will provide sketches of the modal personality profiles of each target group as well as to catalogue the traits of its individual members.

The study of social competence as adduced from peer and teacher ratings has undergone more extensive analysis, and provides partial confirmation of some of the negative social correlates of risk status. Since peer regard is a powerful predictor of adaptation [6,34,42], Rolf's findings are of particular interest.

Summarizing the results of his study, Rolf [56] writes:

> At the *triad* level, the prediction was generally supported that the target groups, in comparison with their respective control groups, obtained lower competence scores. At the target group level, the trend of peer-rated competence score ranks indicated that for both males and females the externalizing children were lowest, followed in order of ascending competence by the children with schizophrenic mothers, the internalizing children, and by the children whose mothers had internalizing symptoms. This order was also obtained for the *teacher-rated* competence scores, but only for the girls and not the boys. In the latter case, teachers rated externalizers lowest, and internalizers next lowest, but teachers generally did not discriminate, with certain exceptions, between the two male target groups with mentally ill mothers and their respective control groups.

Stage Two Studies: Performance Variables in Relation to Competence

The study of attentional factors in children-at-risk holds promise for evaluating precursor signs of deficit functioning. In the long history of experimental research with schizophrenic patients, the area of attention and preparatory set as measured by reaction time has repeatedly been shown to differentiate reliably patients from control subjects [59]. This suggests that the study of such attentional factors in children-at-risk could hold promise as a precursor sign to deficit functioning. To link such attentional components to independent indices of competence would satisfy the structure demanded of Stage Two studies. An ongoing experiment [45] using the basic design of Rolf's study with its four different groups of target children and two groups of matched and random control exemplifies the strategy. Three studies consitute the core of Marcus dissertation research: 1) attention and set are studied as a function of varying preparatory intervals presented in a simple reaction time procedure using regular and irregular RT sequences. (This study essentially replicates the classical RT study of Rodnick and Shakow [54] and those of Shakow and his associates [69].) A second study of attentional mechanisms uses the reaction time procedure with irregular preparatory intervals but with pre-trial information given to reduce temporal uncertainty. The third study also utilizes the basic reaction time task, but modified to focus on decision-making and choice behavior of the children under conditions of risk-taking in which the probability of positive and negative outcomes is varied by experimenter. The purpose of the latter study is to provide pilot data regarding behavior relevant to several competence criteria including success versus failure orientation, impulsive versus reflective cognitive styles, and internal versus external locus of control.

The Marcus study provides an example of the Minnesota strategy.* Reaction time is essentially a Stage Two variable; it has demonstrable significance as a measure of attentional deficit in schizophrenia and may provide in similar fashion a measure of task set in children. But it is necessary that a Stage One variable also be incorporated into the study to provide a measure of competence that can be correlated with RT performance. Such a proviso has been made in three ways: 1) formal evaluation of the children's cumulative school records; 2) teachers' ratings on the Competence Rating Scale; 3) the creation within the experiment of a risk-taking component to measure level of aspiration, locus of control, and cognitive style.

VERBAL ASSOCIATIVE BEHAVIOR OF CHILDREN VARYING IN VULNERABILITY TO PSYCHOPATHOLOGY: THE EFFECT OF REWARD AND CENSURE

In a study now getting underway,[†] we are attempting to test an observation by Mednick and Schulsinger [46,47] that children who are at risk for schizophrenia tend, in a continuous word association test, to show evidence of associative "drift." Thus, instead of maintaining a task set that generates sequential, relevant associations to the original stimulus word, such risk individuals, it is reported, tend to associate to their own previous responses so that subsequent response chains that appear bear no direct and evident relationship to the original stimulus word. Such findings are relevant to the successful maintenance of a task set, association deficits and subtle manifestations of cognitive deviance. As such, they accentuate the potential relevance of data derived from the association procedure as a Stage Two variable in the study of adaptation in children-at-risk.

The present experiment is more than a cross-validating effort. It provides for a test of the hypothesis using carefully matched groups of words which, prior to the continuous association procedure, are exposed to experimental manipulations designed to arouse in the subjects positive and negative affects specific to each

---

*Preliminary evidence from the Marcus study indicates that difficulty in maintaining a task set falls to the children born to schizophrenic mothers and clinic children manifesting externalizing behaviors. These findings parallel Rolf's data on social competence. Since there is overlap in the subjects within the risk groups, these results suggest the viability of the Stage One-Stage Two strategy.

[†] Coparticipants in the research include Roger Lazoff, Paul Sanders, Susan Sanger, and Keith Neuchterlein.

group of words. The methods for testing for drift involve, on the one hand, quantitative scores based upon the use of existent standard lists of the word associations provided by American children and, on the other, clinically sensitive methods designed to elicit the pattern of drift, if it occurs. The competence measures employed in Marcus' study will also be applied to the investigation of associative drift, with responsiveness to the stress imposed by success and failure experiences constituting an additional criterion of adaptation.

Stage Three Studies: Short-Term Longitudinal Studies

For a description of a program of risk research centering on a relatively short-term prospective design, it is necessary that I turn toward research that is now underway in the Department of Psychiatry at the University of Rochester Medical School. Under the coordinating leadership of Lyman Wynne more than 20 professional staff have gathered, drawn from four major universities (Rochester, Cornell, Berkeley, Minnesota) and a diversity of disciplines that include psychiatry, clinical psychology, developmental psychology, experimental psychopathology, epidemiology, sociology and statistics.

The program focus, rooted in the study of children and family members who are *at risk* for schizophrenia, is comprised of three subprograms, all interrelated and all central to an understanding of the problem of predisposition. One set of studies focuses on children and siblings born to schizophrenic and depressive mothers as defining the samples at risk; another core is oriented to studies of the families and the parents of these children with an emphasis on family organization, modes of intrafamilial communication and significant aspects of parenting patterns; a third group of studies is epidemiological in orientation and utilizes as instrument uniquely advantageous to epidemiological research—the Psychiatric Case Register for Monroe County, which is housed in the department's Division of Preventive and Social Psychiatry [3,48].

The program incorporates both a cross-sectional and a short-term longitudinal-prospective strategy which, in combination, may produce within a span of six years a projected developmental picture of the vulnerable child ranging from infancy to early adolescence. Hopefully, this can be achieved through the convergence procedure to which reference has been made earlier. This strategy involves the use of groups of children at ages one,* four, seven and ten who will be followed initially over a span of four years thus providing for retesting of subgroups of children at the convergence age points of four, seven, ten. If the project warrants

---

*The one year olds are children who are being seen in an infant risk project under the direction of professors Arnold Sameroff and Melvin Zax of the University of Rochester with the collaboration of Professor Harouton Babigian.

extension to seven years, it would be possible to strengthen the Ns at each con-vergence point and to extend the group's observations on these children into the period of puberty and beyond.

The variables to be studied are numerous and embrace such diverse compo-nents of adaptation as cognitive and social development employing a Piagetian framework, modes of information processing and sensory integration, psycho-physiological studies of children and parents, learning efficiency under censure and praise, quality of school functioning, formal diagnostic assessment of children and parents, modes of parenting, patterns of communication within the family, mother-child play interaction, styles of conflict resolution, follow-up studies of disturbed adolescents, and the role of birth complications in risk. Since the components of competence have been made an integral part of the research program, the relevance of adaptation to the range of dependent variables to be studied may indicate whether such a relatively short-term longitudinal-developmental program can sug-gest the pattern of outcomes in groups of vulnerable children.

Stage Four—Intervention

There is little to be said about this final stage of risk research, for no data have been brought to bear on the important problem of intervention. What, indeed, can one say about intervention when there exists such restricted information about the etiology of schizophrenia? On the island of Mauritius, Mednick and Schulsinger (see this volume) are readying a program of preschool training with children whose risk status is revealed not by the psychiatric status of parents but by the preschool-ers' "deviant" psychophysiology. The investigators indicate that they hope to pro-vide specific training for the children in learning to cope with frustration, to develop approach rather than avoidant behaviors in the face of environmental stress, and to encourage active play participation with peers. Such an orientation fulfills the specific function I have assigned to Stage Four studies.

A closer approximation to the projected intervention strategy and one whose power has been empirically demonstrated has been provided by one of my Rochester colleagues, Michael Chandler [10]. Reared in a developmental-Piagetian framework and tempered by clinical training at the Menninger Foundation, Chandler has been interested in problems of egocentricism and role playing as sig-nificant manifestations of adaptation within delinquent and abandoned-neglected children. Chandler sees the ability to demonstrate perspective or role-taking skills as a dispositional attribute of the maturing organism. Children with tenuous adaptive qualities presumably would show a lessened skill to assay the role of another and to perceive a self-other differentiation. Devising and adapting a variety of assessment procedures for measuring egocentricism, Chandler applied his pro-cedures to the study of 75 normal public school children between the ages of six and 13. Other groups consisted of children with "long histories of social and inter-personal failures" including 50 children from eight to 13 years of age who were institutionalized for emotional disturbance and 50 chronically delinquent boys. The disturbed children showed a significant deficit in role-taking skills. Most normal children by age 10 can adopt the orientation of another, but many disturbed children had difficulty in doing so and failed to show the maturing of such skills between the ages of six and 10. Chandler reports that the typical disturbed child at 13 was more egocentric than the average seven year old normal child. But variability prevailed in the disturbed group. Nonegocentricity characterized those children who had been brutalized and were "hyperalert to and suspicious of the motives of others."

Chandler's intervention strategy exemplifies best the Stage Four principle. In effect, he asked: "What would be the consequence of training these disturbed children in role-taking skills? His procedure was to take 50 chronically delinquent boys between the ages of 11 and 13 and to divide them into three groups on the basis of their prior egocentricism scores. One group remained untreated and was retested after 10 weeks. Remaining subjects were seen in groups of five for a total of 30 hours over a 10 week period. Half of these Ss were involved in making documentary films during the intervention period. Chandler's comments [10] on the experiences given the remaining half of the Ss warrants description:

Having noted the presence of substantial developmental delays in the role-taking abilities of various populations of disturbed children, an effort was made to formulate and test a program of remedial training intended as a means of testing the modifiability of these observed deficiencies.

There is little information in the research literature that provides useful clues as to how a curriculum for improving faulty role taking skills might be devised. What seemed required was some vehicle for transporting the subjects into the perspectives of others and for encouraging and facilitating their efforts to adopt roles other than their own. Efforts to satisfy these task requirements led to the development of a storefront video film and actors workshop where delinquent adolescents were provided closed-circuit television recording equipment and encouraged to make films about their own role taking efforts. They were helped to develop skits or short plays about persons their own age and required to replay and review these skits until each participant had occupied every role in the plot.

What, then, were the consequences of this training effort to get subjects to relinquish their egocentric biases and take on the more mature social skills commensurate with their age? The documentary film making group and the no treatment group, together with the role-playing groups, were all administered equivalent forms of the pretest role-taking test. Only the group trained in role-taking showed substantial improvement on the tests.

A broader question, of course, relates to the daily real-life adaptation of children exposed to such a specific regimen of social skills training. Chandler, in response to my direct inquiry on this important point, wrote to me of a group of delinquent and a group of abandoned and neglected children with whom he and his associates have worked:

> In both instances the Ss were selected on the grounds that they were egocentric beyond their years and were enrolled in a three-month long intervention effort which relied on the making of videotape movies of dramatic productions as a training vehicle. With the delinquent sample, behavioral ratings were made both before and after the termination of the treatment effort and the delinquent histories of the group members were examined through a search of police and court records. This record search was carried out at the outset of the program and at six, 12 and 18 months intervals after the program's termination. Program participants, in contrast to control Ss, did show some minor reduction in the level of their postprogram delinquent involvement. The population of institutionalized abandoned and neglected children were rated by teachers and custodial staff before and after the intervention project. In addition to changing in more manifest behavioral ways, these observed changes were generally in the direction of increased social competence and comfort in social situations.

Despite my enthusiasm for research of this type, I must confess to a concern regarding this fourth stage of a total research strategy. Can piecemeal treatment via skills training provide for the development of those higher skills necessary for adaptation in our complex society? I am torn between the search for a form of precise intervention and an awareness that the future environmental demands that children must face, whatever their degree of vulnerability to psychopathology, are so great that a broader program of intervention seems necessary. As Bruner [7] has indicated, the development of skill and competence occurs by small accretions on a day-to-day basis throughout infancy and childhood. Such skills lead to new mastery which, in turn, encourages the development of other skills that generate an in-depth competence that effectively inoculated an individual against the ravaging effects of disorder. But this pattern of skill acquisition requires a view of intervention that must be broadly gauged. Bruner's appraisal of the problem of intervention with disadvantaged children provides a fitting final statement to the critical problem of reducing the risk potential of vulnerable children:

. . . little can be done for a human being with a "one-shot" intervention. One has to work at it. Head Start alone does not work, if afterward the child is dumped into a punishing school experience. When we build an expectancy, build a skill, we incur a responsibility for nurturing it. It may, in some instances, be a compounding of evils to open the child's vulnerabilities and then disappoint or dump him. If we are to be effective in helping disadvantaged children cope better, it is their life cycle that must be dealt with not their preschool or their nursery or their street life. That is why we need diverse forms of care and can hardly tolerate quarrels about this form versus that form on ideological grounds rather than evidence. . . . The important thing is to get going. We must surely praise the attitude that though the first programs may not happen to be our preferred ones, nonetheless, we try to make them as good as possible knowing that we shall surely go on from there (pp. 115-116).

REFERENCES

1. Alvarez, R. R. A comparison of the preferences of schizophrenic and normal subjects for reward and punished stimuli. Unpublished Ph.D. dissertation, Duke University, 1957.
2. Atkinson, R. L. and Robinson, N. M. Paired associate learning by schizophrenic and normal subjects under conditions of personal and impersonal reward and punishment. *J. Ab. Soc. Psychol.* 62:322-326, 1961.
3. Babigian, H. M. Schizophrenia in Monroe County. In *Schizophrenia: Implications of research findings for treatment and teaching*, edited by M. M. Katz, A. Littlestone, L. Mosher, M. S. Roath, and A. H. Tuma (in press).
4. Bell, R. Q. Convergence: An accelerated longitudinal approach. *Child Development* 24:142-145, 1953.
5. Bower, E. M. *Early identification of emotionally handicapped children in school*, 2nd ed. Springfield, IL: Charles C. Thomas, 1969.
6. Bradford, N. Comparative perceptions of mothers and maternal roles by schizophrenic patients and their normal siblings. Unpublished Ph.D. dissertation. University of Minnesota, 1965.
7. Bruner, J. S. Discussion: Infant education as viewed by a psychologist. In: *Education of the infant and young child*, edited by V. H. Denenberg. New York: Academic Press, 1970, pp. 109-116.
8. Buss, A. H. and Lang, P. J. Psychological deficit in schizophrenia. I. Affect, reinforcement, and concept attainment. *J. Ab. Psychol.* 70:2-24, 1965.
9. Cavanaugh, D. K. Improvement in the performance of schizophrenics on concept formation tasks as a function of motivational change. *J. Ab. Soc. Psychol.* 57:8-12, 1958.

10. Chandler, M. J. Egocentricism in normal and pathological child development. Paper presented at the First Symposium of the International Society for the Study of Behavioral Development. Nijmegen, Holland, 1971.

11. Cicchetti, D. V. Reported family dynamics and psychopathology. *J. Ab. Psychol. 72*:282-289, 1967.

12. Clarke, A. R. Conformity behavior of schizophrenic subjects with maternal figures. *J. Ab. Soc. Psychol. 68*:45-53, 1964.

13. Coopersmith, S. *The antecedents of self-esteem.* San Francisco: W. H. Freeman Co., 1967.

14. Dunn, W. L., Jr. Visual discrimination of schizophrenic subjects as a function of stimulus meaning. *J. Pers. 23*:48-64, 1954.

15. Englehardt, R. S. Semantic correlates of interpersonal concepts and parental attributes in schizophrenia. Unpublished Ph.D. dissertation, Duke University, 1959.

16. Farina, A. Patterns of role dominance and conflict in parents of schizophrenic patients. *J. Ab. Soc. Psychol. 61*:31-38, 1960.

17. Farina, A. and Dunham, R. M. Measurement of family relationships and their effects. *Arch. Gen. Psych. 9*:64-73, 1963.

18. Farina, A. and Holzberg, J. D. Interaction patterns of parents and hospitalized sons diagnosed as schizophrenic or non-schizophrenic. *J. Ab. Psychol. 73*:114-118, 1968.

19. Garmezy, N. Approach and avoidance behavior of schizophrenic and normal subjects under conditions of reward and punishment. *J. Pers. 20*:253-276, 1952.

20. Garmezy, N. Prediction of performance in schizophrenia. In: *Psychopathology of schizophrenia*, edited by P. Hoch and J. Zubin. New York: Grune & Stratton, 1966, pp. 129-181.

21. Garmezy, N. Process and reactive schizophrenia: Some conceptions and issues. In: *The role and methodology of classification in psychiatry and psychopathology*, edited by M. Katz and J. Cole. U.S. Department of HEW, Government Printing Office. Reprinted (with addendum) in Schizophrenia Bulletin, Washington, D.C., NIMH, Vol. 1, A 2, 1970, pp. 30-74.

22. Garmezy, N. Commentary. *J. Nerv. Ment. Dis. 153*:317-322, 1971.

23. Garmezy, N. Research strategies for the study of children who are at risk for schizophrenia. In: *Schizophrenia: Implications of research findings for treatment and teaching*, edited by M. M. Katz, R. Littlestone, L. Mosher, M. S. Roath, and A. H. Tuma (in press).

24. Garmezy, N. Models of etiology for the study of children who are at risk for schizophrenia. In: *Life history research in psychopathology*, Vol. II, edited by M. Rolf, L. Robins, and M. M. Pollack. Minneapolis: University of Minnesota Press (in press).

25. Garmezy, N. and Neucheterlein, K. H. Vulnerable and invulnerable children: The fact and fiction of competence and disadvantage. *Am. J. Orthopsych.* (abstract) 77, 1972.

26. Garmezy, N. and Rodnick, E. H. Premorbid adjustment and performance in schizophrenia: Implications for interpreting heterogeneity in schizophrenia. *J. Nerv. Ment. Dis. 129*:450-466, 1959.

27. Goldstein, M. J. Premorbid adjustment, paranoid status, and patterns of response to phenothiazine in acute schizophrenia. Washington, D.C.: NIMH *Schizophrenia Bulletin*, No. 3, Winter 1970.

28. Goldstein, M. J., Judd, L. L., Rodnick, E. H., Alkire, A., and Gould, E. A method for studying influence and coping patterns within families of disturbed adolescents. *J. Nerv. Ment. Dis. 147*:233-251, 1968.

29. Goldstein, M. J., Judd, L. L., Rodnick, E. H., and LaPolla, A. Psychophysiological and behavioral effects of phenothiazine administration in acute schizophrenics as a function of premorbid status. *J. Psych. Res. 6*:271-287, 1969.

30. Goodman, D. Performance of good and premorbid male schizophrenics as a function of paternal vs. maternal censure. *J. Ab. Soc. Psychol. 69*:550-555, 1964.

31. Grinker, R. R. Psychiatry and our dangerous world. In: *Psychiatric research in our changing world*. Proceedings of an International Symposium. Montreal, 1968, Excerpta Medica International Congress Series, No. 187.

32. Haggard, E. A., Brekstad, A., and Skard, A. On the reliability of the anamnestic interview. *J. Ab. Soc. Psychol. 61*:311-318, 1960.

33. Harris, J. E. Size estimation of pictures as a function of thematic content for schizophrenic and normal subjects. *J. Pers. 25*:651-671, 1957.

34. Hartup, W. W. Peer interaction and social organization. In: *Manual of child psychology*, 3rd Ed., edited by P. H. Mussen. New York: John Wiley & Sons, 1970, pp. 361-456.

35. Higgins, J. The concept of process-reactive schizophrenia: Criteria and related research. *J. Nerv. Ment. Dis. 138*:9-25, 1964.

36. Higgins, J. Process reactive schizophrenia: Recent developments. *J. Nerv. Ment. Dis. 149*:450-472, 1969.

37. Higgins, J. and Peterson, J. C. Concept of process-reactive schizophrenia: A critique. *Psychol. Bull. 66*:201-206, 1966.

38. Kantor, R. E. and Herron, W. G. *Reactive and process schizophrenia*. Palo Alto, CA: Science & Behavior Books, 1966.

39. Klein, E. B., Cicchetti, D. V., and Spohn, H. E. A test of the censure-deficit model and its relation to premorbidity in the performance of schizophrenics. *J. Ab. Psychol. 72*:174-181, 1967.

40. Knight, R. Effect of punishment on performance of schizophrenics: A review of the literature. Minneapolis: University of Minnesota, 1968.

41. Leventhal, A. M. The effects of diagnostic category and reinforcer on learning and awareness. *J. Ab. Soc. Psychol. 59*:162-166, 1959.

42. Lippitt, R. and Gold, M. Classroom social structure as a mental health problem. *J. Soc. Issues 15*:40-49, 1959.

43. MacFarlane, J. W. Perspectives on personal consistency and change: The guidance study. *Vita Humana 7*:115-126, 1964.

44. MacFarlane, J. W. and Clausen, J. A. Childhood influences upon intelligence, personality and mental health. In: *The mental health of the child*, edited by J. Segal. Rockville, MD: National Institute of Mental Health, 1971, pp. 131-154.

45. Marcus, L. Attention, set and risk-taking in disturbed, vulnerable and normal children. Ph.D. dissertation proposal, Minneapolis University of Minnesota Project Competence, 1971.

46. Mednick, S. Breakdown in individuals at high risk for schizophrenia: Possible predispositional perinatal factors. *Ment. Hygiene 54*:50–63, 1970.
47. Mednick, S. A. and Schulsinger, F. Some premorbid characteristics related to breakdown in children with schizophrenic mothers. In: *The transmission of schizophrenia*, edited by D. Rosenthal and S. S. Kety. Oxford: Pergamon Press, 1968, pp. 267–291.
48. Miles, H. C. and Gardner, E. A. A psychiatric case register. *Arch. Gen. Psych. 14*:571–580, 1966.
49. Nathanson, I. A. A semantic differential analysis of parent–son relationships in schizophrenia. *J. Ab. Psychol 72*:277–281, 1967.
50. Nuechterlein, K. H. Competent disadvantaged children: A review of research. Thesis, University of Minnesota, 1970.
51. Phillips, L. Case history data and prognosis in schizophrenia. *J. Nerv. Ment. Dis. 117*:515–525, 1953.
52. Phillips, L. *Human adaptation and its failures*. New York: Academic Press, 1968.
53. Rodnick, E. H. and Garmezy, N. An experimental approach to the study of motivation in schizophrenia. In: *Nebraska Symposium on Motivation*, edited by M. R. Jones. Lincoln, NE: University of Nebraska Press, 1975, pp. 109–184.
54. Rodnick, E. H. and Shakow, D. Set in the schizophrenic as measured by a composite reaction time index. *Am. J. Psych. 97*:214–225, 1940.
55. Rolf, J. E. The academic and social competence of school children vulnerable to behavior pathology. Unpublished Ph.D. dissertation, University of Minnesota, 1969.
56. Rolf, J. E. The academic and social competence of children vulnerable to schizophrenia and other behavior pathologies. *J. Ab. Psychol.* (in press).
57. Rosenthal, D. *Genetic theory and abnormal behavior*. New York: McGraw-Hill, 1970.
58. Schaie, K. W. and Strother, C. R. A cross-sequential study of age changes in cognitive behavior. *Psychol. Bull 70*:671–680, 1968.
59. Shakow, D. Psychological deficit in schizophrenia. *Behavioral Science 8*:275–305, 1963.
60. Stoller, F. H. The effect of maternal evaluation on schizophrenics and their siblings. Unpublished doctoral dissertation, Los Angeles University of California, 1964.
61. Teele, J. E., Schleifer, M. J., Corman, L., and Larson, K. Teacher ratings, sociometric status, and choice-reciprocity of anti-social and normal boys. *Group Psychotherapy 19*(3–4):183–197, 1966.
62. Watt, N. F., Stolorow, R. D., Lubensky, A. W., and McClelland, D. C. School adjustment and behavior of children hospitalized for schizophrenia as adults. *Am. J. Orthopsych. 40*:637–657, 1970.
63. Wenar, C. Competence at one. *Merrill Palmer Quarterly 10*:329–342, 1964.
64. Wenar, C. *Personality development*. Boston: Houghton Mifflin Co., 1971.
65. White, B. L. An analysis of excellent early educational practices: Preliminary report. *Interchange 2*:71–88, 1971.

66. White, R. W. The experience of efficacy in schizophrenia. *Psychiatry 28*: 199-211, 1965.
67. Yarrow, M. R., Campbell, J. D., and Burton, R. W. Recollections of childhood: A study of the retrospective method. *Monographs of the Society for Research in Child Development 35*, No. 5, 1970.
68. Zahn, T. P., Rosenthal, D., and Shakow, D. Effects of irregular preparatory intervals on reaction time in schizophrenics. *J. Ab. Soc. Psychol. 63*:161-168, 1961.
69. Zahn, T. P., Rosenthal, D., and Shakow, D. Reaction time in schizophrenic and normal subjects in relation to the sequence of series of regular preparatory intervals. *J. Ab. Soc. Psychol 63*:161-168, 1961.

AUTHOR'S COMMENTS

Introduction

We are now removed by 20 to 30 years from the Duke studies in schizophrenia that spanned the decade 1950 to 1960. In 1973, its co-principal investigators wrote independent accounts of their then ongoing research at UCLA and Minnesota. Those post-Duke studies revealed a relationship between the earlier studies of the motivational and performance attributes of adult schizophrenic patients and the subsequent studies of children-at-risk for later psychopathology which were then being conducted independently at the two universities.

Readers interested in research continuities stemming out of the Duke studies may find it of value to turn to these 1973 articles in which Rodnick [25] wrote of *Antecedents and Continuities in Schizophreniform Behavior* and Garmezy [11] traced a lineage linking Duke and Minnesota through research studies focused on *Competence and Adaptation in Adult Schizophrenic Patients and Children at Risk*.

Now, a decade later one can perceive some of the ramifications of these earlier investigations of schizophrenia. In 1973, examining the research program that had been conducted at Duke, I wrote of six areas of research in schizophrenia that characterized the project.

1. The first, and undoubtedly the most significant, transcended any individual experiment and came to serve as the basic framework for subject selection. This was the effort to test the experimental consequences of differentiating patients along a continuum of good or poor premorbid competence;

2. Within the context of patient selection, emphasis was placed upon the creation of experimental-laboratory analogues that made use of constructs that were conceptually significant for psychodynamic formulations of the development of schizophrenia;

3. Specifically, the stimulus focus became the depiction of early family relationships that emphasized the power and punishment aspects of the mother-child and, to a lesser extent, the father-child relationship. Later in the program ways were sought to test for variations in such patient-parent relationships as a function of the premorbid interpersonal competence of the patient;

4. Interest in the reinforcing properties of punishment led to studies designed more broadly to examine the effects of censure *per se* on the level of performance of schizophrenic patients in comparison with other normal control subjects;

5. The study of such effects suggested an ubiquitous pattern of avoidance responding that appeared to characterize patients exposed to censure stress;

6. Finally when the so-called "tranquilizers" began to assume therapeutic ascendancy in the mid and late 1950s, preliminary assessments were begun on the influence of the phenothiazines on the patients' avoidance response systems under conditions of stress (p. 165).

There is a danger in looking back to one's earlier research efforts, for reflections on the past can lead to a justification for studies which the passage of time and the acquisition of new knowledge have rendered obsolete. This awareness creates even greater discomfort when a review of the recent literature on aftercare of schizophrenic patients suggests that the evidence to support the central tenets of the Duke studies seem more viable today than they did two decades ago. Let me attempt to provide the linkages between studies in the present and those of the past.

These new links to the six investigative areas covered in the original Duke studies represent extensions provided by seven different areas of current research:

1. Revisions in the classification of schizophrenia [1]

2. The management of schizophrenia in the community [33]

3. The influence of the family on the course of psychiatric illness [6,9,17,17a,29,30]

4. Laboratory studies of schizophrenic patients living in families that vary in their manifestations of *expressed emotion* [27a,28]

5. Studies of the extent and magnitude of the patient's contact with mother and degree of medication usage in relation to the likelihood of the patient's stay in the community [6,14a]

6. Skills training for patients and family members designed to reduce manifestations of expressed emotion in these families [9a,14,18,20,22a]

7. Studies of children at risk for schizophrenia [12,13,21,24,32]

The Classification of Schizophrenia

The revision of the *Diagnostic and Statistical Manual of Mental Disorders: DSM-III* [1] has brought radical changes in the classification of schizophrenia [10]. The current focus has been placed on: (1) certain specific psychotic symptoms; (2) deterioration from a previous level of functioning in work and social relationships; (3) prolongation of the illness; (4) age of onset; (5) absence of etiological factors related to affective disorders, organic conditions or mental retardation. Schizophreniform disorder, a new classification now removed from schizophrenia, is described as being of shorter duration, involving more emotional turmoil and confusion, a markedly acute onset, a more favorable prognosis with anticipated recovery to the premorbid level of functioning, and an absence of evidence of a heightened prevalence of schizophrenia within the family pedigree.

These differentiations seem to apply to many aspects of the poor premorbid-good premorbid dichotomy used by the Duke investigators to create more homogenous subsamples of patients in an effort to reduce the marked variability that was evident in the total group of schizophrenic patients tested. This heterogeneity had resulted initially in a failure to obtain significant differences between the index group of patients and their normal controls.

Selection criteria in the early Duke studies were based solely on the patient's level of economic, social and sexual functioning in the premorbid period. There were, however, correlated criteria as well that were components of the Phillips Scale [23] although we had not used them. These included brief duration, acute onset, a favorable prognosis, and a return to the prior level of pre-illness functioning.

Although we never investigated the prevalence of schizophrenia in the families of our poor premorbid and good premorbid subgroups, there were behavioral indices evident in parents and patients that suggested a greater degree of disturbed relationships in the nuclear families of the poor premorbid patients.

At the level of the patients' functioning there were redundant findings of deficit behaviors in the poor premorbid group. The deterioration of performance evident solely in the poor premorbid patients when symbolic cues of censure were employed suggested that our recurrent findings of differences led us to infer that we were likely dealing with two different entities. The appearance of DSM-III provides a determinate support for such a differentiation in the psychiatric nomenclature that may have been facilitated by the components of the Phillips Scale and the consequences of its usage in laboratory studies.

Censure and Deficit in Performance in Schizophrenia

As noted, the Duke studies provided reliable evidence of poorer performance in the poor premorbid schizophrenic patients when censure cues were operative.

However, these results frequently went unconfirmed by subsequent studies in other laboratories, while there were reaffirmations of our findings in other investigations. One evaluation of 91 studies reviewed by R. Knight [11] found that "In general, schizophrenics were found to be more reactive to punishment than to reward, remaining unchanged in their performance in fewer instances when punished than rewarded. Also negative reinforcement was found to be the most successful reinforcers for schizophrenics, producing far more positive and fewer negative results than any other reinforcer."

Nevertheless, there remained a considerable number of studies published during the 1960s, that produced inconsistent results and others that failed to replicate our basic findings. Klein and Salzman [16] in discussing these inconsistencies have written of the problems associated with the studies.

> Discrepant findings may have resulted from variations in procedure or from sampling problems which interfere with integration of data from different studies. Social class, chronicity, social competence, sex, and premorbid status, for example, often differ or are uncontrolled in the sample studies; these potentially important differences have received little attention (p. 249).

Perhaps the most significant factor of all may have been omitted from consideration. The difficulty, if not the failure, to remove phenothiazine therapy from patients who were research participants assured a confounding factor on poor premorbid patients once the therapeutic drug revolution (which began in the late 1950s) was extended with great vigor throughout the 1960s and 1970s. Early demonstrations of a reduction in avoidance behavior to censure reinforcement, such as that by Garmezy [11], suggested a mechanism that could have been partially responsible for the reduction of its power to generate deficit behavior in the poor premorbid group.

The Institute of Psychiatry's Studies of the Role of the Family on the Course of Schizophrenic Disorders

In 1958, George Brown and his colleagues published a paper in *Lancet* that was the forerunner of a very significant series of studies that stretched into the 1970s [3,4,7-9,17,27,29,30].

The early observations of these distinguished British researchers stemmed out of a long-term interest on the part of John Wing and George Brown that is reflected in two volumes: *Institutionalism and Schizophrenia* [34] and *Schizophrenia and Social Care* [5].

Three paragraphs extracted from the 1958 *Lancet* paper presaged the direction of the research that was to follow:

Psychiatrists are familiar with the mother who tolerates and perhaps even encourages, an attitude of child-like dependence in her schizophrenic son. This kind of relationship was especially common in the 53% of unemployed schizophrenics living with their parents and was very rarely found in other types of living groups. Perhaps it is not always beneficial for a schizophrenic to return to the *close emotional ties of a parental or marital household.*

There was some evidence that behavior deteriorated more in schizophrenic patients who went to parental and marital groups. For example, the relative incidence of *violent outbursts and displays of temper* was significantly greater in those groups, whereas other psychotic manifestations (e.g., delusions) were more often reported for patients living with other kin and in lodgings.

Where a schizophrenic lived with his parents and was unemployed, with the result that he was brought into *close contact with his mother* all day, he was likely to fail; but if he or his mother was away at work for part of each day, he was more likely to succeed (p. 687).

There are three elements clearly evident in the above passage: first, the heightened emotional reactivity that characterizes some families; second, the reciprocal reactivity of patients to such experiences; third, the therapeutic necessity of reducing the amount of time patient and mother spend together.

These three factors are interrelated components of a construct now called *expressed emotionality* (EE). On the parental side there are three elements in the behavior of family members that contribute to the overall EE index. The first is the number of *critical comments* made by the family about the patient; the second are manifestations of their *hostility*; the third is the family's degree of *emotional overinvolvement* [29]. Of these three, "the single most important measure contributing to the overall index of a relative's expressed emotion proved to be the number of critical remarks made about the patient by the relative when interviewed alone. Hostility and emotional overinvolvement also contributed to the overall index, but hostility appeared to be highly related to criticism, while marked emotional overinvolvement was found only in parents and not in spouses" [30].

The similarity between the current formulation of expressed emotion and the Duke findings that schizophrenic patients proved to be markedly sensitive to the effects of personal criticism seems to be quite apparent. The consistency is even more striking when one considers that the construct of EE was formulated a decade or more after the termination of the Duke research program, and that its investigators were concerned with research, not in a laboratory setting, but rather through observations and interviews conducted in the patient's homes. These observational studies appear to have far more ecological validity than did those many laboratory analogues of the 1950s and 1960s which produced both confirmation and disconfirmation of the central findings of the Duke research program.

There are two other facets to the study of expressed emotion in the families of schizophrenic patients that warrant mention. The first is the recent return to the laboratory to test the construct validity of the EE variable. Tarrier et al. [28] have measured psychophysiological responsiveness in schizophrenic patients both in their homes and in the laboratory. In the home the patient was brought into face to face contact with the key relative who remained during the period of the recording, but who was not present during study in the laboratory setting. Included in the study were schizophrenic patients from high and low EE families. In addition, a group of physically healthy, age, and sex-matched normal subjects with no history of psychiatric illness served as controls. Skin resistance and heart rate were first recorded with only the experimenter present and then was repeated with the key relative in the room. During testing the patients and relatives talked about the illness, the hospitalization of the patient, and psychiatry in general. The controls discussed only the last named topic.

The results indicated that the laboratory setting failed to elicit significant differences, whereas in the home situation, with the key relative present, significantly higher levels of ANS arousal were present in the patient group, with high EE patients significantly more aroused than were their low EE counterparts. Furthermore, the low EE patients showed habituation in their arousal response despite the continued presence of the relatives. By contrast, in the laboratory, with the key relative absent, high and low EE patients failed to differ in their psychophysiological responsiveness. Recently Sturgeon et al. [27a] have essentially confirmed these findings with acutely ill patients.

Zahn et al. [36] recent report of a greater probability of relapse in acutely ill schizophrenic patients who show a high resting state of arousal, slow habituation, and attenuated ANS activity to demanding stimuli and situations adds further support to a network of consequences.

These studies would seem to establish a link between manifest and extensive criticism of some schizophrenic patients by key relatives (high EE), accompanied by high physiological arousal states in the patients which can trigger a return to the hospital. Low EE patients, by contrast, seem to live in calmer environments outside the hospital which can serve to buffer them from the stressors that are inevitably present for former mental patients in their social and work environments.

These data also suggest that the prognoses of the two groups may differ, with the likelihood of a more chronic pattern in one group (high EE) and a greater tendency toward stable remission in the other (low EE).

Preventing Relapse: Phenothiazines and Social Skills

Mention has already been made of the greater resistiveness to the negative effects of censure seen in chronic patients who received antipsychotic medication following its introduction into psychiatric treatment. It is, therefore, of great interest to observe that one way of attenuating the regressive effects of a high EE relative's influence on the patient is to insure maintenance of the drug regimen. A concurrent intervention effort that has been suggested is to reduce contact between patient and relative [29]. If these two factors are operative, they exert an important protective influence on the patient enabling him to achieve a more favorable clinical outcome [30]. A third element in this therapeutic effort is to train the patient in needed social skills so as to encourage participation in the community and to foster efforts to move him satisfactorily into the broader community that exists beyond the familial homestead.

We are now witnessing an upsurge in social skills training aimed at preparing the patient to live more satisfactorily and in a more stable fashion following discharge from the mental hospital. This union of medical and behavioral interventions would seem to bode well for future efforts to contain the movement of the "revolving door" of the mental hospitals.

Children-at-Risk

One can speak only in a general way of the proliferation of studies of children who are perceived to be at risk for schizophrenia and/or other forms of psychopathology. There have been efforts to study the effects of censure and praise on performance in learning tasks of children born of schizophrenic parents [15]. Results indicate that these children performed significantly poorer than did children of nonschizophrenic parents on a nonsense syllable discrimination task administered under conditions of response-contingent social reinforcement (both praise and censure). With the children's mothers providing the reinforcement, schizophrenics' offspring made twice the number of errors as did children of controls, but this was true under both praise and censure conditions.

Recently, these same investigators [16] replicated their earlier study demonstrating that their results held not only with a 10-year-old group of children at risk but with seven-year-olds as well. A control group of nonpsychotic patients' offspring, however, produced results comparable with those obtained with offspring of schizophrenic mothers. Thus, there may be a greater degree of generalization produced by the presence of mothers or the sound of their voices that extends beyond any single age risk group.

Finally, there have been studies of the competence qualities of the children of schizophrenic mothers that parallel to some extent the effort to secure premorbid ratings of social competence as indexed by classroom peers [22,26].

A lengthy treatment of the status of the findings of all of the major risk groups is now available to provide an update on the similarities and the differences that are to be found between the behavior of schizophrenic patients and children who are born to schizophrenic parents when compared with normal children and those to other forms of deviance [32].

The more recent studies represent an effort to provide a developmental account of children vulnerable to schizophrenia. A sturdy prospective literature can undo the shortcomings of the retrospective method that has been used previously to elicit data on the premorbid developmental histories of those who in time may become mentally disordered. It is also likely to provide information about other children similarly at risk who manage to escape disorder. In accomplishing these goals, such studies will move far beyond the purview of the early Duke studies.

REFERENCES

1. American Psychiatric Association. *Diagnostic and statistical manual of mental disorders: DSM-III.* Washington, D.C., 1980.
2. Bellak, A. S., Hersen, M., and Turner, S. M. Generalization effects of social skills training with chronic schizophrenics: An experimental analysis. *Behav. Res. Ther. 14*:391-398, 1976.
3. Birley, J. L. T. and Brown, G. W. Crises and life changes preceding the onset or relapse of acute schizophrenia: Clinical aspects. *Br. J. Psych. 116*:327-333, 1970.
4. Brown, G. W., Birley, J. L. T., and Wing, J. K. Influence of family life on the course of schizophrenic disorders: A replication. *Br. J. Psych. 121*:241-258, 1972.
5. Brown, G. W., Bone, M., Dalison, B., and Wing, J. K. *Schizophrenia and social care.* London: Oxford University Press, 1966.
6. Brown, G. W., Carstairs, G. M., and Topping, G. Post-hospital adjustment of chronic mental patients. *Lancet 2*:685-689, 1958.
7. Brown, G. W., Monck, E. M., Carstairs, G. M., and Wing, J. K. Influence of family life on the course of schizophrenic illness. *Br. J. Prev. Soc. Med. 16*: 55-68, 1962.
8. Brown, G. W. and Rutter, M. The measurement of family activities and relationships: A methodological study. *Human Relations 19*:241-263, 1966.
9. Creer, C. and Wing, J. K. Living with a schizophrenic. *Br. J. Hosp. Med. 14*: 73-82, 1975.
9a. Falloon, I. R. H., Boyd, J. L., McGill, C. W., Razani, J., Moss, H. B., and Gilderman, A. M. Family management in the prevention of exacerbations of schizophrenia. *New Engl. J. Med. 306*:1437-1440, 1982.
10. Fox, H. A. The DSM-III concept of schizophrenia. *Br. J. Psych. 138*:60-63, 1981.
11. Garmezy, N. Competence and adaptation in adult schizophrenic patients and children at risk. In: *Schizophrenia: The first ten Dean Award Lectures,* edited by S. R. Dean. New York: MSS Information Corp., 1973.
12. Garmezy, N. Children at risk: The search for the antecedents to schizophrenia. Part II: Ongoing research programs, issues, and intervention. *Schiz. Bull. 1*(9):55-125, 1974.

13. Garmezy, N. and Streitman, S. Children at risk. The search for the antecedents of schizophrenia. Part I. Conceptual models and research methods. *Schiz. Bull.* *1*(8):14-90, 1974.

14. Hersen, M. and Bellak, A. S. Social skills training for chronic psychiatric patients: Rationale, research findings, and future directions. *Compr. Psych.* 7: 559-580, 1976.

14a. Hogarty, G. E. and Goldberg, S. C. Drugs and social therapy in aftercare of schizophrenic patients. *Arch. Gen. Psych.* *28*:54-64, 1973.

15. Klein, R. H. and Salzman, L. F. Censure-praise learning of children at risk. *J. Nerv. Ment. Dis.* *166*:799-804, 1978.

16. Klein, R. H. and Salzman, L. F. Response-contingent learning in children at risk. *J. Nerv. Ment. Dis.* *169*:249-252, 1981.

17. Leff, J. P. Schizophrenia and sensitivity to the family environment. *Schiz. Bull.* *2*:566-574, 1976.

17a. Leff, J. and Vaughn, C. The role of maintenance therapy and relatives' expressed emotion in relapse of schizophrenia: A two-year follow-up. *Br. J. Psych.* *139*:102-104, 1981.

18. Liberman, R. P., Falloon, L. I., et al. *Social skills training for relapsing schizophrenics: An experimental analysis.* Unpublished manuscript, 1978.

19. Liberman, R. P., Wallace, C., Teigen, J., et al. Behavioral interventions with psychotics. In: *Innovative treatment methods in psychopathology*, edited by K. S. Calhoun, H. E. Adams, and E. M. Mitchell. New York: John Wiley & Sons, 1974.

20. Liberman, R. P., Wallace, C. J., Vaughn, C. E., Snyder, K. S., and Rust, C. Social and family factors in the course of schizophrenia. Towards an interpersonal problem-solving therapy for schizophrenics and their families. Presented at the Conference on Psychotherapy of Schizophrenia. New Haven, CT: Yale University School of Medicine, April 1979.

21. Mednick, S. A., Schulsinger, H., and Schulsinger, F. Schizophrenia in children of schizophrenic mothers. In: *Childhood personality and psychopathology: Current topics Vol. II*, edited by A. Davids. New York: John Wiley & Sons, 1975, pp. 221-252.

22. Nuechterlein, K. Dysfunctions of sustained attention and personality attributes of children vulnerable to schizophrenia and other adult psychopathology. Unpublished Ph.D. dissertation, Minneapolis, University of Minnesota, 1978.

22a. Paul, G. L. and Lentz, R. J. *Psychocial treatment of chronic patients.* Cambridge, MA: Harvard University Press, 1977.

23. Phillips, L. Case history data and prognosis in schizophrenia. *J. Nerv. Ment. Dis.* *117*:515-525, 1953.

24. Rieder, R. O. Children at risk. In: *Disorders of the schizophrenic syndrome*, edited by L. Bellak. New York: Basic Books, 1979, pp. 232-263.

25. Rodnick, E. H. Antecedents and continuities in schizophreniform behavior. In: *Schizophrenia: The first ten Dean Award Lectures*, edited by S. R. Dean. New York: MSS Information Corporation, pp. 139-161, 1973.

26. Rolf, J. E. The academic and social competence of children vulnerable to schizophrenia and other behavior pathologies. *J. Ab. Psych. 80*:225-243, 1972.

27. Rutter, M. and Brown, G. W. The reliability and validity of measures of family life and relationships in families containing a psychiatric patient. *Soc. Psych. 1*:38-53, 1966.

27a. Sturgeon, D., Kuipers, L., Berkowitz, R., Turpin, G., and Leff, J. P. Psychophysiological responses of schizophrenic patients to high and low expressed emotion relatives. *Br. J. Psych. 138*:40-45, 1981.

28. Tarrier, N., Vaughn, C. E., Lader, M. H., and Leff, J. P. Bodily reactions to people and events in schizophrenia. *Arch. Gen. Psych. 36*:311-315, 1979.

29. Vaughn, C. E. and Leff, J. P. The influence of family and social factors on the course of psychiatric illness. *Br. J. Psych. 129*:125-137, 1976.

30. Vaughn, C. E. and Leff, J. P. The measurement of expressed emotion in the families of psychiatric patients. *Br. J. Soc. Clin. Psychol. 15*:157-165, 1976.

31. Wallace, C. J., Nelson, C. J., Liberman, R. P., Aitchison, R. A., Lukoff, D., Elder, J. P., and Ferris, C. A review and critique of social skills training with schizophrenic patients. *Schiz. Bull. 6*:42-63, 1980.

32. Watt, N. J., Anthony, E. J., Wynne, L. C., and Rolf, J. E. (Eds.). *Children at risk for schizophrenia: A longitudinal perspective.* Cambridge: Cambridge University Press, 1984.

33. Wing, J. K. The management of schizophrenia in the community. In: *Psychiatric medicine*, edited by G. Usdin. New York: Brunner/Mazel, 1977, pp. 427-477.

34. Wing, J. K. and Brown, G. W. *Institutionalization and schizophrenia.* Cambridge: Cambridge University Press, 1970.

35. Zahn, T. P., Carpenter, W. T., and McGlashan, T. H. Autonomic nervous system activity in acute schizophrenia. *Arch. Gen. Psych. 38*:260-266, 1981.

The Management of Schizophrenia in the Community

John K. Wing, M.D., Ph.D.

INTRODUCTION

The title of this paper requires explanation. In the early part of the nineteenth century, such care as people with schizophrenia received was given "in the community," not in hospital. Attitudes toward institutional and community psychiatry have changed over the years. The first psychiatric hospitals were set up, both in Britain and in the United States, in reaction against the appalling conditions of "community care" then prevailing [12,22]. They were small, with a high turnover, and in the best of them the staff fostered a family atmosphere, with emphasis on moral treatment, the ideals of which were not altogether dissimilar to those fashionable during the past 20 years, and took pride in the fact that they used no methods of physical restraint. Since then, the balance of advantage to the patient, of receiving care inside hospital compared with receiving care outside it, has swung both ways, but it has never been an absolutely clear-cut decision as to which was preferable; there has always been something to be said on both sides. The psychiatric hospitals, in both our countries, went through a prolonged period in which patients experienced pauperism, neglect, institutionalism, restraint, and occasionally cruelty. This was the "custodial" era. Whether conditions outside would have been any better at that time is rarely discussed.

As late as the 1930s, in England, a patient admitted for the first time with a

Research in the Schizophrenic Disorders: The Stanley R. Dean Award Lectures, vol. 2, edited by R. Cancro and S. R. Dean. Copyright © 1985 by Spectrum Publications.

diagnosis of schizophrenia had only one chance in three of discharge within two years. After two years, the discharge curve reached a plateau, and the only chance most people had of being taken off the books was by dying. There was, of course, a large excess mortality. The English Mental Treatment Act of 1930 was the first substantial movement back of the pendulum for half a century, since it allowed patients to be admitted voluntarily instead of by commitment. When a fresh influx of psychiatrists entered the mental hospital service after the second world war, full of high social ideals and pioneering spirit that was characteristic of that time, they found large numbers of patients whose psychiatric condition did not justify being in hospital. This was the time when new concepts of social treatment, offering alternatives to long-term hospitalization, were introduced: the therapeutic community, social and vocational rehabilitation, sheltered work, after-care, day centers, hostels, domiciliary supervision, crisis intervention, pre-admission screening [57]. In hospitals such as Mapperley, in Nottingham, the numbers of beds were being drastically reduced long before reserpine and chlorpromazine became available [58].

The combination of social and pharmacological treatments was astonishingly successful. Large numbers of patients were discharged and the expectation became that single or multiple admissions to hospital would rarely result in a prolonged stay. Since the mid-1950s, the number of occupied beds in mental hospitals has gradually been decreasing, and it is still going down. Because of the marked improvement often occurring in patients who had previously spent long years in hospital, some psychiatrists began to think most of the symptoms and the disability found in long-stay hospital patients were actually caused by being in the institution. If people could be prevented from coming into hospital, much of this morbidity would be prevented altogether. At the same time, sociologists were showing that, in some areas of the United States, people were being admitted to hospital and given a diagnosis of schizophrenia although their problems appeared to be social rather than medical, and the functions of the hospital seemed custodial rather than therapeutic [39]. Some of the seeds of the anti-psychiatry movement were actually planted by psychiatrists themselves. The very success of psychiatric efforts seemed to show that psychiatry was unnecessary, at least so far as the functional psychoses were concerned. In some parts of the United States it became as hard to get into a hospital as, in former days, it had been hard to get out of one.

We have passed through a similar period in England, though not perhaps quite so extreme as in some states here, but there is general agreement that the time has come for a reappraisal of the situation. This is due to two factors. The first is that many patients quite clearly continue to be disabled, and vulnerable to relapse, even though they have spent only a brief time in hospital, or no time at all. The other is the rediscovery that "community" (that is, non-hospital) *care* also has deficiencies, sometimes quite as great as those of hospitals. The conditions most involved are the major psychoses, dementia and mental retardation, but outstanding among them is schizophrenia, so much so that we can use it as an acid test against which to measure the success of services.

From now on, when I use the term "community care," I shall assume that it includes the services provided in hospitals as well as those provided outside. Three principles of service organizations have evolved in different parts of the world, and are now accepted as being fundamental, although they have been applied with very varied degrees of thoroughness. First, there is the principle of district responsibility. This means that it is possible to identify, in any geographical area, who is responsible for providing the service and who, therefore, is responsible if the service is ineffective. It means that the service is geographically accessible to the local population. The main danger is that rigid insistence on geographical boundaries could limit choice. It is important, therefore, that this principle, like the others, should be applied flexibly and in the interest of the consumer. The second principle is that a sufficiently broad range of services, both medical and social, should be provided to cover all contingencies. The third principle is that any service which provides a wide variety of social and medical agencies needs to be properly coordinated, so that there is adequate communication between agencies and no block or delay when a patient moves from one part of the service to another. These three principles are conveniently summarized as the provision of a responsible, comprehensive, and integrated community service [54].

However, important as these principles are, there is one that is considerably more fundamental. The chief aim of the health services is to decrease or contain disease, disability, or distress—first, in the patient; second, in the patient's immediate family; third, in the community at large. Each service agency has a combination of diagnostic, therapeutic, rehabilitative, and preventive functions. Prevention is better than cure. Primary, secondary, and tertiary preventive methods are used to stop disease occurring in the first place, to detect it at an early stage, to limit development of chronic disabilities following acute disease, and to prevent the accumulation of harmful secondary reactions, if chronic intrinsic impairments are unavoidable. This is what is meant by the "containment" of morbidity.

The concept of "management" is basically the same as that of containment. In spite of the bureaucratic sound of the word, it is a useful one and connotations of officialdom and authoritarianism should not be read into it. In the rest of this presentation, I shall try to summarize the knowledge that has accumulated during the past 30 years that enables us now to help, by treatment, counselling, provision of services, and the fostering of self-help, to minimize and contain disabilities that were once regarded as carrying a virtual sentence of institutionalization for life. I shall need to discuss the definition of the acute and chronic syndromes, their prevalence and course, their social reactivity, the influence of adverse secondary reactions and extrinsic disadvantages, and the attitudes and expectations of people in the immediate and more remote social environment. It is only on the basis of this knowledge that we can begin to plan services rationally and hope to give the most effective individual help to affected individuals and their families.

THE ACUTE SYNDROMES OF SCHIZOPHRENIA

The *concept of schizophrenia* has been expanding and contracting like a concertina ever since it was invented. Kraepelin himself merged the *démence précoce* of Morel, Hecker's *hebephrenia*, and Kahlbaum's *catatonia*, into one concept of *dementia* praecox, leaving the *paraphrenias* and *paranoia* as separate entities. Bleuler ended this uncomfortable separation; however, by insisting that all schizophrenias were based upon only two *fundamental symptoms*, neither of which was easy to define (flatness of affect and loosening of associations), he opened the door so wide that some clinicians were able to diagnose virtually anyone as schizophrenic. Berze's suggestion that a general lowering of "psychic activity" was the key factor was even more vague. Psychodynamic formulations, based on the recognition of primitive or infantile types of thinking, have to be interpreted by experts, and none has ever shown that the judgments involved can be made reliably. Gruhle, Jaspers, and Kurt Schneider did something to redress the balance by emphasizing the processes underlying delusions and hallucinations, but Kleist and Leonhard took the process of differentiation to lengths that bordered on the absurd.

The upshot is that several *diagnostic schools* have evolved in different parts of the world: in the USSR, in France, in the UK, in Germany and other parts of Europe, and in the U.S. The *lifetime expectancy* of schizophrenia calculated from data in the Rochester (New York) *case register* is three per cent, compared with an estimate of just under one per cent derived from the similar register in London, England. This three fold difference must bedevil any attempt at generalization unless we keep it firmly in mind, and it is therefore worthwhile considering in some detail what is involved.

Morton Kramer compared age-adjusted *first admission rates* to public and private mental hospitals in the U.S. with equivalent rates in England and Wales [26]. The American rate in 1960 was 24.7 per 100,000 total population compared with 17.4 for England and Wales. The rates for the major affective psychoses, on the other hand, were 11.0 and 38.5, respectively. Table 1 shows the one-year *prevalence rates* (based on the unduplicated statistics of psychiatric services during 1964) in three urban areas covered by *case registers*—Aberdeen, Scotland; Baltimore, Maryland; and London, England. Schizophrenia was diagnosed markedly more often in Baltimore than in the other two localities, in both sexes, and particularly between the ages of 25 and 64. Depressions were commoner at all ages in Britain but were particularly common in women.

That much of the differences is accounted for by cross-cultural diagnostic biases is suggested by results of the US–UK *Diagnostic Project* in which a standard clinical technique (the *Present State Examination*) was used to interview patients admitted to hospitals in New York and London, usually with acute and severe psychiatric disorders [7]. When hospital diagnoses were compared, the higher frequency of schizophrenia in New York and of affective disorders in London was confirmed. When standardized diagnoses were compared, there was very little

TABLE 1. One-Year Prevalence of Schizophrenic and Depressive Disorders in Three Urban Areas Covered by Case Registers

| | Rates per 100,000 local population, age 15 years and over, in: | | | |
| | Baltimore, USA | | Aberdeen, Scotland | London, England |
Diagnostic category	Non-white	White		
Schizophrenias	722	685	246	317
Manic-depressive psychoses	59	135	225	377
Other depressions	80	134	338	519

Adapted from Wing, Wing, Hailey, Bahn and Baldwin [57].

difference between the two series. In other words, American psychiatrists were using broader criteria to diagnose schizophrenia than their British colleagues. However, the project did not help to determine which of these biases was correct.

Another large-scale international comparison was carried out under the auspices of the *World Health Organization—the International Pilot Study of Schizophrenia (IPSS)* [68]. Nine centers took part, chosen for the cultural diversity of their populations and the divergent schools of psychiatric thought represented. They were: Aarhus (Denmark), Agra (India), Cali (Colombia), Ibadan (Nigeria), London (England), Moscow (USSR), Prague (Czechoslovakia), Taipei (Taiwan), and Washington (USA). In this study, in addition to the standard techniques of examination used in the U.S.-U.K. Diagnostic Project, a standard classification procedure was also used, which produced a reference "diagnosis" based on the standard ratings of symptoms classified according to rules laid down in a computer program (known as CATEGO), thus eliminating variations in subjective judgment [62]. If nine per cent of cases are omitted because insufficient information is available for a reference classification, there is a substantial degree of agreement (in fact, almost an identity) between three of the major CATEGO classes* (S, P and O) and a center diagnosis of schizophrenic or paranoid psychosis. This was true of all nine centers. However, the same was true of concordance between a clinical diagnosis of affective psychosis or neurosis and the equivalent CATEGO classes (M, D, R and N) only in seven of the centers, while in the other two, there was a major amount of discrepancy. The latter two centers were in Moscow and Washington. The data are summarized in Table 2. What this means is that virtually all the cases placed by

---

*For convenience, the major CATEGO classes are listed in Appendix 1, p. 145.

TABLE 2. Concordance between Center Diagnosis and Reference Classification in Two Groups of IPSS Centers

CATEGO classes	Moscow and Washington		Seven other centers	
	N	% concordance	N	% concordance
S+, S?, P+, O+	114	96.5	642	96.0
M+, D+, R+, N+	90	48.9	191	84.8

(Omitted: 94 cases with diagnoses other than schizophrenic, paranoid, or effective disorders, and 71 cases in uncertain CATEGO classes).
    Adapted from WHO [68].

CATEGO into classes S, P and O, purely on the basis of PSE symptoms, were given a diagnosis of schizophrenic or paranoid psychosis, but that psychiatrists in the Moscow and Washington centers also included many of the cases that were regarded elsewhere, and by the reference classification, as affective. This confirms the result of the U.S.-U.K. Project but shows that a broad concept of schizophrenia is not confined to the U.S.A.

The amount of disagreement should not be overemphasized. Out of 801 cases diagnosed as schizophrenic or paranoid psychoses, 726 (90.6 per cent) were also in classes S, P and O. This provides an empirical way to describe the symptoms and syndromes present during the acute psychosis. The syndrome profiles of these three major classes are presented in Appendix 2. By far the largest of the three is class S (67 per cent of all cases diagnosed as schizophrenic or paranoid psychosis in the IPSS). The main symptoms involved are thought insertion, thought broadcast, thought commentary, thought withdrawal, delusions of control and alien penetration, and certain kinds of auditory hallucinations. Most are among the "*symptoms of the first rank*" described by Kurt Schneider [40]. Whenever they are present, there is almost always a wide range of other symptoms as well. The profile in Appendix 2 demonstrates, for example, that 85 per cent had depressed mood, 31 per cent a hypomanic syndrome, 32 per cent grandiose delusions, and so on.

Experiences of this kind are likely to be interpreted differently by different people. The temptation to develop delusional explanations, in terms of ideas which otherwise seem to come straight from science fiction, is obviously great. Suppose, for example, that someone hears his own thoughts being echoed or repeated or spoken aloud in his head, so loud that he feels that anyone standing nearby must be able to overhear them. Suppose that the experience goes further—that some of the thoughts have a distorted quality and do not appear to be his own, or that they seem to come from outside, i.e., are heard as "voices." We are dealing here with a disorder of the most characteristically human experience, "internal language." It is entirely comprehensible that the affected individual will consider all sorts of explanations, including hypnotism, telepathy, radio waves, spirit possession and so on, depending on his cultural background. This effort of the imagination can help us to understand what is happening in the early stages of schizophrenia. In addition, we can see why fear, panic, and depression are so common, and why judgment is so often disorted.

Experiences like these can occur after taking *amphetamine* or in *chronic alcoholism* or as part of the aura in *temporal lobe epilepsy*. Evelyn Waugh gave a vivid description, in *The Ordeal of Gilbert Pinfold*, of a hallucinatory state due to chronic intoxication by *bromide* and alcohol. When there is a known organic cause of this kind, the condition is not usually called "schizophrenia" and is excluded from most types of research project, though they might well give very useful information if studied comparatively.

The second major *CATEGO* class is class P, which contains cases with delusions other than those specific for class S but without predominantly manic or depressive delusions: 17 per cent of all diagnosed schizophrenic or paranoid psychoses fell into this group. Only in one center (Taipei) was there a good concordance between a diagnosis of *paranoid psychosis* (as contrasted with schizophrenic) and class P and it seems that many psychiatrists do not have very clear criteria for making the distinction. It may or may not be useful but, unless the distinction is made, we are unlikely to be able to discover whether it has any value. Conditions in class P included cases characterized by a single "delusion"—for example *morbid jealousy*, or a conviction that the patient's teeth were too protruberant, or that he gave off an unpleasant smell, or that other people thought him homosexual—as well as more florid clinical pictures with widespread persecutory or religious or grandiose delusions. There were also conditions that seemed specifically *subcultural*. There was, however, no case of *"paranoia querulans."*

Finally, there is class O, accounting for only six per cent of the schizophrenic and paranoid psychoses diagnosed in the IPSS. The syndrome profile in Appendix 2 is characterized mainly by behavioral abnormalities: catatonic symptoms, excitement, retardation, bizarre behavior, and so on, in the absence of more specific delusions or hallucinations. These conditions were commoner in the developing countries. Once again, the class is heterogeneous. Some conditions might have had an organic origin, such as encephalitis, some might better be classified as *Asperger's syndrome*, or the later manifestation of *early childhood autism* [65], and some (e.g., acute excitements without other identifying features) might better have been regarded as specific *subcultural* states. Yet others were simply chronic *residual conditions*, following previous more typical episodes of schizophrenic psychosis.

We have seen earlier that yet other conditions than those described in classes S, P and O may also be described as schizophrenic, particularly within certain schools of psychiatry. Anyone who wishes to discuss the causes, treatment, management or prevention of "schizophrenia" must somehow come to terms with this diversity. Every generalization must have so many exceptions that it may seem pointless even to make the attempt. Nevertheless, I shall hope to show, paradoxically, that some useful statements can be made if we first recognize that we must set limits to generalization. For the moment, I shall be concerned mainly with patients who have experienced the central schizophrenic syndrome (class S), and the paranoid syndrome (class P), since these are the largest and the most easily definable groups. Both can, of course, persist for months or years, and can therefore be regarded as chronic as well as acute syndromes. But before considering the environmental factors that precipitate or maintain them (which might provide a basis for rational intervention), it is necessary to give a good deal of attention to the other chronic syndromes associated with schizophrenia.

THE CHRONIC SYNDROMES

Whatever theory of schizophrenia is adopted, it is common ground that acute psychotic attacks characterized by symptoms of the first rank are frequently preceded, accompanied, or followed by more long-lasting impairments, which can be summarized in terms of Bleuler's two *fundamental characteristics*—flatness of affect and loosening of the associations. The first of these chronic syndromes may be called the "chronic poverty syndrome." It is characterized by flatness of affect, poverty of speech, slowness, underactivity, social withdrawal, and lack of motivation. These traits are highly intercorrelated and can be reliably measured, in long-stay patients by means of *behavior scales* [58]. Together they constitute a useful measure of the severity of one kind of chronic impairment. The social withdrawal score is highly correlated with work output at simple industrial tasks, and also with measures of central and peripheral *arousal* [7]. It also represents quantitatively the individual's ability to *communicate* using verbal and non-verbal skills. The most severely impaired person conveys little information through his use of facial expression, voice modulation, or bodily posture, gait or gesture.

In addition to these chronic negative symptoms, there may be *incoherence of speech* and *unpredictability of associations*. The individual does not seem to be able to think to a purpose but goes off on a sidetrack because of some unusual association to a chance stimulus and thus gives the impression of confusion, incoherence, and incompetence. Occasionally this may give the impression of creativity, but usually the syndrome is constricting and handicapping. Most of the creative people who have been afflicted with schizophrenia have had their creativity abolished, not enhanced.

These two kinds of chronic syndrome are not unrelated. Social withdrawal, for example, could in part be a reaction to an individual's experience that his attempts at communicating with other people were received with a more or less polite brush-off. Attempts at communication may actually be painful. However, this is by no means the whole explanation of social withdrawal or slowness. It is

TABLE 3. Severity and Type of Syndrome Present in Three Series of Chronic Schizophrenic Patients

	Series		
Type and severity	A[a] %	B[b] %	C[c] %
No evident syndrome	11	42	20
Moderate severity only	18	22	37
Severe flatness of affect, otherwise only moderate syndromes	5	–	7
Coherently expressed delusions predominant	11	4	22
Severe incoherence of speech	15	–	4
Severe poverty of speech	26	–	7
Mute	15	–	2
Total number in series	(273)	(45)	(54)

[a]Series A. Random sample of long-stay schizophrenic women, aged 18–59, in three mental hospitals [58].

[b]Series B. Long-stay schizophrenic men selected for course at industrial rehabilitation unit [55].

[c]Series C. Schizophrenic men and women who had lived outside hospital at least a year but had been unoccupied [60].

Adapted from Wing, Wing, Griffiths and Stevens [60].

generally accepted that both types of impairment can be manifested in childhood and adolescence long before the first onset of a recognizable syndrome [50]. Certainly, they are often the most characteristic features of the chronic state.

Table 3 shows that the type and severity of chronic syndromes in schizophrenia may vary markedly according to the setting in which the study is being carried out. A serious methodological difficulty in comparing the results of therapeutic trials has been that the composition of the series has not been specified with sufficient clarity to make generalization feasible. The three series described in Table 3 comprised: (a) long-stay schizophrenic women selected at random in three mental hospitals in 1960; (b) long-stay schizophrenic men selected for a trial at an industrial rehabilitation unit [55]; (c) a group of chronic schizophrenic patients who had not been in hospital for at least a year but who had been unemployed throughout that time [60]. The reliability and use of the method used to classify type and severity of syndrome have been described in detail elsewhere [65]. The aims of these three studies were quite different and their conclusions can only be applied to the specific group selected. It is clear that a wide range of type and severity of chronic syndromes can be present and that generalizations about chronic schizophrenia must therefore be applied with great care.

Chronic impairments contain a component that is "intrinsic" and a component that is reactive. From a purely practical point of view, an *intrinsic impairment* is one that will not go away, no matter what biological, psychological, or social factors are brought to bear therapeutically. Of course, we can never be certain, in theory, that we have tested the limits of treatment, and circumstances sometimes change in a way that surprises even the most astute clinician, so that it is never wise to give up hope of improvement. But to deny that severe and chronic intrinsic impairment can exist must do more harm than good. It is only by recognizing and studying disabilities that it is possible to find ways of compensating for them.

THE IMPORTANCE OF DIAGNOSIS

The relationship between the acute and the chronic syndromes has long been a matter of controversy and confusion. Some theorists have thought that specific kinds of delusional experience were primary, while others have regarded the affective flattening and thought disorder as fundamental. A *multigene theory*, together with the action of a variety of environmental precipitants (somatic, psychological and social), could comprehend both points of view, and also link the dimensional with the categorical approach. The main point about a diagnosis, however, is to suggest to the clinician theories that might be helpful when applied in practice, i.e., theories that will predict a useful form of treatment or prevention or management or, at the very least, suggest a prognosis. None of the concepts of schizophrenia at present in vogue is firmly grounded in a set of linked and tested theories in the way that diabetes mellitus, for example, is, although there are many analogies between the two concepts and we may hope that, one day, our knowledge of schizophrenia will equal our knowledge of diabetes. It is therefore necessary to proceed to some extent empirically.

We know that the acute syndromes usually respond to *phenothiazine* treatment whereas the chronic syndromes respond less well. We know that the acute syndromes, particularly when precipitated by organic factors such as amphetamine or alcohol, are less likely to be preceded or followed by the chronic syndromes. We know that social withdrawal often accompanies chronic depression, mental retardation, chronic physical disability, and severe social deprivation. Moreover, mild degrees of affective blunting or of "thought disorder" cannot be recognized reliably. These are good empirical reasons why any theory linking acute and chronic schizophrenic syndromes should not be applied in practice, except with great circumspection. In particular, the supposed presence of chronic syndromes should not lead to a regime of treatment suitable only for the acute syndromes if the latter are absent, and especially if they have never been present. Terms such as simple, pseudoneurotic, pseudopsychopathic, latent, or sluggish "schizophrenia" should not be made to carry too much therapeutic meaning. Still less should the "spectrum" concept be used clinically. Failure to understand this basic distinction has contributed, in my view, to practices that can indeed be criticized. There is a close analogy here with diagnosis of severe mental retardation. The functions of differential diagnosis—for example, between mongolism, phenylketonuria, epiloia, and so on—are different from the delineation of the pattern of impairments that determine the kind of medical, educational, and social help needed by the child and family. There will be a degree of overlap since the diagnosis can, to some extent and in some cases, lead to a prediction of what the pattern of impairments will be. But two essentially different models of "diagnosis" are involved.

Much sterile controversy has stemmed from the fact that this distinction has not been recognized, and absolutist schools have grown up that altogether deny the applicability either of disease concepts or of concepts of chronic impairment. At the most nihilistic, some schools have even denied both.

In the sections that follow, I shall deal separately with the empirical evidence concerning social influences on both acute and chronic syndromes, in order to prepare the ground for a rational discussion of management, which must, of course, include consideration of the problems that occur when both types of syndrome occur, as they often do, together or in alternation.

SOCIAL DISABLEMENT AND THE COURSE OF SCHIZOPHRENIA

So far, we have been considering, in a rather cross-sectional manner, the acute and chronic syndromes that make up the clinical picture of schizophrenia, and have not dealt with their effects upon the individual's level of social functioning. Two other factors need to be dealt with first: extrinsic disadvantages and secondary adverse reactions.

The term "extrinsic" is taken from physical medicine and it cannot be applied to psychiatric disorders without qualification. The basic idea is straightforward—extrinsic factors would be disadvantageous whether or not the individual had become ill. Someone born in a large family in a poor area, with inadequate housing and high unemployment, is socially disadvantaged. If he has few assets in the way of vocational skills, intelligence, social attractiveness, family support, and so on, he is not well prepared to cope with chronic illness of any description, let alone schizophrenia. However, the socially disadvantaged are more at risk of developing all sorts of diseases, schizophrenia being one of them. There has been a long discussion of whether socially disadvantageous conditions are likely to cause schizophrenia, or whether the high rates observed under such conditions are due to a preexisting *drift*. Intrinsic impairments often precede the first acute attack. In such cases the prognosis is less likely to be favorable. The balance of the epidemiological evidence at the moment is in favor of the view that much of the association noted between the onset of acute schizophrenia and factors such as social class, marital status, migration and social isolation is due to selection rather than stress [10]. On the other hand, it is also clear that a disadvantaged background has an adverse effect upon the course because of the lack of social support [20]. This is why an attempt to increase the number of assets, by social skills and vocational training, by the creation of a graded series of supportive social environments, by counseling in methods of self-help, and so on, is of such importance.

The other type of factor is the reaction of the handicapped individual to the combination of intrinsic impairments and extrinsic disadvantages that he or she has experienced. An example from physical medicine may make the distinction clear. Compare the violinist who injures the little finger of his left hand with the laborer who suffers the same injury. The implications and the effects are quite different in the two cases. Compare also an individual who suffers a severe coronary thrombosis, recovers, and is back at work within three months as though nothing had happened, with someone who experiences a mild chest pain and thereafter becomes a "cardiac invalid." We are dealing here with self-attitudes, confidence, expectations, and personal habits, not only in the affected individual but in the people who are important in that individual's social environment, and upon whom his well-being may depend. Whether a man who has had a leg amputated makes a success of his rehabilitation may depend as much upon his wife as on him. Such factors are particularly important in schizophrenia, as we shall see later.

These three types of difficulty—intrinsic impairments, extrinsic disadvantages, and adverse secondary reactions—do not vary independently of each other; how severely each type affects everyday life and social performance depends on a somewhat different group of determining factors. Each individual shows a unique pattern of difficulties. The overall result, however, is that the individual is handicapped from achieving some personal goal; this condition may be called "*social disablement*." *Management* (which includes rehabilitation, training, the realization of potential assets, and the provision of sheltered environments when necessary) is most rational when it is based upon a thorough knowledge of the various types of handicapping factor that produce social disablement, and of the way they can be influenced for better or worse. Having identified the unique pattern of impairments, disadvantages and reactions that is responsible for an individual's disablement, a plan is constructed jointly with him or her, which consists of trying to increase the number of options that are realistically available, so that there is a greater choice of paths forward. The more evident that movement along a particular path will be rewarding, the greater the motivation to take it.

Before coming to a consideration of the social reactivity of the acute and chronic syndromes, and of the adverse secondary reactions, it will be useful to consider the *course of schizophrenia* as it has been determined statistically in various follow-up studies. Mayer-Gross followed up, 16 years later, patients with early schizophrenia admitted to the Heidelberg Clinic in 1912 and 1913 [31]. Harris and his colleagues followed, for five years, patients admitted to the Maudsley Hospital in London for insulin coma treatment, between 1945 and 1948 [18].

TABLE 4. Comparison of Two English Series of Schizophrenic Patients Followed Up for Five Years

	Maudsley 1945–48 (Insulin patients) %	Three hospitals 1956 (First admissions) %
Patient independent: moderate or no symptoms	45	56
Patient dependent: moderate or no symptoms	12	16
Severe disturbance	9	17
In-patient	34	11
No. followed up	123	94
Dead	1	3
Not followed up	1	14

Adapted from Brown, Bone, Dalison and Wing [3].

Brown and his colleagues examined, in 1961, a group of patients first admitted for schizophrenia to three English mental hospitals five years earlier in 1956 [3]. The figures are not strictly comparable because of variations in mortality and numbers followed up, and because standard methods of defining and selecting cohorts were not used, but they do provide a basis for approximate comparison.

The proportions of patients in the three series who were alive, out of hospital, functioning well socially, and without major psychotic symptoms, at the time of follow-up, were 35, 45, and 55 per cent. The proportions who were alive, out of hospital, but unwell or unoccupied, were three, 21, and 32 per cent. The proportions in hospital or dead were 62 (two-thirds of whom were dead), 34, and 13 per cent. A more detailed comparison of the two English series is shown in Table 4.

A considerable improvement in *prognosis* is suggested, part of which seems to have taken place before the new era in pharmacological and social treatment had begun. But even among schizophrenic patients first admitted in 1956, over one-quarter were in hospital (and had been there for more than six months), or were severely disturbed in the community, five years later. Another 16 per cent were outside hospital but unable to work or maintain a home, although their symptoms caused only moderate social disturbance. However, 55 per cent could be said to be "*social recoveries*," if we include a number whose adjustment was rather precarious. Neurotic symptoms were common among this group. Of patients who had been in hospital before 1956, and were readmitted in that year, only 38 per cent were working and reasonably well in 1961.

Perhaps the most remarkable follow-up study of recent years is that carried out by Manfred Bleuler, who had published a detailed account of a personal series of 208 patients admitted in 1942 and 1943 to the Burghölzli clinic in Zurich. He concludes that, after 20 years, between a quarter and a third of the patients are severely handicapped, and that the prognosis has improved during this time [2].

It is a reasonable working hypothesis that the prognosis can be improved still further by the application of knowledge that is already available. The next three sections will therefore deal with the social reactivity of the acute and chronic clinical syndromes, and of the adverse secondary reactions, by way of introduction to a discussion of the ways in which management can be improved.

THE SOCIAL REACTIVITY OF THE ACUTE SYNDROMES

The most convenient way of initiating the scientific study of the social reactivity of the acute syndromes of schizophrenia is to investigate the social concomitants of relapse. Stone and Eldred, for example, described the reemergence of delusions in two patients, shortly after they had been transferred to a special ward for intensive treatment, who had been free of such symptoms for many years while living in the sheltered conditions of a mental hospital [43]. We observed the same phenomenon in six out of 45 moderately disabled patients during the first week of a course at an industrial rehabilitation unit [55]. The course entailed traveling from the mental hospital to the rehabilitation unit by public transport and mixing with a group of people most of whom were physically rather than psychiatrically disabled. It therefore entailed a very considerable change in routine and demanded the exercise of skills that there is not always an opportunity to practice in mental hospitals. In fact, 21 of the 45 patients came from a hospital where social conditions had only recently improved, and there had been very little preparation for the rehabilitation course. Of these 21 patients, five relapsed with delusions and hallucinations that had not been present for years. The other 24 patients came from another hospital which had pioneered methods of social and vocational rehabilitation and only one of these relapsed. More patients from the former hospital were receiving *phenothiazines* (and in higher doses) than patients from the latter, so that this factor cannot explain the difference.

More recently Goldberg and his colleagues found that social casework and vocational rehabilitation hastened relapse in patients who had more severe clinical symptoms at the time of discharge from hospital, while it seemed useful in patients who were asymptomatic at discharge [17].

The significance of these studies will be discussed later, but we can already hypothesize that social pressure can lead to relapse, even when it is applied with the best of intentions by professional helpers. A different kind of social precipitant was described by Brown and Birley who found that there was a marked increase in the frequency of occurrence of certain events, compared with control groups, during the few weeks immediately before the first onset or an acute relapse with florid schizophrenic symptoms [5]. Excluding events that could have been the result rather than the cause of a recrudescence of symptoms did not diminish the extent of the association. Some of the events would be expected to have been experienced as pleasurable, others as unpleasant: becoming engaged to be married or receiving a promotion at work, for example, compared with hearing about the death of a relative or being involved in a traffic accident. Nearly all the events were familiar ones, in the sense that most people would expect to be affected by them during the course of their lives. People who had previously had attacks of schizophrenia seemed more vulnerable to common stresses than most people. The relapse was usually characterized by the same sort of florid schizophrenic symptoms that had been manifested in earlier attacks.

There has long been a theory that the relatives of people with schizophrenia are in some way implicated in the original causation of the disease [28]. Very few empirical studies have been carried out to test this hypothesis and, in any case, the methodological problems are immense. Studies of identical twins brought up together or apart provide little convincing evidence [38]. The best work methodologically is that by Wynne and his colleagues, suggesting that *communication deviances* are common in the parents of patients with schizophrenia, though this has no necessary bearing on origins in childhood [69,70]. However, this clear-cut result has not been replicated in a recent very careful study in England, and it may be that the reason, at least in part, is that the criteria for diagnosis in the two series were quite different [21]. As for the really passionate advocates of various forms of the early environment model of causation, it is difficult to understand where the strong motivation to believe such theories springs from, in the absence of any substantial body of evidence in favor.

There are, however, several recent studies that seem to show that factors in the family might be responsible for precipitating the first onset or relapse of acute schizophrenia. The most recent study, by Vaughn and Leff, used the same design and method as an earlier one and yielded very similar results, thus providing one of the rare samples of replication within the field of social psychiatry [7,46]. Series of schizophrenic and depressed patients were examined shortly after admission to hospital, using a standard technique to describe and classify their symptoms and to ensure that the processes of clinical selection were unambiguous. A key relative was interviewed while the patient was still in hospital and ratings made of hostility, emotional overinvolvement and criticism. Of these three factors, the last proved most important and could be measured very reliably in terms of the number of critical comments made about the patient during the interview [47]. All three factors were combined into one index of *"expressed emotion."* This index was associated with previous work history and with the amount of disturbance in behavior during the three months before admission. It was also highly significantly related, but quite independently, to relapse rate during the nine months after discharge. Two other factors were important in predicting relapse. One was the amount of time that patient and key relative spent in face-to-face contact with each other (itself affected by whether the patient was working or attending a day center); the other was whether the patient was taking *phenothiazine* drugs.

Thus a hierarchy could be set up in terms of risk of relapse. Those at most risk (over 90 per cent) were living at home in constant face-to-face contact with highly involved relatives and not protected by taking phenothiazine drugs. If they were taking medication or if they had little contact with the involved relatives (another protective feature), the risk was much lower (40 to 50 per cent). If both protective features were present, the risk of relapse (15 per cent) was the same as if they had returned to a low-emotion family. Approximately half the families came into the high-emotion and half into the low-emotion category. The association of high emotional expressiveness in the relative with previously disturbed behavior and poor work history in the patient suggests that, at least in part, there was a vicious circle effect of patient and relative on each other. Further light is thrown on this by Vaughn's analysis of the content of the criticisms made about patients by relatives [8]. Less than a third of these were concerned with symptoms of the florid attack; these criticisms were usually not associated with a poor previous relationship between patient and relative. Over two-thirds of the criticisms were directed at longstanding personality traits which had been present before the first onset of florid symptoms. These traits were mainly lack of sociability, communication and affection. So far as could be judged, such earlier indications of negative impairments were not found only in patients whose relatives were critical about them, but the measurement of *previous personality* is notoriously difficult, and the relatives were the main source of information. The question of cause and effect must therefore remain open.

It is difficult to see a single common factor underlying all these situations associated with relapse, except the very vague one of "stress." One possibility is *social intrusiveness*; i.e., an environmental agent (whether an eager therapist, an accidental contact, or an over-involved relative) that does not allow a sufficient degree of protective withdrawal, so that an individual with inadequate equipment for communication is forced into interaction in a social situation and an underlying cognitive disorder becomes evident.

Data concerning the level of psychophysiological *arousal* in schizophrenia are relevant to this explanation, but discussion will be reserved until after the social reactivity of the clinical poverty syndrome has been considered.

THE SOCIAL REACTIVITY OF THE CHRONIC SYNDROMES

A great deal of attention was given by psychologists between the wars to the nature of *"psychological deficit"* in schizophrenia. Jung had pointed out in 1906 that patients with *dementia praecox* showed a "passive registration of events which are enacted . . . but all that which requires an effort of attention passes without heed" [23]. He quoted Stransky's experiment in which subjects were asked to talk at random on a given topic for a minute without giving any attention to what they were saying. The result recalled the incoherence of dementia praecox. These two features, *lack of attention* and *incoherence of speech*, which were regarded by Eugen Bleuler as *fundamental*, were usually investigated separately and few further attempts were made to link them theoretically. Clinicians tended to regard *deterioration* (deficit increasing with time) as inevitable in schizophrenia. Babcock thought that deficit began in childhood and that she could trace it through its course, but her conclusions were based on cross-sectional surveys of verbal and performance tests [1]. Most subsequent workers agreed that her conclusions did not follow from the data. Kendig and Richmond, for example, showed that there was a genuine intellectual impairment, but no tendency for it to increase over time [24]. Foulds and Dixon later confirmed this [13]. Many other psychologists described experiments in which it was demonstrated that the performance of schizophrenic patients could improve with practice and that there were ways of motivating them to perform better [14,15,25,34-37]. This large body of work on psychological deficit succeeded in more precisely describing and measuring the two components of impairment identified by Kraepelin and Bleuler but not in suggesting practical methods of rehabilitation.

Subsequently, a distinction was made between "*paranoid*" and "*nonparanoid*" patients, who tended to react in different ways, and studies were made in realistic workshop conditions using more global measures of *social performance* [6,32,33]. One interesting observation was that the improvement due to practice, which would be expected, by analogy with normal people and even with the severely mentally retarded, to be negatively accelerated, tended in schizophrenia to be slow and linear. One workshop experiment gave more encouraging results than the others, perhaps because patients were encouraged to perform better by well-known and trusted nurses rather than by giving artificial rewards reminiscent of a laboratory setting [52]. This study demonstrated that active social supervision could considerably increase output in an experimental group but that the response was sharp and immediate, rather than taking the expected form of another *learning curve*, and that as soon as the extra *social stimulation* was withdrawn, output fell at once to its former level. Moreover, output in the control group (who were situated in the adjoining workshop and were aware that they were *not* getting extra attention) fell and then rose again in a mirror image of what was happening next door. The improvement in output was reflected in a decrease in aimless wandering and fidgeting, or simply sitting staring into space, but this change in behavior was not generalized to other times of the day. The patients involved were all very severely impaired, but they were living in a resocialization villa in a hospital with an excellent reputation for *social therapy*, so that the improvement was not simply due to the fact that they had previously been neglected. It appeared that the extra social stimulation provided a source of motivation that needed to be kept up continuously, since the improvement disappeared at once when it was withdrawn. It

TABLE 5. Socal Environment of Samples of Long-Stay Schizophrenic Women in Three Mental Hospitals in 1960

Index	Hospital A (N = 100)	Hospital B (N = 73)	Hospital C (N = 100)
Hours spent doing absolutely nothing	2h.48m.	3h.15m.	5h.39m.
Patients possessing a comb	89%	69%	30%
Nurses who thought patient could do useful work in hospital	76%	52%	26%
Mean ward restrictiveness score	27	40	60

Adapted from Wing and Brown [58].

could be said that the supervisors were passively exercising functions which the patients were unable to exercise actively for themselves, in the hope that the functions would eventually return. This is an excellent and fundamental principle of rehabilitation.

A larger-scale survey, in which samples of long-stay schizophrenic women in three mental hospitals were interviewed, was designed to test the hypothesis that the social environment in which the patient lived had a measurable influence on the severity of the *clinical poverty syndrome* [58]. There were marked social differences between the three hospitals which could be summarized in terms of the degree of *"social poverty"* characterizing them. The measures used were the number of personal possessions owned by patients, the attitudes of nurses toward them, the amount of contact with the world outside, the restrictiveness of ward regimes, and the amount of time spent by the patient doing absolutely nothing. This last measure proved particularly crucial. At the first hospital, it was two hours 48 minutes per patient per day, on the average. At the third it was five hours 39 minutes, on the average. All the indices used showed equivalent differences. Thus 79 per cent of the subjects owned a handbag at the first hospital, but only 42 per cent did so at the third. This revealing item shows how useful simple indices can be; the ownership of items such as a handbag tells us a great deal about how the role-performance of hospitalized women is affected by administrative policy. Table 5 illustrates the differences between the hospitals on selected measures.

Having demonstrated differences between the hospitals, it was possible to proceed to a first test of the hypothesis that there should be equivalent clinical differences. In fact, these were also marked and followed the predicted pattern. Symptoms such as blunting of affect and poverty of speech, rated at a standard interview with each patient, and social withdrawal, independently rated by nurses, indicated a much higher level of impairment at the third hospital than at the first.

TABLE 6. Clinical Condition of Sample of Long-Stay Schizophrenic Women in Three Mental Hospitals

Clinical condition	Hospital A (N = 100) %	Hospital B (N = 73) %	Hospital C (N = 100) %
No evident syndrome	10	15	8
Moderate severity only	21	19	13
Severe flattening of affect, otherwise only moderate symptoms	9	5	2
Coherently expressed delusions predominant	17	8	6
Severe incoherence of speech	17	14	15
Severe poverty of speech	20	25	32
Mute	6	14	24

Adapted from Wing and Brown [58].

Table 6 shows the clinical condition of patients at the three hospitals. Of all the social indices used, the one most clearly correlated with clinical condition was length of time doing nothing.

Since the groups were followed up at the three hospitals over an eight-year period, during which time the social conditions improved and then to a varying extent fell back again, it was possible to test this hypothesis fairly rigorously, but it was not possible to disprove it. An increase in social poverty was accompanied by an increase in clinical poverty, and social improvement was accompanied by clinical improvement. Moreover, the least impaired patients were most likely to be discharged, so that improvement did mean something in concrete terms as well.

Thus the conclusions of the earlier experiment were confirmed on a larger scale. A socially rich social environment tends to minimize the development of negative symptoms in schizophrenia. These are symptoms that have been regarded as *fundamental* and immutable. However, in many cases they did not disappear entirely even when active rehabilitation was maintained over many years. This was particularly evident in a series of rehabilitation studies, some of them experimental [51,55,60]. Long-stay schizophrenic patients who appeared very little impaired while in hospital seemed slow, lacking in initiative and unsociable when working alongside physically disabled people in an industrial rehabilitation unit or a sheltered factory. Nevertheless, they were not unfriendly when approached and their work, though plodding, reached an acceptable standard. It was shown that many long-stay schizophrenic patients could achieve a stable resettlement outside hospital, particularly if sheltered working conditions were available.

All the studies so far considered in this section were focused on large mental hospitals. They showed that *"social treatment"* was not merely an empty phrase. It may be questioned, however, whether the results could be applied to conditions outside hospital. In fact, there is considerable evidence that the principles of social treatment can be applied more generally. It has been found that schizophrenic patients are vulnerable to *social understimulation*, whether this is found in hostels, hotels, group homes, day centers, sheltered communities, centers for the destitute, in lodgings, or in a patient's own family. In one five-year follow-up study it was found that, among those who were unemployed at home, the length of time spent doing absolutely nothing was of the same order as in long-stay inpatients in a hospital with a poor social environment [3].

A careful study was made in Camberwell, a predominantly working-class area in southeast London where psychiatric services are reasonably good. The purpose of the study was to find all the patients known to the local case register who had not been in hospital for the previous year although they had been given a diagnosis of functional psychosis, were of employable age (18 to 54), and expected to be in employment, but had been out of work for at least a year [60]. There were 75 such people, most of whom were schizophrenic, a rate of 44 per 100,000 population. Many of these people were socially isolated and, as the results of an experimental evaluation of a rehabilitation workshop demonstrated, in need of protected living environments. Other studies, of the "new long-stay" patients still accumulating in English mental hospitals (though at a much slower rate than hitherto) [29, 30], and of the alternative residential accommodation now available [19], indicate that social withdrawal and social isolation remain a problem even when patients are no longer in hospital. Thus there are two major problems: one the reduction of an excessive level of social withdrawal and associated behaviors by enriching the social environment, without at the same time provoking an acute relapse with florid symptoms; the other, the maintenance of these impairments at a minimal level by providing the appropriate kinds of protected environments.

OPTIMUM CONDITIONS FOR DRUG AND SOCIAL TREATMENT

The work discussed in the previous two sections suggests that many patients who experience an attack of acute schizophrenia remain vulnerable to social stresses of two rather different kinds. On the one hand, too much social stimulation, experienced by the patient as *social intrusiveness*, may lead to an acute relapse. On the other hand, too little stimulation will exacerbate any tendency already present towards social withdrawal, slowness, underactivity, and an apparent lack of motivation. Thus the patient has to walk a tightrope between two different types of danger, and it is easy to become decompensated either way. The difficulty in thinking and the inability to communicate, verbally and non-verbally, will be most evident in interactions, for example with close kin, and will be exacerbated in conditions of anxiety and high arousal. Families might provide protective or provocative settings; there is no general rule. The same is true of non-family environments. However, the first onset usually occurs within a family setting and there is a natural tendency on the part of relatives towards normalization, which can lead to further intrusion, particularly if the patient withdraws in natural reaction. The patient may try to explain the abnormal experiences in terms of hypnotism, thought transfer, witchcraft, or other subculturally acceptable ideas. The more intrusive the environment the less the patient is able to withdraw and the more the symptoms are perpetuated and provoked. Being admitted to hospital, or wandering off alone, may provide relief simply by reducing the degree of social stimulation.

After several attacks, the patient may come to terms with his experiences and find (whether he consciously formulates this or not) that a degree of withdrawal is protective. Relatives, too, may learn for themselves just what degree of stimulation is permissible, so that a working solution is arrived at. Patients usually do not wish to be entirely alone, but they do like to remain in control of the intensity of contact. The *role of the relative*, however, is a very difficult one since it is unnatural to acquire the degree of detachment and neutrality required; it is much easier for professional people to adopt such a stance. The relatives of handicapped people may easily become overprotective and overinvolved. In the case of schizophrenia, this overinvolvement has specific effects in increasing the liability to further breakdown. When this shows itself in symptoms such as violence, noisiness at night, refusal to eat, or delusions involving the relative, a circular effect is set up. The relative expects such reactions on future occasions. This is a much more likely explanation of the findings than is the assumption that the original cause of schizophrenia is somehow familial.

The alternative temptation for the patient is to withdraw altogether into contemplation of private experiences. When a patient is allowed to do so in the understimulating conditions of some large hospital wards, or in badly run hostels, or in reception centers, or even in an attic at home, the negative impairments become more and more obvious.

If *phenothiazine* drugs work, at least in part, by reducing a high level of arousal [61], they should be most effective in conditions of social overstimulation, and there is some evidence that this is so [4,46]. It has been suggested that patients who live in low-emotion families can avoid relapse (except for the operation of accidental stresses) even without taking preventive medication [27]. Tarrier has recently shown that schizophrenic patients, sitting quietly by themselves in their own homes, have a high arousal level as measured by skin conductance [44]. If a "high-emotion" relative enters the room, there is no change in arousal level. If a "low-emotion" relative enters, the high arousal habituates towards normal. The work needs to be repeated but, if it holds true, it suggests that some relatives have a calming and supporting effect, similar to the protective effect of a good hospital or hostel. The evidence that phenothiazines are useful for long-stay patients is conflicting but it may be hypothesized that the more optimal the social conditions provided, the less the need for medication. Most studies of the effects of medication do not give sufficient information to assess how protective, or how *overstimulating* or *understimulating*, the social environments of patients are, but on the whole it appears that medication is less necessary when the patient is living in optimal social surroundings [61].

To generalize: The *optimal environment* is structured, with expectations geared precisely to the level of performance that the patient actually can achieve, or perhaps a little above, but with supervision that is not emotionally involved. Complex decision-making is particularly difficult for patients and tends to rouse their anxiety or their withdrawal. Control of the level of social stimulation must be left to some extent in the patient's hands. A relationship of trust and confidence, whether with relatives or with professional staff, is most likely to develop when these conditions are optimal.

A great deal remains to be explained about the acute and chronic syndromes. The mechanisms whereby *first rank symptoms* such as thought echo or "thoughts spoken aloud," which are direct descriptions of basic experiences with very little elaboration, can be explained, in terms of over-arousal, of phenothiazine action, or of cerebral pathology, remain obscure. Recent advances in pharmacology and biochemistry are so promising that it is clear that links will have to be made. Meanwhile, the formulation put forward here has obvious relevance to problems of management. Before considering practical issues of application, one further area of handicap needs to be explored, one where there is also a useful body of empirical work.

ADVERSE SECONDARY REACTIONS

Relatives, employers, workmates, friends, and professional people reflect back to the person handicapped by *intrinsic impairments* or *external disadvantages*, or a combination of both, an indication of their opinion of his social status and worth. If they think he is less of a man because of his illness, he will tend to think so too. If he is helpless, he may have no choice but to depend upon his environment to help him. A congenitally deaf child must rely on others to teach him to talk and to help him communicate. The more severe the intrinsic impairment, the more inevitable is the development of secondary impairment as well.

Handicapped people have to (or wish to) opt out of certain social responsibilities. Therefore others must take them on. If the others are professionally trained for the job, there tends to develop what Goffman called a staff–client split, each side taking up rather stereotyped attitudes toward the other [16]. The longer the client experiences *dependence*, the more he will tend to prefer it. Staff, in turn, may exaggerate the severity of intrinsic impairments, and thus a

TABLE 7. Attitudes to Discharge of Sample of Long-Stay Schizophrenic Women in Three Mental Hospitals: by Length of Stay

	Indifferent or wish to stay		
Length of stay	Hospital A (N = 100) %	Hospital B (N = 73) %	Hospital C (N = 100) %
2 to 10 years	40	53	50
11 to 20 years	58	58	75
21 years or more	75	82	83

Adapted from Wing and Brown [58].

vicious circle is set up, with unnecessary dependence developing as well as that which is unavoidable. The acceptance by the handicapped person of limitations which are not actually necessary is the essence of secondary reaction.

The most obvious example of adverse *secondary reaction* in schizophrenia is *institutionalism*, at the heart of which is a gradually acquired contentment with life in the institution which culminates in the individual no longer wishing to live any other. Institutionalism is thus caused partly by a reflection back to the handicapped person of his own altered status as a human being. He is seen as a patient, rather than as an employee, a father, a customer, or a companion. The patient role is a constricted one, replacing many others which he might have been able to undertake. Partly, however, institutionalism results from forces in the patient—his own previous experience of illness, his self-confidence, his potential for developing alternative skills, and his determination to achieve independence [53,58].

Table 7 illustrates the process occurring in the three mental hospitals described earlier (see Tables 5 and 6). In spite of the marked social differences between them, patients in all three tend to adopt attitudes of indifference or of desire to stay the longer they have been resident. This remains true when the most severely impaired patients, who were mute or so incoherent that their answers to questions were incomprehensible, were omitted. Other factors, such as age, sex and social environment, are also related, but they do not much reduce the very strong association between attitude to discharge and length of stay.

Other attitudes and personal habits are affected in the same way. The patient loses any ability he might originally have retained to play a wide range of social roles; he does not replenish his stock of useful current information (such as how much a postage stamp costs); he does not practice traveling on public transport, or shopping; he gradually ceases to make plans for the future, or he simply repeats some vague formula if he is asked about them. Visitors, of course, also begin to fall off, and even if he is allowed to go outside the hospital, he will tend to do so less and less.

It was mentioned earlier that hospitals tend to differ markedly in the social conditions they provide. Although exposure to a socially impoverished environment is not necessarily associated with increased negative impairments, with the major exception of an imposed idleness, it is fairly obvious that *social poverty* is likely to encourage the development of adverse secondary reactions. Thus institutionalism, pauperism, and neglect contribute to the *disablement* of hospital patients as well as to their discomfort.

Once attitude change has taken place, as in institutionalism, it is very difficult to remedy. Festinger's early work on reference groups provides a possible theoretical approach. It may be that the success of rehabilitation units is attributable to the improvement in attitude and *self-confidence* which comes about when a handicapped person is exposed to a social group in which confidence and self-reliance are valued and are visibly being acquired by other physically handicapped people. The evidence of a study carried out among 212 entrants to such a unit certainly pointed that way [56]. Moreover, those who acquired confidence were more likely to be employed two months after leaving the unit than those who remained unconfident. Unfortunately, however, there was a group of people with idiosyncratic motivation or with very little motivation at all who did not wish to join the more conforming group. Their self-confidence did not change, and they were not very successful. Most of those with schizophrenia were in this group.

Probably attitudes toward discharge or toward work outside the hospital can be changed in schizophrenic patients only by methods designed specifically to effect such a change. Thus in one of our studies, we successfully changed the *attitude toward work* by getting moderately handicapped long-stay schizophrenic patients to go to an industrial rehabilitation unit outside, after adequate preparation. We found that those who improved did in fact find jobs [51]. What we did not anticipate was that there was no change in their attitude toward discharge. Even those who began working out still wanted to live in. These attitudes are very specific.

Although this analysis of secondary reactions has been based mainly on the problem of institutionalism, it is of course true to say that nowadays far fewer long-stay patients are accumulating in hospitals. Nevertheless, there are *"new"* long-stay patients in hostels, day centers, and sheltered communities, and it is important to remember that the principles of the development of institutionalism are most unlikely to be different in kind for these people than for the *"old"* long-stay patients in mental hospitals. Indeed, in some respects the problems are more difficult. Mann and Sproule, for example, found that after only two or three years in the hospital the "new" long-stay patients already wanted to stay where they were [29]. This is what one would expect as the process of selection focuses to a greater extent on those with more severe intrinsic impairments and extrinsic disadvantages, and as the "hospital" becomes more and more a sheltered community.

Even those who spend most of their time at home and do not have to remain in sheltered environments of various kinds are still at risk for developing secondary handicaps other than institutionalism. Brown and his colleagues found that 20 per cent of their schizophrenic patients admitted to the hospital in 1956 left home during the subsequent five years and did not return [3]. Divorce is much commoner among schizophrenic patients than in the general population. The secondary problems arising from unemployment, solitary living, poverty, and even destitution are important causes of an attitude of indifference or despair. Once you are down, you tend to be ground further down; handicapped schizophrenics are no exception to that rule.

Even among those who stay at home, there are some who have outstayed their welcome. There tends to be a discrepancy between the attitudes of relative and patient, the former seeing the patient as disabled and often a bit of a trial to live with, the patient himself regarding his circumstances with a touch of complacency [54]. We have heard a good deal about the effect of relatives upon patients but much less about the reverse.

THE MANAGEMENT OF SCHIZOPHRENIA

The *management* of schizophrenia needs to be considered from three different points of view: those of the patient; the patient's relatives and others in the immediate social environment on whom his livelihood may depend; and the professional people, administrators and politicians who provide, plan, or control access to services. Each group sees a different set of problems and often fails to appreciate the difficulties faced by the others. It may therefore be useful to look through the eyes of each group separately.

Self-Help

Every individual has to learn to live with himself, warts and all. Everyone is handicapped in some way. But those people who are severely impaired in their *inner language* and in their ability to *communicate* with themselves have very little chance to achieve a satisfactory solution unless others help them. Fortunately, most people with schizophrenia are not as severely impaired as that. The central problem is still, however, one of "*insight*." The acute symptoms of schizophrenia carry a peculiar conviction to most of those who experience them, which cancels out the skepticism of their relatives or professional helpers. Given that many have always been somewhat detached from social opinion and that the first onset often occurs during the rebellious teens anyway, it is not surprising that patients find themselves at odds with their relatives and unwilling to take advice from them. Sometimes it is only after prolonged experience and suffering that patients begin to understand some of the factors that make matters better or worse for them.

The following list of factors that are at least partially under the control of patients contains some of the major keys to self-help:

Whether to take medication
Recognition and avoidance of triggering situations
Specific and restricted social withdrawal
Methods of dealing with primordial symptoms
Finding work within competence
Finding companions who are not intrusive
Helping others to understand the condition

In every case, much depends on others as well. At the moment, misunderstanding of the nature of schizophrenia is common, among specialists as well as among relatives and the general public. Nevertheless, some people do, by trial and error, discover that they can help themselves. They discover, for example, that *medication* controls the acute symptoms and that discontinuing medication brings them back. They also discover the side-effects and then have to balance the advantages against the disadvantages. Three highly intelligent professional people, who contributed to a volume of essays on experiences of schizophrenia, each said that they found phenothiazines dampening and depressing, although each one found the acute symptoms intensely distressing and was grateful for the relief that medication brought [64]. Studies of long-term medication in conditions such as tuberculosis, parkinsonism, epilepsy, and diabetes have shown equivalent problems, and the fact that up to one-third of patients with schizophrenia do not persist with their medication is not extraordinary [20]. Serious attention given to explanations about the type of drugs available, the use of other drugs to minimize side-effects, the effects of varying the dose, the value of doing so in anticipation of stressful occasions, and the consequences of discontinuing medication may prove very helpful, particularly if it matches the patient's own experience.

Some patients learn eventually to recognize situations which make them feel worse and which might trigger episodes of relapse. One said: "There is a sensitivity in myself and I have to try to harden my emotions and cut myself off from potentially dangerous situations. . . . When I get worked up I often experience a slight recurrence of delusional thoughts." He avoided arguments on topics that made him emotionally upset. Another found that sometimes, when sitting in a subway train, he noticed the eyes of another passenger begin "to radiate." Then he would deliberately turn his attention to something else; in fact, he had evolved a relaxation technique in order to deal with such occasions. Another only experienced hallucinatory voices last thing at night, when his attention began to wander before he went to sleep, but he knew that he would not act on them and quite enjoyed them. A very bright girl chose her men friends among people who were not her intellectual equals because she did not get so involved with them and found she could control the situation better.

Social withdrawal is a technique that can be consciously manipulated by patients and used in a specific way to avoid situations they find painful. It is important that they know the dangers of going too far, since understimulation carries its own risk of increasing morbidity. A degree of external social stimulation is necessary for ordinary social functioning. Nevertheless, being withdrawn is often found preferable to being forced into unwanted social interaction. Work that is within the patient's competence and not too socially exacting is a great help but other people do not always recognize how exhausting many patients find a full-time job, even of this kind.

The degree of *insight* required to use these techniques sensibly is uncommon and some patients take them too far. Many never attain this degree of control; severity of illness is an important factor and it can vary independently of the quality of the social environment. Many could achieve more insight but are given very little help. Many meet with difficulties that are outside their competence to deal with: an over-involved relative who uses the emotional relationship to intrude upon the patient and force him into unwanted interaction; a lack of protected environments such as day centers, sheltered workshops or group homes; a critical attitude from friends or employers; an unrealistic professional helper who does not understand the impairments; accidental stresses such as can occur to anyone.

Most patients are not highly intelligent and articulate. Those who are have to speak for their fellows. The less capacity an individual has to recognize and cope with his own intrinsic impairments, extrinsic disadvantages, and personal reactions, the more important become the sympathy and the help of those with whom he or she has to live.

The Problems of Relatives

A study of schizophrenic patients admitted to three English mental hospitals in 1956 showed that, on discharge, 40 per cent went to live with parents, 37 per cent (mostly women) with a spouse, eight per cent with some other relative or friend, and 15 per cent went to lodgings, rooming houses or residential jobs [3]. By the time of follow-up, five years later, only 29 per cent were living with parents; there was less change in the other groups. Few separations from parental homes were due to disturbed behavior, which parents tolerated with remarkable fortitude. A third were due to the death or ill health of a parent and a third to positive reasons for leaving. Parents often made very little complaint even when they felt great distress and some developed very skillful methods of managing disturbed behavior. Three-quarters of the parents were over the age of 60 and 40 per cent over the age of 70.

There was a *high divorce and separation rate*, probably three times that in the general population. It was particularly high among men. Although a much smaller proportion of the men had married, the rate of separation was nearly double that of the women. Disturbed behavior was responsible for nearly all the separations during the follow-up period. Other studies have emphasized the difference in the types of *social problem* experienced by parents and spouses [4,11,48]. If neighbors come to call, it is fairly socially acceptable for a mother to say that her son is sick, by way of explanation of the fact that he dashes for the safety of his room as soon as he hears the doorbell ring, and does not come down until the visitors have gone. Such behavior by a husband or wife is much more difficult for the other partner to explain, particularly if there are children about.

As time goes on, there is no doubt that patient and relative, if they stay together, come to acquire a tolerance which neither might have had earlier. The relative, however, does so at the expense of restricting his or her life [42]. Often the parents of unmarried schizophrenics are elderly widows who are glad to have some companionship, to have someone to do a bit of shopping if they are physically disabled, and who are not too worried by not being able to live a life of their own. Under such circumstances, even a patient with a turbulent history of frequent breakdowns may eventually settle into a routine. It is another kind of *institutionalism*, less expensive, of course, and less demanding of the patient than a good hospital with workshops, leisure activities, and socialization programs would be, and sometimes a good deal more restricting on the activities and interests of relatives. Few, however, complain. The major problem raised by relatives articulate enough to be able to make a point is worry over the patient's future [7]. One father called it the WIAG ("when I am gone") syndrome.

However, this contented, if restricted, outcome for family life, at least for the unmarried patient, is sometimes only reached after what is, in some cases, a lengthy and profoundly distressing time, during which the patient's condition is constantly unstable and the relatives do not know what will happen next. It is not surprising that many patients find themselves homeless and drift to common lodging houses or reception centers. There do not have to be very many patients from each area each year to account for the large numbers found in Salvation Army hostels and shelters for the destitute [45,67].

Since relatives are almost as much in the front line as patients, so far as living with schizophrenia is concerned, it is surprising that there have been so few informed surveys of their views on the subject. Much work appears to have been carried out with the major object of selecting quotations that fit the author's preconceptions of the pathogenesis of schizophrenia. Relatives do, of course, acquire considerable experience of *coping with difficult behavior* but their methods are inevitably trial and error. Some learn not to argue with a deluded patient; others never learn. Some discover just how far they can go in trying to stimulate a rather slow and apathetic individual without arousing resentment. Others push too hard, find their efforts rejected or that they make matters worse, and then retreat into

TABLE 8. Behavior Problems Encountered by the Relatives of Schizophrenic
Patients

Characteristic	Fellowship (N = 50) %	Camberwell (N = 30) %	Total (N = 80) %
Social withdrawal	76	70	74
Underactivity	60	50	56
Lack of conversation	54	53	54
Few leisure interests	54	43	50
Slowness	34	70	48
Overactivity	54	20	41
Odd ideas	44	17	34
Depression	33	27	34
Odd behavior	44	17	34
Neglect of appearance	38	17	30
Odd postures or movements	38	3	25
Threats or violence	32	7	23
Sexually unusual behavior	6	10	8
Suicidal attempts	6	–	4
Incontinence	–	10	4

Adapted from Creer and Wing [11].

inactivity themselves. Some never give up intruding until the patient is driven away from home.

A recent survey of the experience of relatives was undertaken with the deliberate objective of learning from them what could be done to help people with schizophrenia [11]. Fifty patients were living with relatives who had joined a newly-formed voluntary organization, the National Schizophrenia Fellowship, and another 30 were selected from those known to specialists in an area of southeast London where services were reasonably good. The two groups therefore formed a marked contrast, since it was to be expected that relatives who joined the Fellowship would be articulate and responsible, but also that they would have particularly marked problems. Between the two groups it was possible to form a clear impression of the difficulties of families. Table 8 shows the kinds of problem mentioned. It is hardly necessary to illustrate them since they are so obvious. A few examples will suffice.

Altogether two-thirds of the 80 patients were reported as being markedly or somewhat underactive. However, even those who had some activity, tended to adopt some ritual way of spending the time, for example by brewing tea continuously or chain smoking. One relative explained graphically that, "in the evenings you go into the sitting room and it's in darkness. You turn on the light, and there he is, just sitting there, staring in front of him." Some relatives used the word "uncanny" to describe this kind of behavior. One mother said her son spent most of his time closeted in his room, only coming out at night when everyone was in bed. Usually he was talking to himself and moving about, but every few weeks there would be complete silence for a few days. "After that has been going on for a day or two, I sometimes wonder whether he is dead."

Relatives had various theories to explain the periods of total inactivity and the long hours spent in bed. Many felt that it was because ordinary everyday living and contact with people was simply an unbearable effort to the individual suffering from schizophrenia—he had to withdraw and "recharge" himself frequently. As one mother put it, "He just can't bear *people*—even to be in the same room as another person." One patient himself explained that he had to have the time lying on his bed that he did, because he was "all fizzing up inside." Some relatives feared that if they allowed too much underactivity, the patient would get worse, and therefore insisted that the patient should perform certain household tasks, even though they had to stay in the room with him to make sure he did them and even though the pace at which he worked was often painfully slow. Others decided this kind of thing was too exhausting. "I'd sooner not ask him to wash up," one said, "because it only means he uses cold water or forgets to use any washing up liquid, or leaves half the food on the plates." Some people tried to keep the patients active by keeping them entertained as much as possible—taking them out in the car, for walks, etc. But this was usually very draining emotionally.

About a third of the patients had odd ideas of various kinds—for example, that the neighbors were plotting against them or that some particular relative was at fault. The latter could be very distressing for the relative concerned. The odd ideas often concerned agencies or organizations whom the patient believed to have power over him or to be planning to harm him. Relatives found it difficult to know what to do when a patient expressed this kind of idea. If he said he had just been pursued up the road by a secret agent, ought they to accept what he said and pretend to believe it, or should they tell him he was imagining it? Many relatives feared that if they used the former approach they were encouraging the patient to lose touch even more with reality. But if they took the latter course, would the patient lose his confidence in them?

Patients tended to develop sudden irrational fears. They might, for instance, become fearful of a particular room in the house. Maybe they would tell the family the reason for their fear. "There's a poisonous gas leaking into that room" or "There are snakes under the bed in that room." At first relatives are baffled by this. Some admitted they had grown frustrated with a patient's absolute refusal to abandon some idea, despite all their attempts to reason with him, and had lost their temper. But they found this only resulted in the patient becoming very upset, and in any case the idea continued to be held with as much conviction as ever.

Several patients talked or laughed to themselves, but did this only in their own rooms, and not in front of their family. "If you stand outside his room you can hear that he's keeping up a more or less constant monologue in there." "Sometimes I hear shrieks of laughter coming from her room." Others, however, would sit throughout a meal laughing to themselves, regardless of who was there. Some would occasionally cry out in great distress in reaction to hallucinatory voices. A few relatives had been able to persuade a patient that he must not behave in this way in public. If he forgot and started to do it when others were present, a discreet reminder from the relative would silence him, or else he would go off somewhere alone until he had finished his muttering.

Innumerable examples like these could be given. One of the major complaints by relatives was that when they asked for advice from professional people as to the best way to react, they received no answer at all, or the question was simply turned back on them and their own amateur answers received with polite disdain. Perhaps their advisers had no better idea than they and were doing their best to conceal their ignorance?

A particularly important and distressing problem concerned compulsory admission to hospital. Relatives at the end of their tether and concerned for the long-term welfare of the patient are less inclined to argue the ins and outs of the civil rights issue than those who do not have to live in the situation, although I have met very few who are not distressed by the necessity. Relatives very rarely have advocate lawyers to argue their case for them.

Other difficulties concerned the administration of medication, the lack of sheltered work, the nonavailability of hostels or homes where the patients could go when the relatives needed a break, and the difficulty of obtaining welfare support when the patient was unwilling to claim it for himself. Relatives were often under strain themselves—depressed, anxious and guilty. Sometimes there was division within the family as to the best way of proceeding. There was also the agony of not knowing what to tell children.

It is extraordinary that so many relatives do manage to find a way of living with schizophrenia that provides the patient with a supportive and non-threatening home. Some of the factors that are to some extent under their control are as follows:

Creating a noncritical, accepting, environment
Providing the optimal degree of social stimulation
Keeping aims realistic
Learning how to cope with fluctuating insight
Learning how to respond to delusions or bizarre behavior
Making use of whatever social and medical help is available
Learning to use welfare arrangements

Obtaining rewards from the patient's presence
Helping patient's attitudes to self, to relatives, to medication, to work

General Factors in Management of Schizophrenia

Part of the peculiar difficulty in managing schizophrenia is that it lies some-where between conditions like blindness which, though severely handicapping, do not interfere with an individual's capacity to make independent judgments about his own future, and conditions like severe mental retardation, in which it is clear that the individual will never be able to make such independent judgments. There is frequently a fluctuating degree of insight and of severity. In some cases the fluc-tuations appear to follow no obvious pattern. They are then quite unpredictable and environmental circumstances seem unrelated. The fact that the impairments are invisible makes the problems even more difficult, since neighbors and friends, and even unfortunately some professional advisers, deny their existence altogether.

Nevertheless, a good deal is now known about the social and pharmacological reactivity of the acute and chronic syndromes and of the way adverse secondary reactions develop. This means that it is possible to influence the course of schizo-phrenia for the better. Indeed, in a great many areas, some part of the service al-ready operates effectively to minimize impairment, maximize assets or relieve burden. If it were possible to put together a service that was effective in all its parts, instead of in only some of them, a great deal of illness, handicap, and suffering would be prevented. Such a service would have to include a range of sheltered day and residential environments. The elements of a responsible, comprehensive and integrated service are well-known on both sides of the Atlantic.

One further point requires emphasis, because it has been insufficiently recog-nized hitherto. This is that the relatives are nowadays the real primary care agents. They and the patients deserve better *counseling*. Our attitudes to cancer have changed fundamentally during the past decade. Opinion leaders are willing to say that they have been treated for it and that they can live with it. This is not yet true for schizophrenia. Doctors themselves are still frightened of the word. Because the most effective methods of management, and the services to back them up, are not yet universally available, known or applied, it is rare for the condition to be frankly and realistically discussed. Public opinion remains ill informed and prejudiced. But just as the results of treatment have dramatically improved during the past quarter of a century, so the application simply of our present knowledge could produce a further substantial step forward. We would find that public opinion moved forward with us.

Perhaps the most important and most difficult requirement, for patient, relative and therapist alike, is to be realistic. If the therapist expects too much (endeavoring, perhaps, to fit the patient into the straightjacket of some untested theory), he can very easily make the patient worse [17,41,43,55]. It may seem reasonable to submit the patient to an intensive regime of social skills training, or rehabilitation, or social activation, on the theory that invisible impairments need not be considered, but although it will sometimes work, it often will not. It should always be remembered that there is a noteworthy suicide rate in schizophrenia [60]. If the patient has severe intrinsic impairments, they will usually be obvious and it may, for example, be worthwhile to accept work in a sheltered setting where only part of the social role need be adopted. A neutral (not over-emotional) expectation to perform up to *attainable* standards is the ideal. This rule, if difficult for the specialist to adopt, is a thousand times more difficult for relatives. Nevertheless, we should be humbled to recognize that a large portion of relatives, by trial and error, do come to adopt it, without any help from professionals. The patient will have the greatest chance of acquiring insight if those around him, in spite of everything, are realistic. This is why skilled counseling (which means much more willingness by professionals to learn from relatives and patients) is the essence of good management.

I have concentrated attention on the severely handicapped patient with schizophrenia for obvious reasons, but most make a good social recovery. With the application of all that we now know, many more can do so.

APPENDIX 1

The Major CATEGO Classes

Class	*Diagnostic Equivalent* (approximate only)
Class S	Central schizophrenia
Class P	Paranoid psychoses not classified as S, M or D
Class O	Non-paranoid and borderline psychoses not classified as S, P, M or D
Class M	Manic psychoses
Class D	Depressive psychosis
Class R	Retarded depressions not classified as D
Class N	Neurotic depressions
Class A	Anxiety states
Class B	Obsessional neuroses

Each of these classes has a more certain form (S+, P+, etc.) and a less certain form (S?, P?, etc.).

APPENDIX 2

Syndrome Profiles of Three Major CATEGO Classes
(IPSS Data)

	Syndrome	Class S %	Class P %	Class O %
1.	Nuclear syndrome	82	–	–
2.	Catatonic syndrome	10	–	58
3.	Incoherence of speech	31	19	53
4.	Residual syndrome	22	7	55
5.	Depressive delusions	19	1	3
6.	Simple depression	85	77	39
7.	Obsessional symptoms	15	8	2
8.	General anxiety	66	51	42
9.	Situational anxiety	19	17	6
11.	Affective flattening	51	32	65
12.	Hypomanic syndrome	31	11	29
13.	Non-affective auditory hallucinations	74	1	–
14.	Delusions of persecution	64	69	3
15.	Delusions of reference	64	52	5
16.	Grandiose and religious delusions	32	13	11
17.	Fantastic delusions	78	72	16
18.	Visual hallucinations	51	26	5
19.	Olfactory hallucinations	17	7	3
20.	Overactivity	10	2	24
21.	Slowness and underactivity	36	25	76
22.	Nonspecific psychosis	69	50	71
23.	Depersonalization	41	17	11
24.	Guilt and self-depreciation	42	24	8
25.	Agitation	20	11	32
26.	Self-neglect	19	6	40
27.	Ideas of reference	68	52	10
28.	Muscular tension	72	58	23
29.	Lack of energy	42	29	19
30.	Worrying	88	87	37

Syndrome 10 and syndromes 31–38 omitted.
For details of the symptoms making up the syndromes, and the glossary of definitions, see
Wing, Cooper, and Sartorius [62].

REFERENCES

1. Babcock, H. *Dementia praecox: A psychological study.* New York, 1933.
2. Bleuler, M. *Die schizophrenen Geislesslörungen im Lichte langjähriger Kranken und Familiengeschichte.* Stuttgart: Thieme, 1972.
3. Brown, G. W., Bone, M., Dalison, B., and Wing, J. K. *Schizophrenia and social care.* London: Oxford University Press, 1966.
4. Brown, G. W. Birley, J. L. T., and Wing, J. K. Influence of family life on the course of schizophrenic disorders: A replication. *Br. J. Psych. 121*:241-258, 1972.
5. Brown, G. W. and Birley, J. L. T. Social precipitants of severe psychiatric disorders. In: *Psychiatric epidemiology*, edited by E. H. Hare and J. K. Wing. London: Oxford University Press, 1970.
6. Carstairs, G. M., O'Connor, N., and Rawnsley, K. The organization of a hospital workshop for chronic psychotic patients. *Br. J. Prev. Soc. Med. 10*:136, 1956.
7. Catterson, A., Bennett, D. H., and Freundunberg, R. L. A survey of longstay schizophrenic patients. *Br. J. Psych. 109*:750, 1963.
8. Cooper, B. Social class prognosis in schizophrenia. *Br. J. Prev. Soc. Med. 15*: 17-41, 1961.
9. Cooper, J. E., Kendell, R. E., Gurland, B. J., Sharpe, L., Copeland, J. R. M., and Simon, R. *Psychiatric diagnosis in New York and London.* Maudsley Monograph No. 20. London: Oxford University Press, 1972.
10. Cooper, B. and Morgan, H. G. *Epidemiological psychiatry.* Springfield, IL: Charles C. Thomas, 1973.
11. Creer, C. and Wing, J. K. *Schizophrenia at home.* London: National Schizophrenia Fellowship, 1974.
12. Deutsch, A. *The mentally ill in America.* New York: Columbia University Press, 1949.
13. Foulds, G. A. and Dixon, P. The nature of intellectual deficit in schizophrenia. *Br. J. Clin. Soc. Psychol. 1*:199, 1962.
14. Garmezy, N. Stimulus differentiation by schizophrenic and normal subjects under conditions of reward and punishment. *J. Pers. 20*:253, 1952.
15. Gatewood, L. C. An experimental study of dementia praecox. *Psychol. Monogr. 11*:2, 1909.
16. Goffman, E. *Asylums: Essays on the social situation of mental patients and other inmates.* New York: Doubleday, 1961.
17. Goldberg, S. C., Schooler, N. R., Hogarty, G. E., and Roper M. Prediction of relapse in schizophrenic outpatients treated by drug and sociotherapy. *Arch. Gen. Psych.* (in press).
18. Harris, A., Linker, I., Norris, V., and Shepherd, M. Schizophrenia: A social and prognostic study. *Br. J. Prev. Soc. Med. 10*:107-114, 1956.
19. Hewett, S., Ryan, P., and Wing, J. K. Living without the mental hospitals. *J. Soc. Policy 4*:391-404, 1975.
20. Hirsch, S. R., Gaind, R., Rohde, P. D., Stevens, B. C., and Wing, J. K. Outpatient maintenance of chronic schizophrenic patients with long-acting fluphenazine: Double-blind placebo trial. *Br. Med. J. 1*:633-637, 1973.

21. Hirsch, S. R. and Leff, J. P. *Abnormality in parents of schizophrenics: A review of the literature and an investigation of communication defects and deviances.* London: Oxford University Press, 1975.
22. Jones, K. *A history of the mental health services.* London: Routledge, 1972.
23. Jung, C. G. *The psychology of dementia praecox,* translated by A. A. Brill. Nervous and Mental Diseases Monographs, 1936, 1960.
24. Kendig, I. and Richmond, W. V. *Psychological studies in dementia praecox.* Ann Arbor, MI, 1940.
25. Kent, G. H. Experiments in habit formation in dementia praecox. *Psychol. Rev. 18:*375-410, 1911.
26. Kramer, M. Some problems for international research suggested by observations on differences in first admission rates to mental hospitals of England and Wales and of the United States. *Proceedings of the 3rd World Cong. Psychiat. 3.* Montreal: McGill University Press, 1963.
27. Leff, J. P. and Wing, J. K. Trial of maintenance therapy in schizophrenia. *Br. Med. J. 3:*599-604, 1971.
28. Lidz, T., Fleck, S., and Cornelison, A. R. *Schizophrenia and the family.* New York: International Universities Press, 1965.
29. Mann. S. and Sprogle, J. Reasons for a six months stay. In: *Evaluating a community psychiatric service,* edited by J. K. Wing and A. M. Hailey. London: Oxford University Press, 1972.
30. Mann, S. and Cree, W. New long-stay psychiatric patients: A national sample of 15 mental hospitals in England and Wales 1972/73. *Psychol. Med. 6:*603-616, 1976.
31. Mayer-Gross, W. Die Schizophrenie. In: *Handbuch der Geisteskrankheiten. Band IX,* edited by O. Burnke. Berlin: Springer, 1932.
32. O'Connor, N., Heron, A., and Carstairs, G. M. Work performance of chronic schizophrenics. *Occup. Psychol. 30:*1-12, 1956.
33. O'Connor, N. and Rawnsley, K. Incentives with paranoid and non-paranoid schizophrenics in a workshop. *Br. J. Med. Psychol. 32:*133-143, 1959.
34. Peffer, P. A. Money: A rehabilitation incentive for mental patients. *Am. J. Psych. 110:*84-92, 1953.
35. Peffer, P. A. Motivation of the chronic mental patient. *Am. J. Psych. 113:*55-59, 1956.
36. Peters, H.N. and Jenkins, R. L. Improvement of chronic schizophrenic patients with guided problem-solving motivated by hunger. *Psych. Quart. Sup. 28:*84, 1954.
37. Peters, H. N. and Murphree, O. D. The conditional reflex in the chronic schizophrenic. *J. Clin. Psychol. 10:*126, 1954.
38. Rosenthal, D. and Kety S. S. *The transmission of schizophrenia.* New York: Pergamon, 1968.
39. Scheft, T. J. *Being mentally ill.* Chicago: Aldine, 1966.
40. Schneider, K. *Clinical psychopathology,* Fifth edition, translated by M. W. Hamilton. New York: Grune and Stratton, 1959.
41. Stevens, B. Evaluation of rehabilitation for psychotic patients in the community. *Acta Psych. Scand. 49:*169-180, 1973.

42. Stevens, B. C. Dependence of schizophrenic patients on elderly relatives. *Psychol. Med. 2:*17-32, 1972.
43. Stone, A. A. and Eldred, S. H. Delusion formation during the activation of chronic schizophrenic patients. *Arch. Gen. Psych. 1:*177-179, 1959.
44. Tarrier, N., Vaughn, C., Lader, M. H., and Leff, J. P. Bodily reactions to people and events in schizophrenia (to be published), 1977.
45. Tidmarsh, D. and Wood, S. Psychiatric aspects of destitution. In: *Evaluating a community psychiatric service*, edited by J. K. Wing and A. M. Hailey. London: Oxford University Press, 1972.
46. Vaughn, C. E. and Leff, J. P. The influence of family and social factors on the course of psychiatric illness. *Br. J. Psych. 120:*125-137, 1976.
47. Vaughn, C. E. and Leff, J. P. The measurement of expressed emotion in the families of psychiatric patients. *Br. J. Clin. Soc. Psychol. 15:*157-165, 1976.
48. Vaughn, C. E. Patterns of interaction in families of schizophrenic patients. In: *Schizophrenia: The other side*, edited by H. Katsehnig. Vienna: Urban and Schwarzenberg, 1977.
49. Venables, P. H. and Wing, J. K. Level of arousal and the subclassification of schizophrenia. *Arch. Gen. Psych. 7:*111-119, 1962.
50. Watt N. F. and Lubensky, A. W. Childhood roots of schizophrenia. *J. Cons. Clin. Psychol. 44:*363-375, 1976.
51. Wing, J. K. A pilot experiment on the rehabilitation of long-hospitalized male schizophrenic patients. *Br. J. Prev. Soc. Med. 14:*173, 1960.
52. Wing, J. K. and Freudenberg, R. K. The response of severely ill chronic schizophrenic patients to social stimulation. *Am. J. Psych. 118:*311, 1961.
53. Wing, J. K. Institutionalism in mental hospitals. *Br. J. Soc. Clin. Psychol. 1:* 38, 1962.
54. Wing, J. K., Monck, E., Brown, G. W., and Carstairs, G. M. Morbidity in the community of schizophrenic patients discharged from London mental hospitals in 1959. *Br. J. Psych. 110:*10, 1964.
55. Wing, J. K., Bennett, D. H., and Denham, J. *The industrial rehabilitation of long-stay schizophrenic patients.* Med. Res. Council Memo. No. 42. London: H.M.S.O., 1964.
56. Wing, J. K. Social and psychological changes in a rehabilitation unit. *Soc. Psych. 1:*21-28, 1966.
57. Wing, L., Wing, J. K., Hailey, A. M., Bahn, A. K., Smith, A. E., and Baldwin, J. A. The use of psychiatric services in three urban areas: An international case register study. *Soc. Psych. 2:*158-167, 1967.
58. Wing, J. K. and Brown, G. W. *Institutionalism and schizophrenia.* London: Cambridge University Press, 1970.
59. Wing, J. P., and Hailey, A. M. (Eds.). *Evaluating a community psychiatric service: The Camberwell Register, 1964-1971.* London: Oxford University Press, 1972.
60. Wing, L., Wing J. K., Griffiths, D., and Stevens, B. An epidemiological and experimental evaluation of industrial rehabilitation of chronic psychotic patients in the community. In: *Evaluating a community psychiatric service*, edited by J. K. Wing and A. M. Hailey. London: Oxford University Press, 1972.

61. Wing, J. K., Leff, J. P., and Hirsch, S. R. Preventive treatment of schizophrenia: Some theoretical and methodological issues. In: *Psychopathology and psychopharmacology*, edited by J. O. Cole, A. M. Freedman, and A. J. Friedhoff. Baltimore: Johns Hopkins University Press, 1973.
62. Wing, J. K., Cooper, J. E., and Sartorius, N. *The description and classification of psychiatric symptoms: An instruction manual for the PSE and catego system*. London: Cambridge University Press, 1974.
63. Wing, J. K. Institutional influences on mental disorders. *Psychiatric der Gegenwart, Band III 2 Aufl*. Berlin: Springer Verlag, 1975.
64. Wing, J. K. (Ed.). *Schizophrenia from within*. London: National Schizophrenia Fellowship, 1975.
65. Wing, L. (Ed.). *Early childhood autism: Clinical, educational and social aspects*. New York: Pergamon, 1976.
66. Wing, J. K. *Reasoning about madness*. London and New York: Oxford University Press, 1977.
67. Wood, S. M. Camberwell Reception Center: A consideration of the need for health and social services of homeless single men. *J. Soc. Pol.* 5:389–399, 1976.
68. World Health Organization. *The international pilot study of schizophrenia*. Geneva: W.H.O., 1973.
69. Wynne, L. C. Methodologic and conceptual issues in the study of schizophrenics and their families. In: *The transmission of schizophrenia*, edited by D. Rosenthal and S. S. Kety. London and New York: Pergamon, 1968.
70. Wynne, L. C. Family research on the pathogenesis of schizophrenia. In: *Problems of psychosis*, edited by P. Doncet and C. Laurin. Excerpta Medica International Congress Series, No. 194, 1971.

AUTHOR'S COMMENTS

Family Problems

This is a uniquely well-informed audience, impossible to satisfy with platitudes, fantastic theories, or idealized descriptions of services that manifestly do not exist anywhere in the world. Nor does one have to remind such a group that schizophrenia often entails disability, disappointment, and suffering. Talking to groups of members throughout the U.K., and to the sister organizations that have been set up in Eire, Australia, and New Zealand, I have found the same realistic questions being put each time. There is, of course, keen interest in the progress being made in genetics, biochemistry, pathophysiology, and epidemiology. But discussion always returns to problems of treatment, management, and services and to difficulties in relationships with professionals.

In his introduction to the reprint the President of the NSF, John Pringle, was kind enough to say that laymen would be able to follow the text in spite of the fact that it was written for professionals. "Laymen who are also relatives may be in for a pleasant surprise at the appreciative things the author occasionally has to say about them. They are so used to being targets for hostile criticism either for allegedly being implicated in the causation of the condition in the first place, or for their clumsy handling of it afterwards, that it comes as a slight shock to be told that they may have things to teach as well as to learn."

This point deserves further emphasis. I suggested in the lecture that there was little satisfactory evidence to substantiate theories that inadequate child-rearing practices play a significant part in the subsequent development of schizophrenia. The best empirical studies, by Wynne and his colleagues, are remote from the theme and, in any case, the results could not be reproduced in a careful replication in London. Rutter [7] has reviewed the discrepancies between these two studies and concluded that they could not be explained by methodological differences but were most likely accounted for by variations in diagnostic practice. There may perhaps be a high frequency of communication deviances in the parents of certain psychiatric patients but many would be diagnosed in the U.K. as having personality disorders, not schizophrenia. Moreover, even in this group, the evidence could equally well be explained on a genetic basis, or as a reaction to unusual behavior in the patient.

It is unfortunate that ideas put forward on a purely theoretical basis (a legitimate and necessary part of scientific activity) should have become accepted as gospel by many people who have not themselves studied the evidence, so that they are taken from the air by administrators, professional people, medical students, interested laymen and, of course, by patients and relatives. Whatever the intentions of the authors of the theories, relatives feel themselves blamed.

That differences of interest must sometimes develop between patients and relatives, particularly when the course of schizophrenia is prolonged is, of course, obvious. The fact that, nevertheless, more than half the relatives in the London studies provided a warm, accepting, and protective environment was emphasized in the lecture; the relapse rate was low. Since then, a trial of intervention has been mounted by Leff, with a group of patients living in "high emotion" families whose risk of relapse is high. All patients receive medication. A randomly selected experimental group is offered a package of measures—daytime occupation (to reduce face-to-face contact), health education for relatives, and group sessions in which the family problems listed by Creer [2] are discussed and in which "low emotion" relatives take part—and a control group receives the usual care from local services. Both groups are followed up to determine relapse rates.

One of the chief activities of the National Schizophrenia Fellowship is to set up local groups in which relatives and patients can discuss their problems and invite, as they feel the need, psychiatrists, social workers, and other professional people to join them. These sessions are useful learning experiences, as much for professional staff as for the regular participants. An account of the first such group, although it does not shirk the real difficulties that arise, suggests that the method is helpful [5,6].

An issue frequently commented upon in these groups, but not discussed in my lecture, concerns civil rights. There has been little research on the subject but what comment is available generally concerns the rights of the patient. At the extreme, it is suggested that no legal restriction should be applied unless the patient commits an offence and that relatives, since they are partial, should not act as witnesses. Everyday practice is more moderate than this but relatives are frequently at a disadvantage; being required to act as primary care and welfare agents and to accept the burden of coping with abnormal behavior and yet not being taken seriously when they asked for advice or, occasionally, for action. Research which considers the human rights of relatives as well as those of patients should receive high priority.

Course

The results of a two year follow-up study of patients included in the International Pilot Study of Schizophrenia (IPSS) have now been published [10]. Overall, they confirm the pattern established by earlier studies. Manfred Bleuler's formulation [1], that the proportion who recover completely and the proportion who remain totally disabled all their lives has not changed since the introduction of the phenothiazines, but that the large intermediate group now runs a fluctuating course much influenced by the quality of the social environment, provides an adequate summary.

A point of new interest in the IPSS is that the two-year course was found to be most favorable in Ibadan and Agra, with Cali, Taipei, Moscow, London, and Prague (in that order) intermediate, and least favorable in Aarhus. There are many explanations other than that environmental factors in developing countries have a beneficial influence, since the IPSS did not have an epidemiological base. Moreover, other results have been conflicting [3]. One possible explanation, if the result should be confirmed in further work now being undertaken, is that any harmful effects due to high expressed emotion in some relatives in the west might be diluted in extended families in developing countries. This hypothesis is being examined in current studies in Aarhus and Chandigarh.

A recent study of destitute men has underlined yet again the importance of homelessness as one of the outcomes in schizophrenia [4] and the difficulties of resocialization once this nadir has been reached.

Institution and Community

The movement for community rather than institutional care was never based on a careful conceptual analysis. Dissatisfaction with total institutions was the starting point and observation of the results of early rehabilitation programs, which succeeded because they corrected the ill-effects of institutionalism, gave credence to the idea that most psychiatric disability was iatrogenic. Some of the optimism of the 1950s and 1960s was based on these doubtful premises and has proved to be illusory. The community has disadvantages and advantages for disabled people, just as institutions have [9].

A recent project to look after new long-stay patients (mostly with schizophrenia) in a hostel-ward (a house with the same staffing ratio as a ward but a domestic scale and quality of living) has been successful, in the sense that no patient has had to be transferred back to a hospital ward during the first three years of operation [11]. The only new long-stay patients from a defined geographical district who could not be accommodated in the hostel-ward were two under legal restriction in a special hospital and one whose disturbed behavior made it impossible. Costs are no different from those in a general ward. Further units will be opened as necessary, some of them with a lower staff ratio. Hostel-wards will also be opened in other districts and comparable evaluation techniques applied.

Very little research has been carried out into the advantages and disadvantages of grouping several units on one campus and into methods of administration that would prevent the undoubtedly harmful effects of a single authoritarian hierarchy within closed boundaries. Studies of scattered nonhospital units suggest that residents do not necessarily make use of community facilities (that is, shops, recreation, work) or interact with local people; they can be as isolated and inactive as in a poorly organized hospital [9].

There is now a substantial body of knowledge about the disabilities that occur in schizophrenia and about ways in which these can be minimized by environmental influence. One of the chief areas for research during the next decade should be into methods of translating this knowledge into clinical and administrative practice in order to provide a realistic and humane service for people with schizophrenia and their relatives.

REFERENCES

1. Bleuler, M. The long-term course of schizophrenic psychoses. In: *The nature of schizophrenia: New approaches to research and treatment*, edited by L. C. Wynne, R. L. Cromwell, and S. Matthysse. New York: John Wiley & Sons, 1978.
2. Creer, C. Social work with patients and their families. In: *Schizophrenia: Towards a new synthesis*, edited by J. K. Wing. London: Academic Press, 1978.
3. Kulhara, P. and Wig, N. N. The chronicity of schizophrenia in North West India: Results of a follow-up study. *Br. J. Psych. 132*:186–190, 1978.
4. Leach, J. and Wing, J. K. *Helping destitute men.* London, Tavistock, 1980.
5. Priestley, D. Helping a self-help group. In: *Community care for the mentally disabled*, edited by J. K. Wing and R. Olsen. London: Oxford University Press, 1979.

6. Priestley, D. *Tied together with string*. National Schizophrenia Fellowship, 79, Victoria Road, Surbiton, Surrey, United Kingdom, 1979.
7. Rutter M. Communication deviances and diagnostic differences. In: *The nature of schizophrenia: New approaches to research and treatment*, edited by L. C. Wynne, R. L. Cromwell, and S. Matthysse. New York: John Wiley & Sons, 1978, pp. 512–516.
8. Wing, J. K. (Ed.). *Schizophrenia: Towards a new synthesis*. London: Academic Press, 1978.
9. Wing, J. K. and Olsen, R. (Eds.). *Social care for the mentally disabled*. London: Oxford University Press, 1979.
10. World Health Organization. *Schizophrenia: An international follow-up study*. New York: John Wiley & Sons, 1979.
11. Wykes, T. and Wink, J. K. A ward in a house: Alternative accommodation for "new" long-stay patients (in press).

BIOLOGIC FACTORS

Biochemical Studies in Schizophrenia

Seymour S. Kety, M.D.

The concept of a chemical etiology in schizophrenia is not new. The Hippocratic school attributed certain mental aberrations to changes in the composition of the blood, but it was Thudichum, the founder of modern neurochemistry, who in 1884 expressed the concept most cogently: "Many forms of insanity are unquestionably the external manifestations of the effects upon the brain substance of poisons fermented within the body, just as mental aberrations accompanying chronic alcoholic intoxication are the accumulated effects of a relatively simple poison fermented out of the body. These poisons we shall, I have no doubt, be able to isolate after we know the normal chemistry to its uttermost detail. And then will come, in their turn, the crowning discoveries to which our efforts must ultimately be directed, namely, the discoveries of the antidotes to the poisons and to the fermenting causes and processes which produce them" In these few words were anticipated and encompassed most of the current chemical formulations regarding schizophrenia.

It may be of value to pause in the midst of the present era of psychochemical activity to ask how far we have advanced along the course plotted by Thudichum. Have we merely substituted "enzymes" for "ferments" and the names of specific agents for "poisons" without altering the completely theoretical nature of the concept? Or, on the other hand, are there some well-substantiated findings to support the prevalent belief that this old and stubborn disorder which has resisted all previous attempts to expose its etiology is about to yield its secrets to the biochemist?

An examination of the experience of another and older discipline may be of

Research in the Schizophrenic Disorders: The Stanley R. Dean Award Lectures, vol. 2, edited by R. Cancro and S. R. Dean. Copyright © 1985 by Spectrum Publications.

help in the design, interpretation, and evaluation of biochemical studies. The pathological concepts of schizophrenia have been well reviewed recently. Prompted by the definite histological changes in the cerebral cortex, described by Alzheimer and confirmed by a number of others, an early enthusiasm developed which penetrated into the thinking of Kraepelin and Bleuler. This was followed by a period of questioning and the design and execution of more critically controlled studies leading to the present consensus that a pathological lesion characteristic of schizophrenia or of any of its subgroups remains to be demonstrated.

Because of the chronicity of the disease, the prolonged periods of institutionalization associated with its management, and the comparatively few objective criteria available for its diagnosis and the evaluation of its progress, schizophrenia presents to the biochemical investigator a large number of variables and sources of error which he must recognize and attempt to control before he may attribute to any of his findings a primary or characteristic significance.

Despite the phenomenological similarities which permitted the concept of schizophrenia as a fairly well defined symptom-complex to emerge, there is little evidence that all of its forms have a common etiology or pathogenesis. The likelihood that one is dealing with a number of different disorders with a common symptomatology must be recognized and included in one's experimental design. Errors involved in sampling from heterogeneous populations may help to explain the high frequency with which findings of one group fail to be confirmed by another. Recognition of the probability that any sample of schizophrenia is a heterogeneous one should seem to emphasize the importance of analyzing data not only for mean values but also for significant deviations of individual values from the group. The biochemical characteristics of phenylketonuria would hardly have been detected in an average value for phenylalanine blood levels in a large group of mentally retarded patients. Most biochemical research in schizophrenia has been carried out in patients with a long history of hospitalization in institutions where overcrowding is difficult to avoid and hygienic standards cannot always be maintained. It is easy to imagine the spread of chronic infections, especially of the digestive tract, among such patients. The presence of amebiasis in a majority of patients at one large institution has been reported, and one wonders how often this condition or a former infectious hepatitis has accounted for the various disturbances in hepatic function found in schizophrenia. Even in the absence of previous or current infection, the development of a characteristic pattern of intestinal flora in a population of schizophrenic patients living together for long periods and fed from the same kitchen is a possibility which cannot be dismissed in interpreting what appear to be deviant metabolic pathways.

The variety and quality of the diet of the institutionalized schizophrenic is rarely comparable to that of the nonhospitalized normal control. Whatever homeostatic function the process of free dietary selection may serve is often lost between the rigors of the kitchen or the budget and the overriding emotional or obsessive features of the disease. In the case of the "acute" schizophrenic, the weeks and months of emotional turmoil which precede the recognition and diagnosis of the disease are hardly conducive to a normal dietary intake. Certain abnormalities in thyroid function previously reported in schizophrenia have been shown to result from a dietary deficiency of iodine, correctable by the introduction of iodized salt into the hospital diet. It is not surprising that a dietary vitamin deficiency has been found to explain at least two of the biochemical abnormalities recently attributed to schizophrenia. ·It is more surprising that the vitamins and other dietary constituents, whose role in metabolism has become so clearly established, should so often be relegated to a position of unimportance in the intermediary metabolism of schizophrenics. Horwitt has found signs of liver dysfunction during ingestion of a diet containing borderline levels of protein while nonspecific vitamin therapy accompanieq by a high protein and carbohydrate diet has been reported to reverse the impairment of hepatic function in schizophrenic patients.

Another incidental feature of the schizophrenic which sets him apart from the normal control is the long list of therapies to which he may have been exposed. Hypnotic and ataractic drugs and their metabolis products or effects produce changes which have sometimes been attributed to the disease. Less obvious is the possibility of residual electrophysiological or biochemical changes resulting from repeated electroshock or insulin comas.

Emotional stress is known to cause profound changes in man, in adrenocortical and thyroid function, in the excretion of epinephrine and norepinephrine, of water, electrolytes or creatinine, to mention only a few recently reported findings. Schizophrenic illness is often characterized by marked emotional disturbance even in what is called the basal state and frequently exaggerated anxiety in response to routine and research procedures. The disturbances in behavior and activity which mark the schizophrenic process would also be expected to cause deviations from the normal in many biochemical and metabolic measures: in urine volume and concentration, in energy and nitrogen metabolism, in the size and function of numerous organic systems. The physiological and biochemical changes which are secondary to the psychological and behavioral state of the patient are of interest in themselves as part of a total understanding of the schizophrenic process; it is important, however, not to attribute to them a primary or etiological role.

An additional source of error which must be recognized is one which is common to all of science and which it is the very purpose of scientific method, tradition, and training to minimize—the subjective bias. There are reasons why this bias should operate to a greater extent in this field than in many others. Not only is the motivation heightened by the tragedy of this problem and the social implications of findings whch may contribute to its solution, but the measurements themselves, especially of the changes in mental state or behavior, are so highly subjective, the symptoms so variable and responsive to nonspecific factors in the milieu that only the most scrupulous attention to controlled design will permit the conclusion that a drug, or a diet, or a protein fraction of the blood, or an extract of the brain is capable of causing or ameliorating some of the manifestations of the disease. This is not to suggest that the results of purely chemical determinations are immune to subjective bias, and the same vigilance is required to prevent the hypothesis from contaminating the data. In a field with as many variables as this one, it is difficult to avoid the subconscious tendency to reject for good reason data which weaken an hypothesis while uncritically accepting those data which strengthen it. Carefully controlled and "double-blind" experimental designs which are becoming more widely utilized in this area can help to minimize this bias.

Obvious as many of these sources of error are, it is expensive and difficult, if not impossible, to prevent some of them from affecting results obtained in this field, especially in the preliminary testing of interesting hypotheses. It is in the interpretation of these results, however, and in the formulating of conclusions, that the investigator has the opportunity and indeed the responsibility to recognize and evaluate his uncontrolled variables rather than to ignore them, for no one knows better than the investigator himself the possible sources of error in his particular experiment. There are enough unknowns in our guessing game with nature to make it unnecessary to indulge in such a sport with one another.

The senior members of this laboratory: Julius Axelrod, Phillippe Carden, Edward Evarts, Marian Kies, Roger McDonald, Seymour Perline, and Louis Sokoloff, pursue fundamental research in the biological sciences and psychiatry. They have also participated in the development of the schizophrenia program of the Laboratory of Clinical Science, a program of biological research in schizophrenia in which they have been joined by Irwin Kopin, Elwood LaBrosse, Jay Mann, and William Pollin. This program is designed to minimize many of the sources of error previously discussed while increasing the opportunity for true biological characteristics, if they exist, to be detected. One of the wards houses a group of approximately 14 clearly diagnosed schizophrenic patients, representative of as many clinical subgroups as possible, chosen from a patient population of 14,000 with an attempt to minimize the nondisease variables of age, sex, race, and physical illness, and on the basis of careful family surveys to maximize the likelihood of including within the group whatever genetic subgroups of the disease may exist. They are maintained for an indefinite period of time on a good diet and under excellent hygienic, nursing, medical, and psychiatric care. Drugs or dietary changes are introduced only for research purposes or, when clinically necessary, for short periods of time. The other ward houses a comparable number of normal controls who volunteer to remain for protracted periods of time exposed to the same diet and in a reasonably similar milieu. We recognize, of course, that only a few of the variables are thus controlled and any positive difference which emerges in this preliminary experiment between some or all of the schizophrenics and the normal population will have to be subjected to much more rigorous examination before its significance can be evaluated.

A decrease in basal metabolism was found in schizophrenia by earlier workers although more recent work has not confirmed this and theories attributing the disease to disturbances in the fundamental mechanisms of energy supply or conversion in the brain have enjoyed some popularity, but on the basis of extremely inadequate evidence, such as spectroscopic oximetry of the ear lobe or nail bed. Our finding of a normal rate of cerebral circulation and oxygen consumption in schizophrenic patients was confirmed by Wilson, Schieve, and Scheinberg, and more recently in our laboratory by Sokoloff and associates, who also found a normal rate of cerebral glucose consumption in this condition. These studies do not, of course, rule out a highly localized change in energy metabolism somewhere in the brain, but cogent evidence for such a hypothesis has yet to be presented.

Richter has pointed out the uncontrolled factors in earlier work which implicated a defect in carbohydrate metabolism as a characteristic of the schizophrenic disease process. The finding in schizophrenia of an abnormal glucose tolerance in conjunction with considerable other evidence of hepatic dysfunction, or evidence of a retarded metabolism of lactate by the schizophrenic does not completely exclude incidental hepatic disease or nutritional deficiencies as possible sources of error. Horwitt and associates were able to demonstrate and correct similar abnormalities by altering the dietary intake of the B group of vitamins.

Evidence for higher than normal anti-insulin or hyperglycemic activity in the blood or urine of a significant segment of schizophrenic patients was reported in 1942 by Meduna, Gerty, and Urse, and as recently as 1958 by Moya and associates. Some progress has been made in concentrating or characterizing such factors in normal urine as well as that from schizophrenics. Harris has thrown some doubt on the importance of such anti-insulin mechanisms in the pathogenesis of schizophrenia and it is hoped that further investigation may clarify the nature of the substance or substances involved and their relevance to schizophrenia.

Defects in oxidative phosphorylation have been thought to occur in this disease. Reports of alterations in the phosphorus metabolism of the erythrocyte, await further definition and independent confirmation.

The well controlled studies of the Gjessings on nitrogen metabolism in periodic catatonia arouse considerable interest in the possible relationship of intermediary protein metabolism to schizophrenia, although earlier workers had postulated defects in amino acid metabolism in this disease. The hallucinogenic properties of some compounds related directly or indirectly to biological amines reawakened this interest and the techniques of paper chromatography offered new and almost unlimited opportunity for its pursuit.

The first group to report chromatographic studies of the urine of schizophrenic and control groups found certain differences in the amino acid pattern and in addition the presence of certain unidentified imidazoles in the urine of schizophrenics. Although a normal group of comparable age was used for comparison, there is no indication of the extent to which dietary and other variables were controlled, and the authors were properly cautious in their conclusions. In a more extensive series of studies, another group has reported a significantly higher than normal concentration of aromatic compounds in the urine of schizophrenic patients, and has suggested certain qualitative differences in the excretion of such compounds. Others have reported the abnormal presence of unidentified amines or indoles, and one group, the absence of a normally occurring indole in the urine of schizophrenic patients. Not all of these studies appear to have controlled possible drug therapy, urinary volume or concentration, and few have controlled the diet. There are numerous mechanisms whereby vitamin deficiencies may cause substantial changes in the complex patterns of the intermediary metabolism of amino acids. In addition, the large number of aromatic compounds in the urine which have recently been shown to be of dietary origin suggest considerably more caution than has usually been employed with regard to this variable. Another point which has not been emphasized sufficiently is that chromatographic procedures which make possible the determination of scores of substances simultaneously, many of them unknown beforehand, also require somewhat different statistical analyses than those which were developed for the testing of single, well-defined hypotheses. It is merely a restatement of statistical theory that in an analysis for 100 different compounds simultaneously in two samples of the same population, five would be expected to show a difference significant at the 0.05 level! It is interesting to note that a more recent study was able to demonstrate considerably fewer differences between the urines of normal and schizophrenic populations and drew very limited and guarded conclusions. In our own laboratory, Mann and LaBrosse undertook a search for urinary phenolic acids in terms of quantity excreted rather than concentration, which disclosed four compounds significantly higher in the urine from the schizophrenic than that from the normals; two of these were found to be known metabolites of substances in coffee as were probably the other two as well; the presence of these four compounds in the urine was, in fact, better correlated with the ingestion of this beverage than with schizophrenia.

The hypothesis that a disordered amino acid metabolism is a fundamental component of some forms of schizophrenia remains an attractive though fairly general one, the chromatographic search for its evidence is interesting and valuable, and the preliminary indications of differences certainly provocative. Proof that any of these differences are characteristic of even a segment of the disease rather than artifactual or incidental has not yet been obtained.

The theory which relates the pathogenesis of schizophrenia to faulty metabolism of epinephrine is imaginative, ingenious, and plausible. It postulates that the symptoms of this disease are caused by the action of abnormal, hallucinogenic derivatives of epinephrine, presumably adrenochrome or adrenolutin. By including the concept of an enzymatic, possibly genetic defect with another factor, epinephrine release, which may be activated by stressful life situations, it encompasses the evidence for sociological as well as constitutional factors in the etiology of the schizophrenias.

The possibility that some of the oxidation products of epinephrine are psychotomimetic received support from anecdotal reports of psychological disturbances associated with the therapeutic use of the compound, especially when it was discolored, and from some early experiments in which the administration of adrenochrome or adrenolutin in appreciable dosage was followed by certain unusual mental manifestations. A number of investigators failed to demonstrate any hallucinogenic properties in adrenochrome, and the original authors were not always able to confirm their earlier results.

Meanwhile, reports were emerging from the group at Tulane University, suggesting a gross disturbance in epinephrine metabolism in schizophrenic patients. Five years previously, Holmberg and Laurell had demonstrated a more rapid oxidation of epinephrine *in vitro* in the presence of pregnancy serum than with serum from the umbilical cord and had suggested that this was due to higher concentrations of ceruloplasmin in the former. There had also been a few reports of an increase in this protein in the blood of schizophrenics. Leach and Heath reported a striking acceleration in the *in vitro* oxidation of epinephrine in the presence of plasma from schizophrenic patients as compared with normals, and shortly thereafter implicated ceruloplasmin or some variant of ceruloplasmin as the oxidizing substance. Hoffer and Kenyon reported evidence that the substance formed from epinephrine by blood serum *in vitro* was adrenolutin.

All of the evidence does not, however, support the epinephrine theory. Despite the considerable new information regarding the metabolism of epinephrine *in vivo* which has been acquired in this laboratory and elsewhere in the past few years in animals and in normal and schizophrenic man, no evidence has been found for the oxidation of epinephrine via adrenochrome and adrenolutin in any of these populations. On the basis of Axelrod's delineation of the normal metabolic pathways of epinephrine and the availability of the triatiated catecholamine with high specific activity, an examination of its metabolism in 12 normals and an equal number of schizophrenics was completed. No significant qualitative or quantitative differences were revealed between these groups either in the rate of disappearance of labelled epinephrine from the blood or its metabolism. Although appreciable levels of adrenochrome have been reported to occur in the blood of normal subjects and to increase considerably following administration of lysergic acid diethylamide, Szara, Axelrod, and Perlin, using techniques of high sensitivity, have been unable to detect it in the blood of normals or of acute or chronic schizophrenic patients. A recent ingenious study of the rate of destruction of epinephrine *in vivo* found no difference between normals and schizophrenic patients in this regard. Finally, it has been shown in our laboratory by McDonald, and by the Tulane group themselves, that the low level of ascorbic acid in the blood was an important and uncontrolled variable in the rapid *in vitro* oxidation of epinephrine by plasma from schizophrenic patients. The fact that McDonald has been able to produce wide fluctuations in the epinephrine oxidation phenomenon from normal to highly abnormal rates in both normals and schizophrenics merely by dietary alterations in blood ascorbic acid level without any effect on the mental processes of either group is quite convincing evidence of the dietary and secondary nature of the phenomenon.

It should be pointed out that none of this negative evidence invalidates the theory that some abnormal product of epinephrine metabolism produces the major symptoms of schizophrenia; it does, however, considerably weaken the evidence which has been used to support it. In addition, there is the bothersome observation of numerous workers, and our own experience, that the administration of epinephrine to schizophrenics which, according to theory, should aggravate the psychotic symptoms, is usually accompanied by considerably less mental disturbance than occurs in normal subjects.

The recent upsurge of interest in ceruloplasmin can be ascribed to the report which depended upon the oxidation of N,N-dimethyl-p-phenylenediamine by ceruloplasmin. Holmberg and Laurell had demonstrated previously that ceruloplasmin was capable of oxidizing a number of substances including phenylenediamine and epinephrine, but that this could be inhibited by ascorbic acid and Leach and Heath had reported a more rapid oxidation of epinephrine in the plasma of schizophrenics which they attributed to ceruloplasmin. All of these observations were compatible with earlier reports in the German literature of increased serum copper in schizophrenia, the demonstration that practically all of the serum copper was in the form of ceruloplasmin, and that this compound was elevated in blood during pregnancy and in a large number of diseases including schizophrenia. Following the announcement of the phenylenediamine test, however, interest in copper and ceruloplasmin rose and very soon, a number of investigators reported this reaction or some modification of it to be positive in a high percentage of schizophrenics, although its value as a diagnostic test was discredited because of the large number of diseases, besides schizophrenia, in which it was positive.

McDonald reported his findings on three groups of schizophrenics, one from the wards of the National Institute of Mental Health, where they had been maintained on a more than adequate diet, and two groups from state hospitals. In none of the schizophrenic groups was there an increase in serum copper or other evidence of increased ceruloplasmin. The state hospital patients and one group of controls, however, showed low ascorbic acid levels and positive phenylenediamine tests which could be reversed by addition of ascorbic acid to the diet; the Institute schizophrenic patients had normal levels of ascorbic acid and negative phenylenediamine tests. It was clear that a high ceruloplasmin was not characteristic of schizophrenia and that the positive phenylenediamine test, where it occurred, could be completely explained by a dietary insufficiency of ascorbic acid.

In the Tulane group, the mean values for serum copper in schizophrenia decreased from a high of 216 ug/100 ml in 1956 to 145 ug/100 ml at the end of 1957, mean normal values remaining at 122 and 124 ug/100 ml during the same time. Other groups have found slight differences or no differences at all in ceruloplasmin or copper blood levels between schizophrenic and normal subjects, and no support for the ability of the phenylenediamine reaction to distinguish between schizophrenic and nonschizophrenic patients. It is not clear why some schizophrenics apparently show an elevated blood ceruloplasmin level; among the possibilities are dietary factors, hepatic damage, chronic infection, or the possible tendency of excitement to raise the blood ceruloplasmin as preliminary experiments appear to suggest.

Quite early in their studies, the Tulane group recognized that the potent oxidant effects of the serum of schizophrenics on epinephrine *in vitro* could not be satisfactorily explained by the ceruloplasmin levels alone. Leaving aside the importance of ascorbic acid deficiency on this reaction, they had postulated the presence in the blood of schizophrenics of a qualitatively different form of ceruloplasmin which they proceeded to isolate, to test in monkey and man, and to which they gave the name taraxein. They have reported that when certain batches of this material were tested in monkeys, marked behavioral and electroencephalographic changes occurred. When samples of these active batches were injected intravenously at a rapid rate into carefully selected prisoner volunteers, all of the subjects developed symptoms which have been described for schizophrenia, including disorganization and fragmentation of thought, autism, feelings of depersonalization, paranoid ideas, auditory hallucinations, and catatonic behavior.

A highly significant decrease in rope climbing speed in rats injected with sera from psychotic patients as opposed to that from nonpsychotic controls has been reported by Winter and Flataker. Their later finding that the phenomenon occurs with sera of patients with a wide variety of mental disorders, including mental retardation and alcoholism, and that there is a considerable variation in this index between similar groups at different hospitals, coupled with the inability of at least one other investigator to demonstrate this phenomenon in the small group of schizophrenic patients under investigation in this laboratory, suggests that the quite real and statistically significant phenomenon originally observed may be related to variables other than those specific for or fundamental to schizophrenia.

It has been reported that rabbits pretreated with serum from schizophrenics do not exhibit a pressor response following the local application of an epinephrine solution to the cerebral cortex. This procedure failed to differentiate between the sera of a small series of normals and schizophrenics on our wards.

One attempt by Robins, Smith, and Lowe to confirm the Tulane findings using comparable numbers and types of subjects, and at least as rigorous controls, was quite unsuccessful. In 20 subjects, who at different times received saline or extracts of blood from normal or schizophrenic donors prepared according to the method for preparing taraxein, there were only five instances of mental or behavioral disturbances resembling those in the original report on taraxein and these occurred as often following the administration of saline, extracts of normal plasma, or taraxein. It is easy to dismiss these negative findings with taraxein on the basis of the difficulty of reproducing exactly the 29 steps described in its preparation; it is considerably more difficult to dismiss the observation that a few subjects who received only saline or normal blood extract developed psychotic manifestations similar to those reported from taraxein.

During the preliminary investigations it was stated that taraxein was qualitatively different from ceruloplasmin on the basis of unpublished studies. A physicochemical or other objective characterization of taraxein would do much to dispel some of the confusion regarding its nature. It is possible, for example, that taraxein is, in fact, ceruloplasmin, but derives its special properties from the psychosocial characteristics of the situation in which it has been tested.

Serotonin, the important derivative of tryptophan, was first shown to exist in the brain in high concentration by Amin, Crawford, and Gaddum. Interest in its possible function in the central nervous system, and even its relationship to schizophrenia, was inspired by the finding that certain hallucinogens, notably lysergic acid diethylamide, could, in extremely low concentration, block the effects of serotonin on smooth muscle. Thus, Woolley and Shaw in 1954 wrote: "The demonstrated ability of such agents to antagonize the action of serotonin in smooth muscle and the finding of serotonin in the brain suggest that the mental changes caused by the drugs are the result of a serotonin deficiency which they induce in the brain. If this be true, then the naturally occurring mental disorders—for example, schizophrenia—which are mimicked by these drugs, may be pictured as being the result of a cerebral serotonin deficiency arising from a metabolic failure . . ."; while simultaneously in England, Gaddum was speculating, ". . . it is possible that the HT in our brains plays an essential part in keeping us sane and that the effect of LSD is due to its inhibitory action on the HT in the brain." Since that time additional evidence has appeared to strengthen these hypotheses.

Serotonin has been found to be considerably higher in the limbic system and other areas of the brain which appear to be associated with emotional states. Bufotenine, or dimethyl serotonin, extracted from an hallucinogenic snuff of West Indian tribes, was found to have some properties similar to those of lysergic acid diethylamide. A major discovery was the finding that the ataractic agent, reserpine, causes a profound and persistent fall in the level of brain serotonin, a process which more closely parallels the mental effects of reserpine than does its own concentration in the brain. By administration of its precursor, 5-hydroxytryptophan, the levels of serotonin can be markedly elevated in the brain with behavioral effects described as resembling those of lysergic acid diethylamide, a finding quite at odds with the original hypotheses. On the other hand, administration of this precursor to mental patients, along with a benzyl analog of serotonin to block the peripheral effects of the amine was reported in preliminary trials to suppress the disease, although the benzyl analog alone is apparently an effective tranquilizing drug in chronically psychotic patients.

Still another bit of evidence supporting the hypotheses of a central function for serotonin was the accidental discovery of toxic psychoses in a certain fraction of tuberculosis patients with iproniazid, which led to the therapeutic use of this drug in psychic depression. Iproniazid is known to inhibit the action of monoamine oxidase, an enzyme which destroys serotonin, and has been shown to increase the levels of this amine in the brain.

There are certain inconsistencies in the information which has accumulated which argue against a simple or singular role for serotonin in schizophrenia. Although the ability of the hallucinogen, lysergic acid diethylamide, to block effects of serotonin on smooth muscle prompted the development of the hypotheses relating serotonin to mental function or disease, a number of lysergic acid derivatives have since been studied, and the correlation between mental effects and anti-serotonin activity in the series as a whole is quite poor. One of these compounds is 2-brom-LSD, which has 1.5 times the antiserotonin activity of LSD, and which can be demonstrated by this property in the brain after systemic administration, but which in doses more than 15 times as great produces none of the mental effects of LSD.

In addition to serotonin, norepinephrine is also markedly reduced in the brain following reserpine. In fact, the brain concentrations of these two amines follow each other so closely in their response to reserpine as to suggest some mechanism common to both and perhaps obtaining as well for other active amines in the brain. In one study, L-dopa, a precursor of norepinephrine, was capable of counteracting the behavioral effects of reserpine whereas the precursor of serotonin was ineffective. Nor are the effects of iproniazid limited to brain serotonin; a comparable effect on norepinephrine has been reported, and it is possible that other amines or substances still to be discovered in the brain may be affected by what may be a nonspecific inhibitor of a relatively nonspecific enzyme. Chlorpromazine, which has the same therapeutic efficacy as reserpine in disturbed behavior, is apparently able to achieve this action without any known effect on serotonin. In addition, the provocative observation that iproniazid which elevates serotonin levels in the brain can cause a toxic psychosis loses some of its impact when one realizes that isoniazid, which does not inhibit monoamine oxidase and can hardly raise the brain serotonin concentration, produces a similar psychosis.

It seems reasonable that the serotonin as well as the norepinephrine in the brain have some important functions there and the evidence in general supports this thesis even though it also suggests that their roles still remain to be defined.

The urinary excretion of 5-hydroxyindoleacetic acid has been used as an indicator of the portion of ingested tryptophan which is metabolized through serotonin to that end product. Although the excretion of 5-hydroxyindoleacetic acid is normal in schizophrenic patients under ordinary circumstances, it may be altered by challenging the metabolic system with large doses of tryptophan. Under these circumstances, failure on the part of schizophrenics to increase their output of 5-hydroxyindoleacetic acid has been reported while nonpsychotic controls double it.

Kopin of our laboratory has had the opportunity to perform a similar study on schizophrenics and normal controls maintained on a good and reasonably controlled diet and in the absence of drugs. In each group, there was a slightly greater

than two-fold increase in output of 5-hydroxyindoleacetic acid following a tryptophan load and no significant deviation from this pattern by any single case.

That the heuristic speculations of Woolley and Shaw and of Gaddum have not
yet been established does not mean that they are invalid. The widespread experimental activity which they stimulated has broadened and deepened our knowledge
of the metabolism and pharmacology of serotonin and its effects on behavior and
may lead the way to their definitive evaluation in normal and pathological states.

GENETIC AND SCHIZOPHRENIC DISORDERS

Many of the current hypotheses concerning the schizophrenia complex are
original and attractive even though, to this time, evidence directly implicating any
one of them in the disease itself is hardly compelling. There is, nevertheless, cogent
evidence responsible to a large extent for the present reawakening of the long
dormant biochemical thinking in this area and sufficiently convincing to promote
its continued development. Genetic studies have recently assumed such a role and
it appears worthwhile briefly to review them in the present context.

Earlier studies on large populations have reported a remarkable correlation
between the incidence of schizophrenia and the degree of consanguinity in relatives
of known schizophrenics. These findings were not conclusive, however, since the
influence of socioenvironmental factors was not controlled. Better evidence is
obtained from the examination of the co-twins and siblings of schizophrenics. The
concordance rate for schizophrenia is extremely high for monozygotic twins in
such studies, while that for dizygotic twins is low and not significantly different
from that in siblings to which, of course, they are quite comparable genetically.
Even these studies, however, are not completely free from possible sources of error,
which prevent a definitive conclusion regarding the role of genetic factors in this
disease. One cannot assume that environmental similarities and mutual interactions
in identical twins, who are always of the same sex and whose striking physical congruence is often accentuated by parental attitudes, play an insignificant role in the
high concordance rate of schizophrenia in this group. This factor could be controlled by a study of twins separated at birth of which no statistically valid series
has yet been compiled, or by a comparison of the concordance rates in monozygotic and dizygotic twins whose zygosity had been mistakenly evaluated by the
twins, their parents, and associates. Another possible means of better controlling
the environmental variables would be a careful study of schizophrenia in adopted
children with comparison of its incidence in blood and foster relatives.* Perhaps

---

*When such a study was carried out in collaboration with Rosenthal, Wender, and Schulsinger,
a significantly evaluated prevalence of schizophrenia-related disorders was found in the biological relatives of adopted individuals who became schizophrenic, but not in their adopted
relatives.

only a survey on a national scale would provide the requisite numbers of cases for any of these studies.

A less satisfactory resolution of this problem can be obtained by an appraisal of environmental similarities in normal fraternal and identical twins. A study of over 100 specific aspects of the environment of normal twins has been made from which it is possible to derive some general impressions. Although there is a difference in this crude measure of environmental congruence between identical and fraternal twins of like sex, it is not statistically significant and can account for only a small fraction of the large difference in the concordance of schizophrenia between these types of twins. On the other hand, there is a highly significant difference in environmental similarity between fraternal twins of like and unlike sex which is sufficient to account for the difference in concordance rate of schizophrenia between them, for which, of course, there is no tenable genetic explanation.

Another possible source of error in the twin studies which have been reported is the personal bias of the investigators who made the judgment of zygocity and the diagnosis of schizophrenia in the co-twins. Until a more definitive study is done in which these judgments are made independently, a rough evaluation is possible at least for the diagnosis of schizophrenia, if not of zygocity, based on diagnoses arrived at in the various hospitals to which the co-twins may have been admitted before or irrespective of their involvement in the study—diagnoses which are not likely to have been contaminated by knowledge of zygocity. Kallmann was good enough to review the material collected in his 1946 to 1949 survey from that point of view. Of 174 monozygotic co-twins of schizophrenic index cases, 103 or 59 per cent had been diagnosed schizophrenic by Kallmann, while 87 or 50 per cent had received a psychiatric hospital diagnosis of schizophrenia before any examination by him. On the other hand, he had made the diagnosis of schizophrenia in 47 or 9.1 per cent of 517 dizygotic co-twins as compared to a hospital diagnosis in 31 or 6 per cent. Although the concordance rates based only on hospital diagnoses are decreased in both types of twins for obvious reasons, the striking difference between the two concordance rates remains. Slater has published individual protocols of his cases from which judgments of zygocity and schizophrenia can be made. Of 21 twin pairs who could be considered definitely uniovular, 15 or 75 per cent were concordant with respect to the simple criterion of admission to a mental hospital, whereas in only 12 or 10.3 per cent of 116 binovular or questionably binovular pairs was there a history of the co-twins having been admitted to a mental hospital for any psychosis. On the basis of this analysis of the two most recent series, it seems that only a small component of the great difference in concordance rates reported for schizophrenia between uniovular and binovular twins can be attributed to the operation of personal bias in the diagnosis of the disease in the co-twin.

Even the most uncritical acceptance of all of the genetic data, however, cannot lead to the conclusion that the schizophrenic illnesses are the result of genetic factors alone. In 14 to 30 per cent of cases where schizophrenia occurs in one of monozygotic twins, the genetically identical partner will be free of the disorder. Attention has already been called to the higher concordance with respect to schizophrenia and the greater environmental similarities in like sexed fraternal twins or siblings than in those of unlike sex, and from the same source a difference in concordance is reported between monozygotic twins separated some years before (77.6 per cent) as opposed to those not separated (91.5 per cent). Neither of these observations is compatible with a purely genetic etiology of the disease, and both suggest the operation of environmental factors. Rosenthal and Jackson have pointed out the striking preponderance of female over male pairs concordant for schizophrenia in all of the reported series whether they be monozygotic or dizygotic twin, sibling, or parent-child pairs. If sampling errors on the basis of a greater mobility of males can be excluded and the observations taken as a reflection of the true incidence of this phenomenon there are several explanations for it on the basis of social interaction but none on purely genetic grounds.

One cannot at this time review the extensive literature supporting the importance of environmental factors in the etiology of schizophrenic disorders. To this reviewer, at least, it is quite as suggestive as the genetic evidence but by no means more conclusive, since few studies in either field have been completely objective or adequately controlled.

It is both interesting and important to note that even if the conclusions of both the genetic and environmental approaches to the etiology of schizophrenic psychoses are accepted uncritically, they are not mutually exclusive. Both are compatible with the hypothesis that this group of diseases results from the operation of socioenvironmental factors on some hereditary predisposition or by an interaction of the two so that each is necessary, but neither alone is sufficient. An excellent example of such a relationship is seen in tuberculosis where the environmental microbial factor is indusputed and where Lurie has shown the importance of genetic susceptibility, so that a population sufficiently heterogeneous with respect to susceptibility and exposure to tuberculosis, yields results in contingency and twin studies which, before the discovery of the tubercle bacillus, could easily have been used to prove a primary genetic cause and almost as convincingly, as the results of similar studies in schizophrenia. Interestingly enough, studies of tuberculosis from the socioenvironmental point of view would obviously secure data equally convincing for the operation of exogenous, social, and economic factors. One hypothesis with respect to the schizophrenic psychoses which remains compatible with all the evidence from the genetic as well as the psychosocial disciplines is that these disorders, like tuberculosis, require the operation of environmental factors upon a genetically determined predisposition.

RESUME

Although the evidence for genetic and therefore biological factors as important and necessary components in the etiology of many or all the schizophrenias is quite compelling, the signposts pointing the way to their discovery are at present quite blurred and, to me at least, illegible.

Genetic factors may operate through some ubiquitous enzyme system to effect general changes in one or another metabolic pathway detectable by studies on blood or urine, and it is to be hoped that the currently active search in these areas will continue.

It is at least equally possible, however, that these genetic factors may operate only through enzymes or metabolic processes peculiar to or confined within the brain or even within extremely localized areas of the brain. We are in need of new hypotheses such as those of Elkes, and many already discussed. In this connection, gamma-aminobutyric acid appears to be just as interesting a substance about which to construct working hypotheses as are the catechols or the indoles. Not only has it been isolated only from nervous tissue, and its metabolism there been investigated in some detail, but its neurophysiological properties appear to be better defined than are those of the other two groups, in addition to which its inhibitory properties may have special relevance to diseases where a failure in central inhibition seems to be involved.

Amphetamine possesses remarkable psychotomimetic properties which should not be overlooked. Its ability to produce a clinical syndrome often indistinguishable from schizophrenia and a possible relation to the naturally occurring catecholamines make it at least as interesting as lysergic acid diethylamide.

In addition to techniques at present available in neurochemistry, neurophysiology, and behavioral pharmacology, the development of others designed to yield information on processes occurring within the psychotic brain will be needed before our explorations in this field have been exhausted.

But the biochemist must not lose sight of the possibility which is certainly as great as any of the others: that the genetic factors in schizophrenia operate to determine inappropriate interconnections or interaction between chemically normal components of the brain, in which case the physiological psychologist, the neurophysiologist, or the anatomist is more likely to find meaningful information long before the biochemist. It would take many biochemists a long time to find a noisy circuit in a radio receiver if they restricted themselves to chemical techniques.

These possibilities are mentioned only to indicate how large is the haystack in which we are searching for the needle; one cannot avoid a feeling of humility when one regards the chance that any one of us has found or will find it in a relatively short time.

That is no cause for discouragement, however. It is not necessary that one be convinced of the truth of a particular hypothesis to justify devoting one's energies to testing it. It is enough that one regard it as worth testing and that the tools with which to do so be adequate. Modern biochemistry, with its wealth of new knowledge of intermediary metabolism and its array of new techniques for the separation and identification of compounds and the tracing of their metabolic pathways, has provided the biologist interested in mental illness with an armamentarium which his predecessor of only a generation ago could hardly have dreamed possible. If he chooses among the approaches which may lead to a definition of the biological factors in schizophrenia, those which will in any case lead to a better understanding of the nervous system and of thought processes and behavior, the present surge of enthusiasm will not have been misdirected.

Pharmacological Approach to Schizophrenia

Arvid Carlsson, M.D., Ph.D.

INTRODUCTION

Schizophrenic symptomatology can be profoundly influenced by drugs: while certain drugs alleviate, others produce or aggravate schizophrenic symptoms. Studies on the mode and site of action of such drugs may yield important clues. The potentiality of this type of approach has been demonstrated, that is, in the case of Parkinson's disease [27].

I hope you will bear with me if I describe the development in this area more or less in chronological order as I have witnessed it myself.

My active research interest in the antipsychotic agents began just 20 years ago. In 1955 to 1956, I had the privilege of working in the famous Laboratory of Chemical Pharmacology, headed by Dr. B. B. Brodie, at the National Institutes of Health in Bethesda, Md. This was only a few years after the discovery of the remarkable therapeutic effect of chlorpromazine and reserpine in psychotic conditions such as schizophrenia. Shortly before my arrival Drs. Brodie and Shore had made an important discovery, namely that reserpine is capable of releasing serotonin from its various stores in the body and causing depletion of these stores [50]. They speculated that the antipsychotic action of reserpine was due to continuous release of serotonin onto receptors.

I am greatly indebted to Drs. Brodie and Shore for introducing me to this fascinating field of research. After returning to Sweden I found, together with the late Dr. N.-A. Hillarp, that the catecholamines are also released by reserpine [17].

Research in the Schizophrenic Disorders: The Stanley R. Dean Award Lectures, vol. 2, edited by R. Cancro and S. R. Dean. Copyright © 1985 by Spectrum Publications.

This resulted in depletion of the adrenergic transmitter noradrenaline and in failure of adrenergic transmission [19].

RESERPINE, DOPA AND DOPAMINE

We felt that deficiency of amines at receptor sites was the most likely explanation of the pharmacological actions of reserpine. To test this hypothesis we gave the catecholamine precursor dopa to reserpine-treated animals and discovered the central activity of this amino acid, presumably mediated via its decarboxylation products [18]. The dramatic reversal of the reserpine syndrome by dopa supported our deficiency theory, and the simultaneously observed inefficiency of 5-hydroxy-tryptophan directed our attention to the catecholamines.

At this time the primary decarboxylation product of dopa, i.e. dopamine, had not yet been detected with certainty in the brain, owing to lack of a specific and sensitive method for assaying this compound. We developed such a method and found that dopamine is stored by a reserpine-sensitive mechanism in the brain in even greater amounts than noradrenaline, suggesting that it is not just an intermediary in the biosynthesis of noradrenaline and adrenaline [20]. In support of this, the distribution of dopamine was found to differ greatly from that of noradrenaline, the largest amounts being found in the basal ganglia rather than the brain stem, where noradrenaline occurs in the highest concentration [14].

BRAIN MONOAMINES: NEUROHUMORAL TRANSMITTERS

The actions of reserpine and dopa, as well as the regional distribution data, suggested to us that the catecholamines are important agonists in the brain and that they (especially dopamine) are involved in the control of extrapyramidal motor functions as well as in higher integrative functions such as wakefulness [21,22]. However, many investigators expressed doubts at this time (around 1960), especially in the case of dopamine, which was known by pharmacologists only as a poor adrenergic agonist. The skepticism was but partly dissipated by Hornykiewicz's discovery of reduced dopamine levels in the brains of Parkinsonian patients [35], or by the subsequent demonstration of the therapeutic properties of L-dopa [16]. Probably more instrumental in this respect was the accumulation of evidence demonstrating the role of the brain monoamines as neurohumoral transmitters. Particularly strong support came from the demonstration, by means of fluorescence histochemistry, of the neuronal localization of the brain monoamines [23] and the subsequent mapping out of the monoamine-carrying neuronal pathways [6,33,34, 37]. Electron-microscopical, biochemical and physiological techniques also contributed to make a strong case for the brain monoamines as neurohumoral transmitters: it was established that the monoamines (dopamine, noradrenaline and

5-hydroxytryptamine) are formed in nerve terminals and stored in synaptic vesicles; they are released by nerve stimulation; after release they cause physiological effects; efficient inactivation mechanisms to terminate the action exist; transmitter turnover is markedly dependent on the nerve impulses (for review, see [7]).

RECEPTOR–BLOCKING ANTIPSYCHOTIC AGENTS

I should like to come back now to the mode of action of the antipsychotic drugs. In the case of reserpine we were rather satisfied with the evidence. Reserpine appears to act by causing transmission failure in monoaminergic nerves, and this is due to depletion of transmitter, which can no longer be concentrated in the synaptic vesicles (or storage granules) because of blockade of a specific uptake mechanism in these organelles [25]. In fact, more recent studies, utilizing the principle of selective protection of individual transmitter stores, have confirmed our original suggestions and clearly demonstrated the dominating role of the catecholamines, notably dopamine, for the characteristic syndrome induced by reserpine, even though 5-HT appears to contribute [30].

We were puzzled, however, by the virtual absence of an effect on monoamine levels exerted by some clinically important groups of antipsychotic agents, i.e., the phenothiazines, the thiaxanthenes, the butyrophenones, etc. As is well known, these drugs are remarkably similar to reserpine with respect to the whole spectrum of psychiatric, extrapyramidal, and endocrinological actions, and yet they must have an entirely different mode of action at the molecular level.

In 1963, Margit Lindqvist and I found that small doses of chlorpromazine and haloperidol specifically stimulated the metabolism of dopamine and noradrenaline in mouse brain, and since this took place without any decrease in catecholamine levels, we inferred that the synthesis of the catecholamines was also stimulated by these agents [24]. It was known at this time that the central effects of catecholamines (from administered dopa) could be antagonized by these agents, and thus we proposed the following mechnism to account for the observations made. The antipsychotic phenothiazines and butyrophenones act by blocking central dopamine and noradrenaline receptors. This blockade activates a negative feedback mechanism which leads to an increased physiological activity of catecholamine neurons, with an increased release, metabolism, and synthesis of the transmitter (Figure 1).

This concept of receptor-mediated feedback control of neuronal activity has been amply confirmed and extended by numerous investigators [31]. Also, in man, antipsychotic agents have been found to stimulate central catecholamine metabolism [49]. Moreover, the hypothesis that phenothiazines and butyrophenones act by blocking catecholamine receptors has received support through the work of Greengard and his colleagues [40], who have demonstrated what appears to be dopaminergic receptor responsiveness of a cell-free adenylate cyclase system,

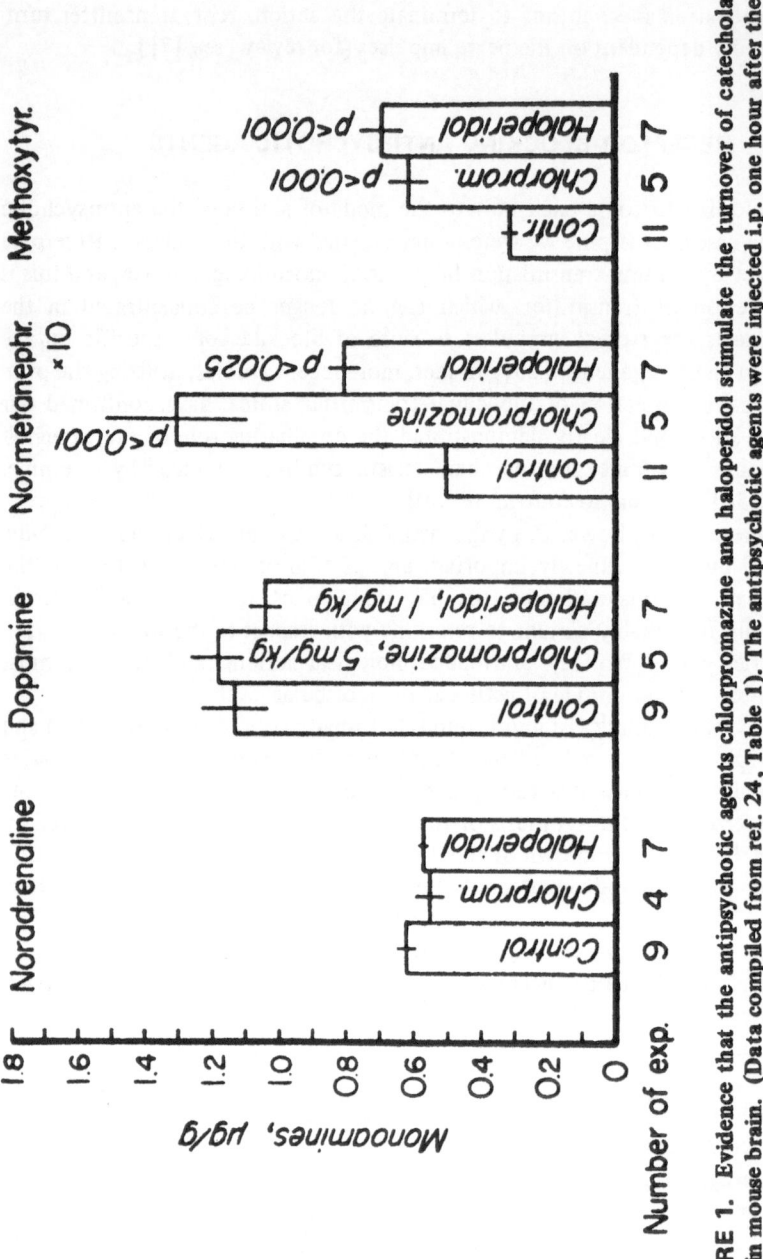

FIGURE 1. Evidence that the antipsychotic agents chlorpromazine and haloperidol stimulate the turnover of catecholamines in mouse brain. (Data compiled from ref. 24, Table 1). The antipsychotic agents were injected i.p. one hour after the monoamine oxidase inhibitor nialamide (100 mg/kg) and were killed after another 3 hours. Controls received nialamide only. NOTE: The antipsychotic agents enhanced the accumulation of the 3-0-methylated catecholamine metabolites normetanephrine and 3-methoxy-tyramine but did not influence the levels of noradrenaline or dopamine.

obtained from striatal tissue. This system can be activated by dopamine, and the effect is blocked by phenothiazines, butyrophenones, etc.

Central noradrenaline receptors, blocked by antipsychotic agents, appear to be similar to the peripheral α-adrenergic receptors. However, central dopamine receptors are different, since either type of receptor can be selectively activated or blocked [10]. A survey of the receptor-blocking properties of antipsychotic drugs indicates that all of them, except for those causing monoamine depletion, possess dopamine receptor-blocking activity. In addition, some of them are capable of blocking central α-adrenergic receptors, but certain agents appear to be devoid of such activity while maintaining antipsychotic properties [9,45]. Besides, no antipsychotic action of pure α-adrenergic blocking agents has thus far been reported. From these data the conclusion seems justified that dopamine receptor-blocking activity is essential for the antipsychotic effect. This does not, however, exclude the possibility that noradrenaline-receptor blockade may play a contributory role.

ROLE OF DIFFERENT DOPAMINERGIC PATHWAYS

It would thus appear that dopamine neurons are not only involved in the extrapyramidal, but also in the antipsychotic actions of the neuroleptic agents. Are we dealing with one and the same dopaminergic pathway? This is unlikely. It is well known that the ratio of extrapyramidal to antipsychotic potency varies for different agents and that for a given agent this ratio can be modified by an anticholinergic drug. Moreover, choreatic side effects of dopa, which are probably due to excessive activation of dopamine receptors, are not strictly correlated to mental side effects. Therefore, we are probably dealing with at least two different systems.

Dopamine occurs in many different parts of the central nervous system. The two quantitatively dominating locations are the striatum, where by far the largest amount occurs, and certain regions belonging to the so-called limbic (or mesolimbic) system, e.g. the olfactory tubercule, the nucleus accumbens, the central nucleus of the amygdala, and certain parts of the paleocortex. Recent observations suggest that the antipsychotic action of neuroleptic drugs largely occurs in the limbic system, whereas the extrapyramidal actions occur in the striatum. Two observations point in this direction. First, anticholinergic agents are known to antagonize extrapyramidal side effects of neuroleptic drugs, while leaving the antipsychotic activity entirely or at least largely unaffected. It has been found, in different sets of experiments, that anticholinergics have a similar differential activity in antagonizing the effect of neuroleptic agents on dopamine metabolism in the two regions [8,32] (Figure 2). Second, neuroleptics with a strong liability to produce extrapyramidal side effects in general cause a stronger effect on the dopamine metabolism in the striatum relative to the limbic system than neuroleptics with little or no extrapyramidal side effects [13,31] (Figure 3, Table 1).

If the antipsychotic action is located in limbic dopaminergic synapses, the question arises as to which of the various dopaminergic regions are involved. This question cannot be answered as yet. Local application of dopamine in the nucleus accumbens has been found to induce hyperkinesia and to stimulate food-reinforced lever pressing, and in the striatum, stereotyped behavior (gnawing, licking, etc.) [41]. Further studies are necessary to clarify this point.

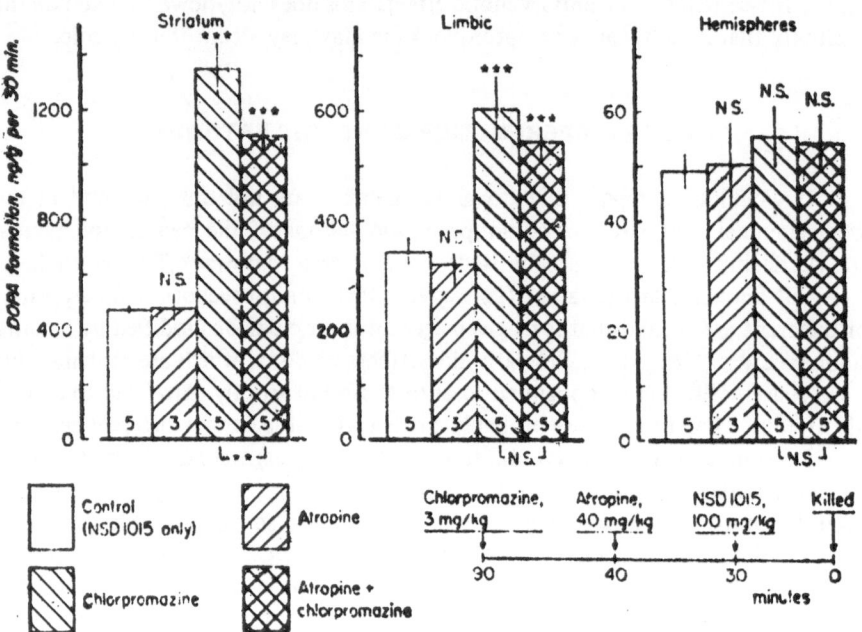

FIGURE 2. Effect of atropine and chlorpromazine, and both drugs combined, on dopa formation in rat brain regions. All animals received the centrally as well as peripherally active inhibitor of the aromatic amino acid decarboxylase, NSD 1015 (3-hydroxybenzylhydrazine HCl) i.p. 30 min before death. Injections of chlorpromazine and atropine were given i.p. at time intervals indicated in the figure. The striatum, limbic dopamine-rich areas and the remaining predominantly noradrenaline-containing cerebral hemisphere portion were analyzed for dopa. NOTE: The stimulating action of chlorpromazine on dopa formation was partly counteracted by atropine in the striatum but not, or less so, in the limbic regions. Atropine *per se* did not influence dopa formation. ***$P < 0.001$; **$P < 0.01$; *$P < 0.05$; (N.S.) $P > 0.05$ (comparison with control, if not otherwise indicated).

TABLE 1. Differential Action of Antipsychotic Drugs on DOPA Formation in Striatum vs. Limbic Regions

Drug	Dose, mg/kg	Diff. in % DOPA incr., striatum–limbic	Significance	Extrapyramidal side effects[a]
Pimozide	0.5 –1.5	153 ± 18.3 (4)		?
Haloperidol	0.25–0.5	117 ± 26.1 (4)	$p < 0.025$	1
Chlorpromazine	3	104 ± 11.1 (8)		3
Chlorpromazine + Atropine	3	73 ± 9.2 (5)	$p < 0.005$	—
Thioridazine	5-10	51 ± 9.3 (8)	$p < 0.05$	4
Clozapine	50–100	37 ± 8.8 (10)		5

[a]From [51]; rank by class; 1 indicates the most side effects.

The differences in % DOPA increase between the striatum and the limbic areas were obtained from the data represented in Fig. 3 and from similar data of an experiment with chlorpromazine and atropine (Atr. 40 mg/kg i.p. 10 min before NSD 1015). The data refer to the doses indicated in the Table. Statistics: t-test.

NOTE: There seems to exist a correlation between a high striatal vs. limbic response and liability to extrapyramidal side effects, which may possibly also be inversely related to anticholinergic activity [51].

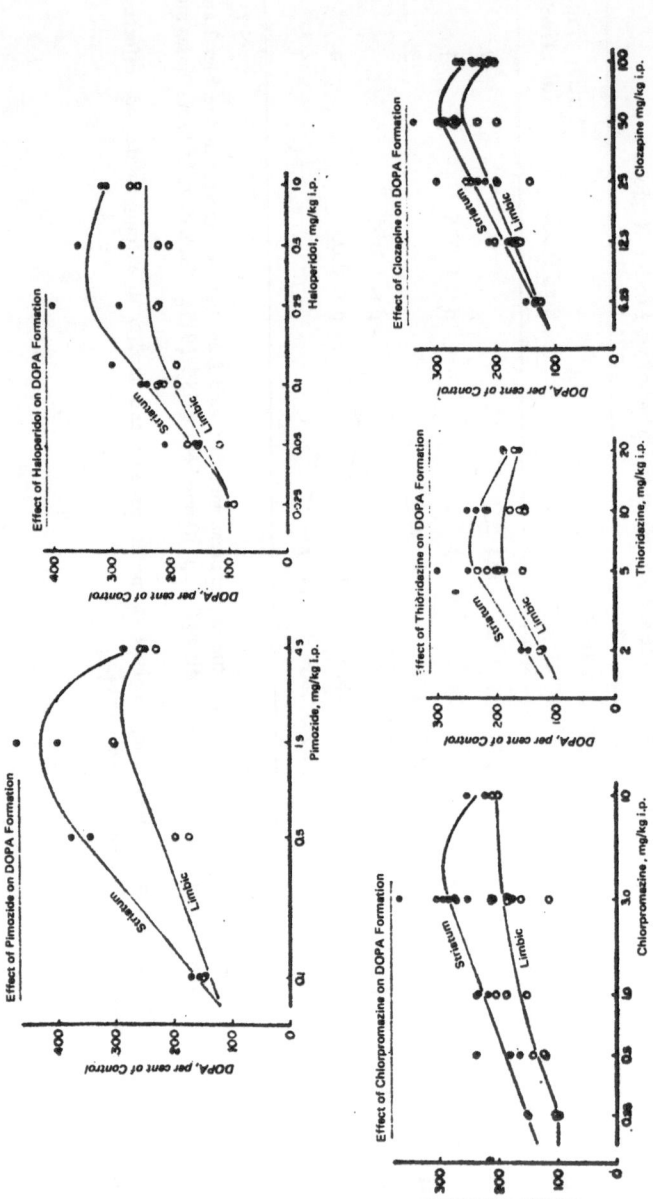

FIGURE 3. Effect of neuroleptics on dopa formation in rat brain regions rich in dopamine. The neuroleptics were given 60 min before NSD 1015 (100 mg/kg i.p.) and the animals were killed after another 30 min. (Unpublished data of this laboratory.)

ANTIPSYCHOTIC ACTION OF A TYROSINE HYDROXYLASE INHIBITOR

As indicated by the evidence presented above, antipsychotic activity can be induced by interfering with central catecholamines, notably dopamine, in two different ways: (1) by blocking the function of the presynaptic storage organelles, leading to neurotransmission failure, and (2) by blocking postsynaptic catecholamine receptors. The question then arises of whether inhibition of catecholamine synthesis will alleviate psychotic symptoms.

α-Methyltyrosine is a relatively specific inhibitor of tyrosine hydroylase, the first enzyme involved in catecholamine biosynthesis. This agent has been administered in fairly large doses to psychotic patients, but no antipsychotic action could be detected [38]. A possible explanation for this failure could be that the degree of enzyme inhibition was insufficient, clinical dosage being limited by renal toxicity. If insufficient dosage is the explanation, it might be possible to demonstrate an antipsychotic action of α-methyltyrosine by utilizing the ability of receptor-blocking neuroleptics to potentiate the action of this enzyme inhibitor. Such potentiation, of a marked degree, has been demonstrated in animal experiments [2,3] (Figure 4).

We have investigated the effect of combined treatment with a phenothiazine and α-methyltyrosine in eight schizophrenic patients with a stationary symptomatology [28,29,53]. In four of these patients, two or three trials have been performed, using a double-blind crossover design in one trial for each patient [53]. Before each trial the patient had been treated with a single antipsychotic agent, usually thioridazine, in constant dosage for several months. The trial was started by rating the symptomatology by means of two different rating scales (Figure 5).

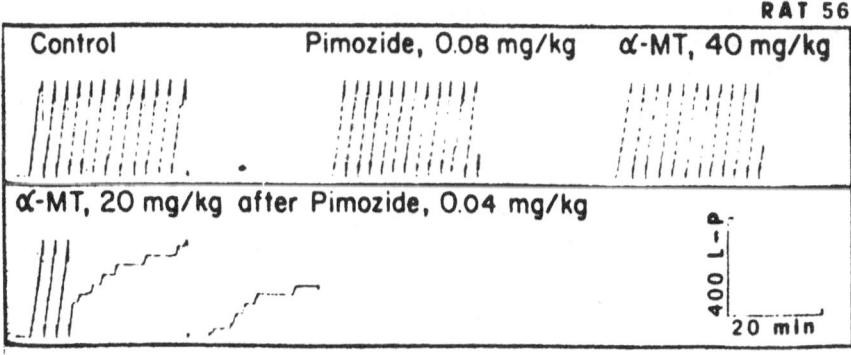

FIGURE 4. Potentiation by α-methyltyrosine of the pimozide-induced inhibition of food-reinforced lever pressing. (Reproduced from ref. 33.) The cumulative records show that pimozide, 0.08 mg/kg, and α-methyltyrosine (α-MT, H 44/68) 40 mg/kg, when given separately to a rat, have no significant influence on the lever pressing. However, a pronounced inhibition is seen if the two drugs are combined, the doses being only half of those given separately.

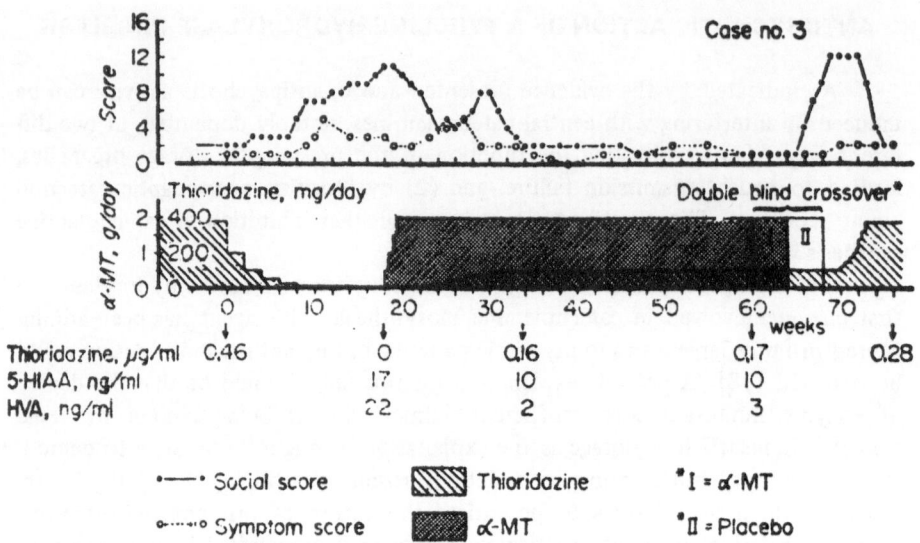

FIGURE 5. Potentiation of the antipsychotic action of thioridazine by α-methyl-tyrosine (α-MT) in a schizophrenic patient. (Data from ref. 53.)

The patient, a 44-year old male, had suffered from schizophrenia with stationary symptomatology for more than six months. He had been hospitalized and treated with thioridazine as the only antipsychotic drug in constant dosage for more than six months. The trial was started by rating the symptoms verbally by a psychiatrist ("symptom score") and behaviorally by the head nurse ("social score"). Then the dosage of thioridazine was reduced stepwise until the care of the patient became difficult. At this time the scores were markedly elevated. α-Methyl-tyrosine therapy was now started; the dosage being gradually increased to 2 g/day, while keeping the thioridazine dosage at a constant low level. This dose regimen was insufficient to control the condition, and thus the dose of thioridazine was gradually increased until the pretrial symptomatology level was attained. This occurred at a thioridazine dosage of 100 as compared to the pretrial dosage of 375 mg/day. The dose regimen was kept for 30 weeks. During this period the patient tended to be in an even better mental condition than before the trial.

At the termination of the trial α-methyltyrosine was replaced by placebo or α-methyltyrosine, using a double-blind crossover design. During the blind α-methyl-tyrosine period, which in this case came first, there was no change in the ratings, but during the placebo period a rapid deterioration occurred. The dosage of thioridazine was then gradually increased to the original level, and the symptomatology level was re-attained.

Similar results were obtained in three additional schizophrenic patients.

Then the dosage of the phenothiazine was slowly reduced stepwise over several weeks and the ratings were repeated at weekly intervals. After several weeks, when the phenothiazine dosage had been drastically reduced and the care of the patients started to be difficult because of a marked deterioration of the mental condition, α-methyltyrosine was given in a dose which was increased to 2 g daily. This was followed by a certain improvement in a few cases. The dose of phenothiazine was then slowly increased in order to titrate the dose necessary to attain the pre-trial level of ratings. This dose was found to be lower than the pre-trial dose in all cases, the average reduction being about 70 per cent (range 33 to 98.5 per cent). Plasma thioridazine levels were similarly reduced. This combined treatment regimen was maintained for periods varying between four weeks and six months without any signs of tolerance. During the treatment the concentration of homovanillic acid in the cerebrospinal fluid was reduced by about 80 per cent. The treatment period was ended by stopping α-methyltyrosine medication or replacing it by placebo, while keeping the phenothiazine dosage unchanged. In all cases a rapid deterioration of the mental status occurred. The phenothiazine dosage was now increased to the same level as before the trial, and the pre-trial level of symptomatology was reattained.

THE NATURE OF ANTIPSYCHOTIC DRUG EFFECT

Thus, in the presence of a phenothiazine, given in a dosage near the threshold for detectable antipsychotic activity, schizophrenic symptomatology appears to be profoundly influenced by alterations in tyrosine hydroxylase activity. Strong additional support is thus provided for the view that this symptomatology is closely related to central catecholamine neurotransmission. The question then arises as to what inference we can draw from this circumstance with respect to the pathogenesis of schizophrenia. The answer to this question is largely dependent on the specificity of the antipsychotic action. In other words, the crucial question is whether the so-called antipsychotic agents possess true antipsychotic activity. If the answer is yes, we have strong reasons to believe that psychotic symptoms are due to a disturbance in catecholamine neurotransmission or in a functionally closely related mechanism. If on the other hand, the so-called antipsychotic action is only a manifestation of a general depression of mental activity, we are not justified in implicating the catecholamines in the pathogenesis of schizophrenia. Of course, we have to also consider the possibility that a true specificity of the antipsychotic agents exists, but is limited to certain components of the schizophrenic process or to a certain group of schizophrenic patients. The implication of the catecholamines in the pathogenesis of schizophrenia may then be at least partiy justified.

Several clinical reports suggest that antipsychotic agents do not simply suppress abnormal behavior but may, in addition, favor normal behavioral components

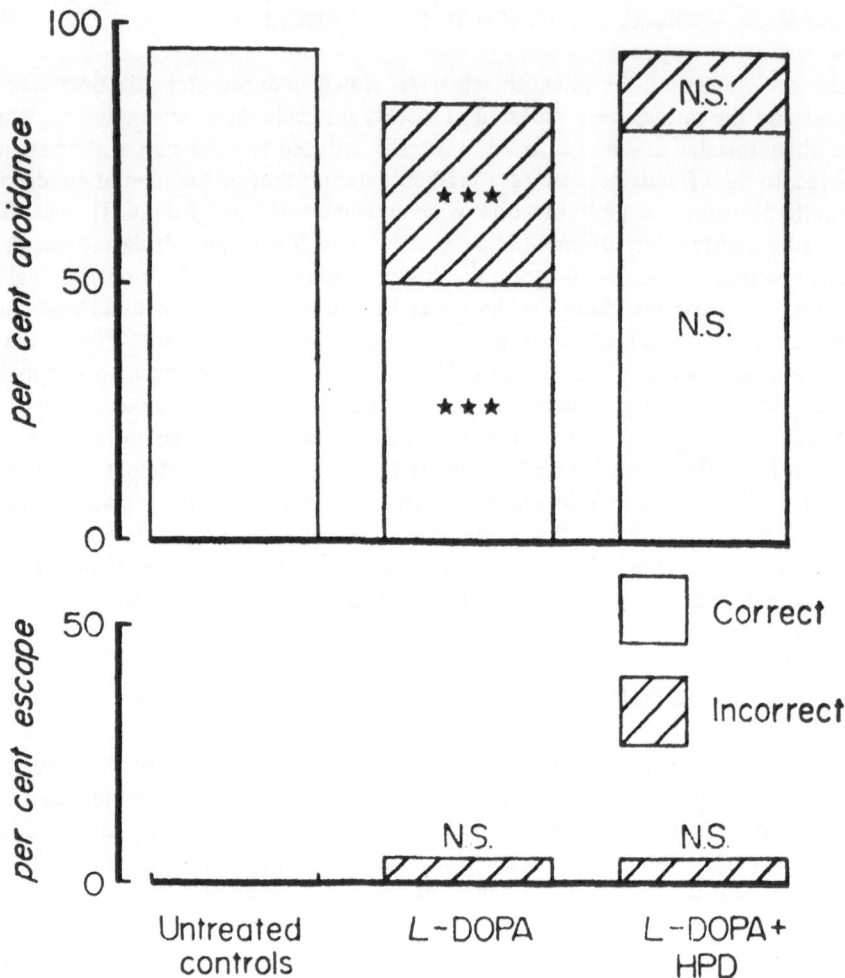

FIGURE 6. Loss of discrimination ability after treatment with a large dose of L-dopa and restoration of correct performance by haloperidol. (Reproduced from ref. 4.)

Rats were trained to pass through one of two passages in a shuttle box. If cue lights above the passage are lit when an auditory warning signal is given the animals were trained to pass through the right passage in order to avoid a grid shock. If the auditory warning stimulus was givens without any light signal, avoidance was obtained by passing through the left passage.

NOTE: Untreated animals were trained to almost 100 per cent avoidance by using the correct passage exclusively. After treatment with L-dopa, 100 mg/kg i.p. 1 hour before the test session (together with the inhibitor of the aromatic amino acid decarboxylase, Ro 4-4602 = N^1, (DL-seryl)-N^2, (2,3,4-trihydroxybenzyl) hydrazine HCl, 50 mg/kg i.p., to inhibit the enzyme in peripheral but no longer discriminated between the correct and the incorrect passage. The addition of haloperidol (HPD), 0.25 mg/kg i.p. 20 min before the test session, restored the performance almost completely to normal.

***$P < 0.001$. N.S. $P > 0.05$ (comparisons with untreated controls, Wilcoxon T-test; the values shown are medians, N = 8).

[38,51]. Also, in animals in which an inadequate behavior has been induced by such means as a large dose of L-dopa, antipsychotic agents have been shown to restore adequate responding [4] (Figure 6).

AMPHETAMINES AND SCHIZOPHRENIA

If drug-induced suppression of central catecholamine functions alleviates schizophrenic symptoms, a psychotomimetic effect would be expected from agents causing excessive stimulation of these functions.

The psychotomimetic agents form a heterogeneous group both from a chemical and pharmacological point of view. Many of them have attracted quite a lot of interest over the years, mainly in view of the possibility that they might provide useful schizophrenia models. The psychotic condition induced by most of these agents can, however, be clearly distinguished from schizophrenia. A striking exception is the amphetamine group of agents, because they are capable of reproducing rather faithfully the picture of paranoid schizophrenia [51]. Therefore, the mode of action of amphetamine is of special interest in the present context and, fortunately, it is at least partly understood. It has been shown that amphetamine, even in low dosage, is capable of releasing central (and peripheral) catecholamines and that its central stimulant action is prevented by pretreatment with α-methyltyrosine. The data indicate that amphetamine acts by releasing catecholamines from a small pool which is immediately dependent upon the synthesis of new catecholamine molecules. The failure of an inhibitor of dopamine-β-hydroxylase to prevent amphetamine-induced excitation suggests that dopamine is involved in this action, although a contributory role of noradrenaline cannot be ruled out (for review, see [26]). It has been shown in experiments on human subjects that the euphoriant action of amphetamines can be prevented by α-methyltyrosine pretreatment. Whether this enzyme inhibitor is capable of preventing amphetamine psychosis remains to be elucidated. However, it seems probable that the psychotic action is related to central stimulation and thus dependent upon catecholamine release.

Paranoid delusions during treatment of Parkinsonian patients with L-dopa have also been reported, although this side effect appears to be less frequent than the confusion-delirium type of mental disturbance [39].

These observations on amphetamine and L-dopa support an involvement of catecholamines in at least a certain type of schizophrenia.

THE POSSIBLE ROLE OF GABA IN SCHIZOPHRENIA

A couple of months ago a man named Per K. Frederiksen came to my office and told me that he was a psychiatrist in one of the mental hospitals near Gothenburg and that he believed he could cure schizophrenia. I couldn't help wondering

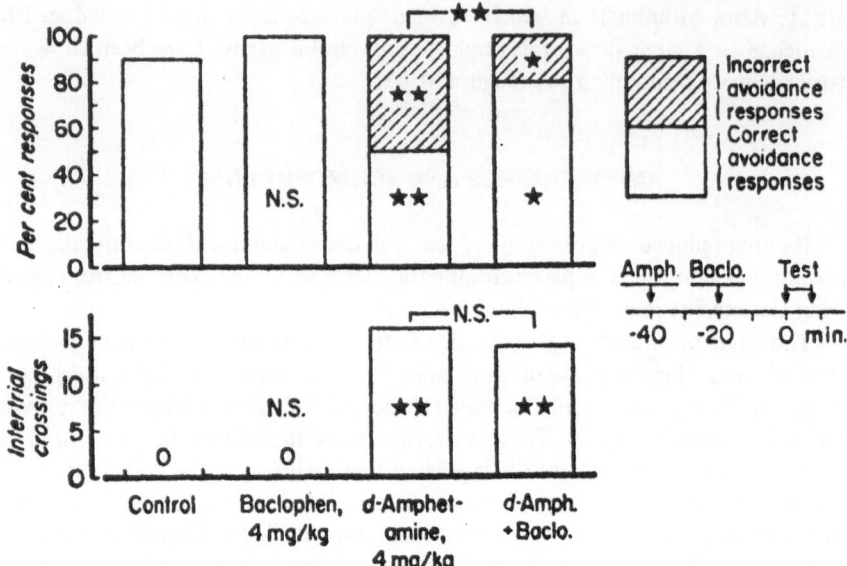

FIGURE 7. Loss of discrimination ability after treatment with dexamphetamine and restoration of correct performance by baclophen. (Reproduced from ref. 5.)
Rats were trained as described in the legend of Figure 6.

NOTE: Amphetamine appeared to block the discrimination ability completely, while preserving the responsiveness to the conditioned stimulus. It has acted similarly to L-dopa as shown in Figure 6. Baclophen restored this ability almost completely, as did haloperidol in the experiment shown in Figure 6. There are, however, two distinct differences between baclophen and haloperidol: the former drug, unlike the latter, does not antagonize the hypermotility induced by amphetamine, and does not inhibit responding when given alone.

**p < 0.01, *p < 0.05, N.S. p > 0.05, Wilcoxon T-test; the values shown are medians, N = 9.

for a moment whether he was a patient rather than a doctor, but then he told me about some remarkable preliminary observations. About six weeks earlier he had started to treat schizophrenic patients with baclophen (Lioresal, Ciba: β-(-4-chlorophenyl)-γ-aminobutyric acid), a derivative of GABA capable of penetrating through the blood-brain barrier. The new drug was superimposed on the previous treatment with neuroleptics, which had been only partially successful in these cases. In most of the patients a marked improvement was seen setting in within a few days of baclophen treatment. In particular, the autism and the thought disorder improved with a partial apparent re-integration of the personality. Many of the patients said they could think and read more easily, and they became more communicative. Hallucinations did not disappear, but were less intense and appeared more remote. In some patients psychotic symptoms disappeared altogether. Phenothiazine medication could be reduced or discontinued. In fact, the tolerance to phenothiazines seemed to be reduced by the drug, causing complaints of rigidity, which disappeared after reducing the phenothiazine dosage [36].

The idea that deficiency in GABA may play a role in schizophrenia has been put forward [48,52]. Similar considerations formed the basis of Frederiksen's trial, even though the mode of action of baclophen has not yet been established.

These speculations and observations on the possible role of GABA in schizophrenia are very interesting and will no doubt stimulate basic and clinical research in this area. It may be recalled that an intimate, mutually antagonistic relationship appears to exist between dopamine- and GABA-carrying neuronal pathways, at least in the striatum, and that deficiency of the latter transmitter might well have the same effect as hyperactivity of the former [12,43,46]. It may also be recalled that paranoid delusions and other schizophrenia-like symptoms may occur in Huntington's chorea, in which a deficiency of a striatal GABA pathway has been detected [15,47]. Baclophen has been tried in Huntington's chorea, and some slight improvement was reported [11].

In rats we have observed actions of baclophen on behavior, somewhat similar to those of a neuroleptic agent such as haloperidol (Figure 7) [5]. Thus amphetamine-induced loss of discrimination was restored by baclophen. However, baclophen, unlike haloperidol, did not antagonize amphetamine-induced hypermotility and did not *per se* disrupt conditioned behavior.

Thus the profile of baclophen appears to be clearly different from that of classical antipsychotic agents, as assessed both clinically and experimentally. In addition, differences in chemical structure suggest a different point of attack. Future work will decide whether baclophen interacts with GABA or some other neurohumoral transmitter.

REFERENCES

1. Aghajanian, G. K. and Bunney, B. S. Pre- and postsynaptic feedback mechanisms in central dopaminergic neurons. In: *Frontiers in neurology and neuroscience research 1974*, edited by P. Seeman and G. M. Brown. Toronto: The University of Toronto Press, 1974, pp. 4-11.
2. Ahlenius, S. and Engel, J. Behavioral effects of haloperidol after tyrosine hydroxylase inhibition. *European J. Pharmacol. 15*:187-192, 1971.
3. Ahlenius, S. and Engel, J. On the interaction between pimozide and α-methyltyrosine. *J. Pharm. Pharmac. 25*:172-174, 1973.
4. Ahlenius, S. and Engel, J. Antagonism by haloperidol of the L-DOPA-induced disruption of a successive discrimination in the rat. *J. Neural Transmission 36*: 43-49, 1975.
5. Ahlenius, S. and Engel, J. Antagonism by baclophen of the d-amphetamine-induced disruption of a successive discrimination in the rat. *J. Neural Transmission*, 1975.

6. Anden, N. E. et al. Demonstration and mapping out of nigro-neostriatal dopamine neurons. *Life Sci. 3*:523–530, 1964.

7. Anden, N. E., Carlsson, A., and Haggendal, J. Adrenergic mechanisms. *An. Rev. Pharmacol. 9*:119–134, 1969.

8. Anden, N. E. Dopamine turnover in the corpus striatum and the limbic system after treatment with neuroleptic and antiacetylcholine drugs. *J. Pharm. Pharmac. 24*:905–906, 1972.

9. Anden, N. E. et al. Receptor activity and turnover of dopamine and noradrenaline after neuroleptics. *European J. Pharmacol. 11*:303–314, 1970.

10. Anden, N. E. Catecholamine receptor mechanisms in vertebrates. In: *Frontiers in catecholamine research*, edited by E. Usdin and S. Snyder. Oxford: Pergamon Press, 1973, pp. 661–665.

11. Anden, N. E., Dalen, P., and Johansson, B. Baclofen and lithium in Huntington's chorea. *Lancet 93*: 1973.

12. Anden, N. E. and Stock, G. Inhibitory effect of gammahydroxybutyric acid and gammaaminobutyric acid on the dopamine cells in the substantia nigra. *Naunyn-Schmiedeberg's Arch. Pharmacol. 279*:89–92, 1973.

13. Anden, N. E. and Stock, G. Effect of clozapine on the turnover of dopamine in the corpus striatum and in the limbic system. *J. Pharm. Pharmac. 25*: 346–348, 1973.

14. Bertler, A. and Rosengren, E. Occurrence and distribution of dopamine in brain and other tissues. *Experientia 15*:10, 1959.

15. Bird, E. D. et al. Reduced glutamic-acid-decarboxylase activity of postmortem brain in Huntington's chorea. *Lancet*, 1090–1092, 1973.

16. Birkmayer, W. and Hornykiewicz, O. Der L-3,4-Dioxyphenylalanin (=DOPA)-Effekt bei der Parkinson-Akinese. *Wien. Klin. Wschr. 73*:787–788, 1961.

17. Carlsson, A. and Hillarp, N. A. Release of adrenaline from the adrenal medulla of rabbits produced by reserpine. *Kgl. Fysiogr. Süllsk Lund Förh 26*: No. 8, 1956.

18. Carlsson, A., Lindqvist, M., and Magnusson, T. 3,4-Dihydroxyphenylalanine and 5-hydroxytryptophan as reserpine antagonists. *Nature* (Lond.) *180*: 1200, 1957.

19. Carlsson, A. et al. Effect of reserpine on the metabolism of catecholamines. In: *Psychotropic drugs*, edited by S. Garattini and V. Ghetti. Amsterdam: Elsevier Publ. Co., 1957, pp. 363–370.

20. Carlsson, A. et al. On the presence of 3-hydroxytyramine in brain. *Science 127*:471, 1958.

21. Carlsson, A. The occurrence, distribution and physiological role of catecholamines in the nervous system. *Pharmacol. Rev. 11*:490–493, 1959.

22. Carlsson, A., Lindqvist, M., and Magnusson, T. On the biochemistry and possible functions of dopamine and noradrenaline in brain. In: *Ciba Symposium on Adrenergic Mechanisms*, edited by J. R. Vane, G. E. W. Wolstenholme, and M. O'Connor. London: J. & A. Churchill, Ltd., 1960, pp. 432–439.

23. Carlsson, A., Falck, B., and Hillarp, N. A. Cellular localization of brain monoamines. *Acta Physiol. Scand. 56*(Suppl.) 196:1–28, 1962.

24. Carlsson, A. and Lindqvist, M. Effect of chlorpromazine or haloperidol on formation of 3-methoxytyramine and normetanephrine in mouse brain. *Acta Pharmacol. et Toxicol. 20*:140–144, 1963.

25. Carlsson, A. Drugs which block the storage of 5-hydroxytryptamine and related amines. In: *5-Hydroxytryptamine and related indolealkylamines*, edited by V. Erspamer. Heidelberg: Springer Verlag, 1965, pp. 529–592.

26. Carlsson, A. Amphetamine and brain catecholamines. In: *Amphetamines and related compounds, proceedings of the Mario Negri Institute for Pharmacological Research, Milan, Italy*, edited by E. Costa and S. Garattini. New York: Raven Press, 1970, pp. 289–300.

27. Carlsson, A. Biochemical and pharmacological aspects of parkinsonism. From Proceedings of the Twentieth Congress of Scandinavian Neurologists, Oslo, 1972. *Acta Neurol. Scandinav.* 48(Suppl.) *51*:11–42, 1972.

28. Carlsson, A. et al. Potentiation of phenothiazines by α-methyltyrosine in treatment of chronic schizophrenia. *J. Neural Transmission 33*:83–90, 1972.

29. Carlsson, A. et al. Further studies on the mechanism of antipsychotic action: Potentiation by α-methyltyrosine of thioridazine effects in chronic schizophrenics. *J. Neural Transmission 34*:125–132, 1973.

30. Carlsson, A. Antipsychotic drugs and catecholamine synapses. *J. Psychiat. Res. 11*:57–64, 1974.

31. Carlsson, A. Receptor-mediated control of dopamine metabolism. In: *Pre- and postsynaptic receptors*, edited by E. Usdin and W. E. Bunney. New York: M. Dekker, Inc., 1975, pp. 49–63.

32. Carlsson, A. The effect of neuroleptic drugs on brain catecholamine metabolism. In: *Antipsychotic drugs, pharmacodynamics and pharmacokinetics*, edited by G. Sedvall, B. Uvnäs, and Y. Zotterman. New York: Pergamon Press, 1975 (in press).

33. Dahlstrom, A. and Fuxe, K. Evidence for the existence of monoamine-containing neurons in the central nervous system. *Acta Physiol. Scand. 62* (Suppl.) *232*:1–55, 1964.

34. Dahlstrom, A. and Fuxe, K. Evidence for the existence of monoamine neurons in the central nervous system. *Acta Physiol. Scand. 64*(Suppl.) *247*: 1–85, 1965.

35. Ehringer, H. and Hornykiewicz, O. Verteilung von noradrenalin und dopamin (3-hydroxytyramin) im Gehirn des Menschen und ihr Verhalten bei Erkrankungen des extrapyramidal systems. *Klin. Wschr. 38*:1236–1239, 1960.

36. Frederiksen, P. K. Preliminärt meddelande angaende Lioresal (Baclofen) vid behandling av schizofreni. *Läkartidningen 72*:456–458, 1975.

37. Fuxe, K. and Anden, N. E. Studies on central monoamine neurons with special reference to the nigro-neostriatal dopamine neuron system. In: *Biochemistry and pharmacology of the basal ganglia*, edited by E. Costa, I. J. Coté, and M. D. Yahr. New York: Raven Press, 1966, pp. 123–129.

38. Gershon, S. et al. Methyl-p-tyrosine (AMT) in schizophrenia. *Psychopharmacologin 11*:189–194, 1967.

39. Goodwin, F. K. et al. Levodopa: Alterations in behavior. *Clin. Pharm. Ther. 12*:383–396, 1971.

40. Greengard, P. Molecular studies on the nature of the dopamine receptor in the caudate nucleus of the mammalian brain. In: *Frontiers in neurology and neuroscience research 1974*, edited by P. Seeman and G. M. Brown. Toronto: The University of Toronto Press, 1974, pp. 12–15.

41. Jackson, D. M., Anden, N. E., and Dahlstrom, A. A functional effect of dopa-
    mine in the nucleus accumbens and in some other dopamine-rich parts of the
    rat brain. *Psychopharmacologin* (Berl.), 1975 (in press).
42. Jonsson, L. E., Ånggard, E., and Gunne, L. M. Blockade of intravenous am-
    phetamine euphoria in man. *Clin. Pharm. Ther. 12*:889–896, 1971.
43. Kim, J. S. et al. Role of γ-aminobutyric acid (GABA) in the extrapyramidal
    motor system. 2. Some evidence for the existence of a type of GABA-rich
    strio-nigral neurons. *Exp. Brain Res. 14*:95–104, 1971.
44. May, P. R. A. Antipsychotic drugs and other forms of therapy. In: *Psycho-
    pharmacology: A review of progress 1957-1967*, edited by D. H. Efron.
    Washington: U.S. Government Printing Office, 1968, pp. 1155–1176.
45. Nyback, H. and Sedvall, C. Further studies on the accumulation and disap-
    pearance of catecholamines formed from tyrosine-^{14}C in mouse brain. Effect
    of some phenothiazine analogues. *European J. Pharmacol. 10*:193–205, 1970.
46. Okada, Y. et al. Role of γ-aminobutyric acid (GABA) in the extrapyramidal
    motor system. 1. Regional distribution of GABA in rabbit, rat, guinea pig
    and baboon CNS. *Exp. Brain Res. 13*:514–518, 1971.
47. Perry, T. L., Hansen, S., and Kloster, M. Huntington's chorea. *New Engl. J.
    Med. 288*:337-342, 1973.
48. Roberts, E. An hypothesis suggesting that there is a defect in the GABA sys-
    tem in schizophrenia. *Neurosci. Res. Progr. Bull. 10*:468–480, 1972.
49. Sedvall, G. et al. Mass fragmentometric determination of homovanillic acid in
    lumbar cerebrospinal fluid of schizophrenic patients during treatment with
    antipsychotic drugs. *J. Psych. Res. 11*:75–80, 1974.
50. Shore, P. A., Silver, S. L., and Brodie, B. B. Interaction of reserpine, sero-
    tonin, and lysergic acid diethylamide in brain. *Science 122*:284–285, 1955.
51. Snyder, S. H. et al. Drugs, neurotransmitters, and schizophrenia. *Science
    184*:1243-1253, 1974.
52. Stevens, J., Wilson, K., and Foote, W. GABA blockade, dopamine and schizo-
    phrenia: Experimental studies in the cat. *Psychopharmacologia* (Berl.) *39*:
    105-119, 1974.
53. Walinder, J. et al. Unpublished data, 1974.

AUTHOR'S COMMENTS

This lecture was held at a time when the possible role of catecholamines in
schizophrenia had just started to attract widespread interest. During the subsequent
years this interest became even more intense, and the "dopamine hypothesis of
schizophrenia" was vividly discussed at many meetings, e.g., at the Fourth International Catecholamine Symposium [1]. For a long time, the evidence was entirely
pharmacological, but more recently some preliminary data suggesting disturbances
in catecholamine metabolism and in dopamine-receptor sensitivity in the brains of
schizophrenic patients have been reported [1-3]. From the therapeutic point of
view this research may help to find better antipsychotic agents (that is, with weaker

extrapyramidal side effects and thus hopefully lacking tardive dyskinesia-inducing effects). Such agents may still be receptor blockers like the classical antipsychotic agents or receptor agonists acting selectively on dopamine autoreceptors, leading to inhibition of presynaptic dopaminergic activity. One such agent has been described recently [4]. It is still in the experimental, preclinical stage.

Future work will also aim at demonstrating the role of noncatecholaminergic neurons in schizophrenic symptomatology and to develop new therapeutic agents acting on different sites such as GABA-ergic and peptidergic synapses. Unfortunately, the preliminary observations of a favorable action of baclophen in schizophrenia mentioned in the lecture could not be confirmed in controlled trials. Similarly, investigations into the possible role of endorphins in schizophrenia are as yet in a preliminary stage. Continued work along these lines may, however, prove very fruitful.

REFERENCES

1. Carlsson, A. The impact of catecholamine research on medical science and practice. In: *Catecholamines: Basic and clinical frontiers*, edited by E. Usdin, I. J. Kopin, and J. Barchas. New York: Pergamon Press, 1979, pp. 4–19.
2. Crow, T. J. et al. Neurotransmitter enzymes and receptors in postmortem brain in schizophrenia. In: *Abstracts of the 12th CINP Congress.* Supplement to Progress in Neuropsychopharmacology, 1980, p. 118.
3. Farley, I. J. et al. Norepinephrine in chronic paranoid schizophrenia. Above-normal levels in limbic forebrain. *Science 200*:456–457, 1978.
4. Hjorth, S., Carlsson, A., et al. A new centrally acting DA-receptor agonist with selectivity for autoreceptors. *Psychopharmacol. Bull. 16*:85, 1980.
5. Seeman, P. and Lee, T. Brain dopamine receptors in schizophrenia. In: *Abstracts of the 12th CINP Congress.* Supplement to Progress in Neuropsychopharmacology, 1980, p. 318.

Neuroleptic Drugs and Neurotransmitter Receptors

Solomon H. Snyder, M.D.

The neuroleptic drugs are probably the most important agents in psychiatry. Not only have they vastly improved the management of schizophrenia, but an abundance of evidence, which we will not review here, indicates that theirs is a fundamental anti-schizophrenic action. The drugs appear to act upon primary symptoms of schizophrenia and induce unique changes not apparent in the effects of the drugs upon other types of psychiatric patients. This unique property of the neuroleptic drugs was apparent even from the earliest clinical trials. When Jean Delay and Pierre Deniker first utilized chlorpromazine in 1952, they were struck by the fact that though the drug was a sedating agent and hyperactive patients were quieted, withdrawn patients were also activated. Reserpine, which differs markedly from chlorpromazine in chemical structure, produced similar therapeutic effects. It was notable that both chlorpromazine and reserpine elicited side effects resembling Parkinson's disease. It seemed that therapeutic improvement tended to go hand in hand with the appearance of these extrapyramidal side effects, giving rise to the concept that the drugs acted "neurologically" in schizophrenics, hence the term "neuroleptic" meaning "to grasp the neuron" [16].

Extensive well controlled clinical studies in the United States established that sedatives such as phenobarbital were no more effective than placebo in relieving schizophrenic symptoms, while all the major phenothiazines were uniformly

Research in the Schizophrenic Disorders: The Stanley R. Dean Award Lectures, vol. 2, edited by R. Cancro and S. R. Dean. Copyright © 1985 by Spectrum Publications.

efficacious. Similarly, antianxiety agents such as the benzodiazepines were not effective in schizophrenia. Moreover, when various classes of symptoms were compared for their response to drug treatment, it became apparent that the fundamental symptoms of schizophrenia responded selectively.

For all these reasons researchers have felt that neuroleptics might act at sites associated with the specific schizophrenic disturbance so that if one understood the therapeutic mechanism of action one might attain insight into the pathophysiology of schizophrenia. Of course, it is also possible that the antischizophrenic effects of the drugs are produced at sites several steps removed from the specific schizophrenic disturbance but somehow linked to it.

INITIAL BIOCHEMICAL INVESTIGATIONS

Because of the substantial number of phenothiazines, butyrophenones, and related neuroleptic drugs which have been utilized therapeutically, one can approach the study of drug action by determining whether a given biochemical effect is exerted by various neuroleptics in some relationship to their clinical potencies. For instance, promethazine (Figure 1) differs from chlorpromazine only in a single carbon in the side chain and by the absence of the chlorine. Though a potent antihistamine, promethazine lacks antischizophrenic efficacy. Promazine, which is merely dechlorinated chlorpromazine, is also markedly less effective in patients than chlorpromazine. The butyrophenones differ markedly from the phenothiazines in chemical structure yet exert essentially the same effects in patients.

Phenothiazines are quite effective membrane-active agents. In early studies they were observed to cause mitochrondrial swelling [29], lysosomal leakage, and expansion of red cell membranes [23]. They also inhibited the sodium-potassium ATPase [15]. All of these effects required relatively high concentrations of the drugs. Moreover, only fairly weak correlations with clinical activity were observed.

Substantially greater success has resulted from investigations of neuroleptic actions on neurotransmitters, especially dopamine. Reserpine was the neuroleptic whose biochemical influences on neurotransmitters were established first. Reserpine profoundly depleted the brain of all its biogenic amines, norepinephrine, serotonin, and dopamine. Soon thereafter it was found that dopamine was deficient in the corpus striatum of patients with Parkinson's disease. The striking success of L-dopa in alleviating Parkinsonian symptoms established that this depletion was related to the Parkinsonian symptoms. Thus it would seem likely that the extrapyramidal side effects elicited by reserpine may well relate to dopamine depletion. According to the "neuroleptic" concept of Delay and Deniker that neurologic and antischizophrenic actions of the drugs are associated, one might go on to speculate that the antischizophrenic actions of reserpine also derive from dopamine depletion. Since phenothiazines and butyrophenones clinically act similarly to reserpine in schizophrenia, one might anticipate a similar mechanism of action. However,

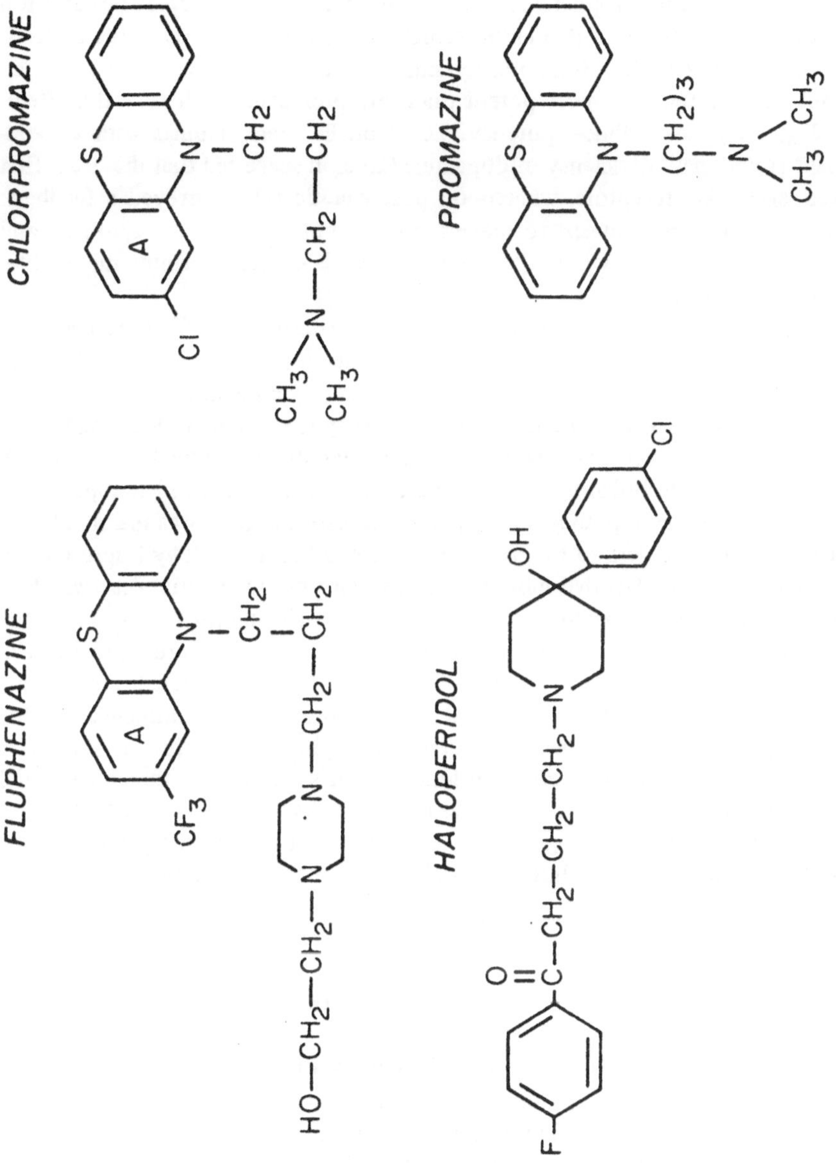

FIGURE 1. Structures of some neuroleptics.

phenothiazines and butyrophenones do not alter brain levels of dopamine. Carlsson and Lindqvist [6] did note that chlorpromazine and haloperidol increased levels of the dopamine metabolite, 3-methoxytyramine, while promethazine did not. Haloperidol was substantially more potent than chlorpromazine in eliciting this effect, paralleling its greater clinical potency. To reconcile these findings with a hypothetical "functional" deficiency of dopamine Carlsson suggested that the drugs first blocked dopamine receptors, whereupon postsynaptic cells conveyed a feedback message to dopamine neurons to make more dopamine. With more synthesis and release of dopamine one would observe higher levels of the dopamine metabolite, 3-methoxytyramine.

The first direct biochemical evidence that neuroleptics affect postsynaptic dopamine receptors derived from studies of the dopamine sensitive adenylate cyclase. Greengard and associates [18] and later Iversen and colleagues [20] demonstrated an enzyme in homgenates of rat corpus striatum which could cause an accumulation of cyclic AMP which was stimulated by dopamine. Unlike the well known effect of catecholamines on β-noradrenergic receptors in which dopamine is weak and isoproterenol potent, the dopamine sensitive cyclase was maximally enhanced by dopamine, less so by norepinephrine and hardly at all by isoproterenol. Phenothiazines were effective inhibitors of the enzyme with influences which in general paralleled pharmacological potencies in animals and man. However, there were marked discrepancies in the case of butyrophenones. Haloperidol, which clinically and pharmacologically is about 100 times more potent than chlorpromazine, appeared weaker than, or at best equal to, chlorpromazine in influences on the cyclase. The most potent known butyrophenone, spiroperidol, which is about five times more potent than haloperidol in intact animals, was weaker than haloperidol and chlorpromazine on the cyclase. These discrepancies suggested either that neuroleptics do not act by blocking dopamine receptors or that phenothiazines might block dopamine receptors, but butyrophenones must act through some other mechanism.

LABELING THE DOPAMINE RECEPTOR

Neurotransmitter receptors generally consist of two portions. The binding site "recognizes" the transmitter, while some "second messenger" site elicits a change in ion permeability or in the formation of some cyclic nucleotide. The most extensive studies of transmitter receptor binding have been performed on the nicotinic acetylcholine receptor of the electric organ of various invertebrate fishes. Studies of neurotransmitter receptor binding in mammalian brain in our laboratory derived from initial investigations of the opiate receptor. The general approach is to measure the binding of an appropriate drug with high affinity for the receptor which is labeled to high specific radioactivity. The most important technical hurdle to overcome in demonstrating specific receptor binding is to minimize the amount of nonspecific association with membrane components, since such nonspecific sites generally exceed by a large margin the number of true receptors. Using low concentrations of relatively selectively drugs and by washing rapidly but thoroughly to remove nonspecific binding while preserving specific interactions, it was possible to demonstrate binding of radioactive opiates to pharmacologically relative opiate receptors. Subsequently numerous other neurotransmitter receptors in the brain could be demonstrated including receptor sites for serotonin, α-noradrenergic, β-noradrenergic effects, glycine, GABA, dopamine, and various peptides such as angiotensin, neurotensin, and thyrotropin-releasing hormone [27].

The dopamine receptor can be labeled either by agonists such as ^{3}H-dopamine or ^{3}H-apomorphine or by the antagonists ^{3}H-haloperidol and ^{3}H-spiroperidol. The receptor can also be labeled with apparent mixed agonist–antagonists such as ^{3}H-LSD and ^{3}H-dihydroergokryptine. Evidence that these ligands in fact label the physiologically relevant dopamine receptors derives from several sources [2-4,8, 24]. First of all, the regional distribution of binding parallels closely regional variations in endogenous dopamine concentration, with highest levels of binding in the corpus striatum, somewhat lower levels in areas of the limbic system that possess dopamine pathways, such as the nucleus accumbens and olfactory tubercle, and very little binding in other regions. Drug specificity also provides important evidence. Among agonists, dopamine is most potent, norepinephrine less so and isoproterenol weakest. Relative potencies of neuroleptics in competing for binding sites in general parallel their pharmacological potencies.

DOPAMINE RECEPTORS IN THE PITUITARY

A dopamine pathway with cell bodies in the arcuate nucleus of the hypothalamus projects axons with terminals about the portal capillary plexus in the median eminence. An abundance of endocrine data indicates that dopamine released from this pathway regulates the subsequent release of peptide hormones from the pituitary. In general dopamine inhibits the release of these substances of which prolactin is one of the best studied. This action has been put to therapeutic use by the use of the dopamine mimicking drug bromokryptine, which now finds widespread utility in treating menstrual abnormalities associated with excessive prolactin release.

By blocking dopamine receptors, all clinically employed neuroleptic drugs elevate plasma levels of prolactin. Since chronic treatment with prolactin is well known to induce mammary tumors in animals, the United States Food and Drug Administration is concerned about the prolactin-elevating effects of neuroleptics, though there is as yet no evidence that long term ingestion of neuroleptics causes breast cancer. Besides the carcinogenic potential, physicians have long known that menstrual abnormalities occur as side effects of neuroleptic treatment, perhaps because of prolactin release.

If the drug specificity of pituitary dopamine receptors differed from receptor sites in the brain, one could design agents to block brain but not pituitary dopamine receptors. Conversely, to treat endocrine disorders with deficiency of prolactin, one might seek agents to block pituitary but not brain dopamine receptors. Labeling pituitary dopamine receptors and comparing their properties with those of the brain is accordingly an important experimental goal. In initial binding studies with ^{3}H-haloperidol, ^{3}H-dopamine, and ^{3}H-apomorphine pituitary dopamine receptor binding could not be detected. However, the more potent ligand, ^{3}H-spiroperidol [13], and the ergot, ^{3}H-dihydroergokryptine [7], do label dopamine receptors. In studies conducted to date properties of pituitary and brain dopamine receptors do not differ markedly except for a lesser potency of agonists in the pituitary than in the brain. Conceivably, more detailed evaluations of a wide range of drugs might reveal agents which are pituitary or brain specific.

TWO STATE MODEL OF THE DOPAMINE RECEPTOR

One of the most striking properties of the opiate receptor is the differential binding of agonists and antagonists. Low concentrations of sodium ion selectively enhance the binding of antagonists and diminish the binding of agonists [22]. These findings indicate that the opiate receptor may exist in two interconvertible states which favor the binding of agonists and antagonists, respectively. Similar observations have emerged in studies of the dopamine receptor.

Though ^{3}H-agonists and ^{3}H-antagonists both label dopamine receptors, there are differences in substrate specificity. Agonists display substantially greater affinity for sites labeled with ^{3}H-agonists than do antagonists, while the reverse situation holds for sites labeled with ^{3}H-antagonists.

The simplest explanation for these discrepancies involves a two state model similar to that described for the opiate receptor. Agonist and antagonist states of the receptor presumably exist in equilibrium so that pharmacological effects of antagonists occur as binding of the drug to the antagonist state makes fewer agonist sites available to the neurotransmitter. This model also explains how neurotransmitter recognition is translated into an alteration in ion conductance or other changes. The appropriate ion or other effecter molecule is postulated to have affinity for one or the other of the receptor states. In its resting condition, the receptor is largely in the antagonist binding sites. When the neurotransmitter binds to the receptor and transforms a portion of the receptors into the agonist state, the binding of the crucial ion or effecter also changes, eliciting the appropriate conductance or other lateration. Besides the opiate receptor, direct evidence supporting this model has been obtained also for glycine, muscarinic cholinergic, and serotonin receptors in the brain.

By comparing the binding of the drug at sites labeled by ^{3}H-agonists or ^{3}H-antagonists it should be possible to predict whether a drug is an agonist, antagonist or mixed agent. LSD is a striking instance of a mixed agonist–antagonist at dopamine receptors. Behavioral studies indicate that LSD possesses dopamine agonist actions. Additionally it stimulates the dopamine sensitive adenylate cyclase but displays antagonist activities in this system by inhibiting the effects of dopamine [31]. In support of mixed agonist–antagonist functions for LSD, the drug displays similar high affinities, about 20 to 30 nM for both ^{3}H-dopamine and ^{3}H-haloperidol sites, very different from the profile of pure agonists or antagonists [3,9]. Apomorphine also seems to possess both agonist and antagonist properties. In studies of the dopamine sensitive adenylate cyclase apomorphine does not elicit as great a maximal enhancement of cyclase activity as does dopamine [18] and thus does not seem to be as "pure" an agonist as dopamine. Interestingly, while dopamine has 30 times greater affinity for ^{3}H-dopamine than ^{3}H-haloperidol sites, apomorphine displays only about a seven fold greater affinity for dopamine than haloperidol binding [4].

The actions of LSD on the dopamine receptor do not appear related to the psychedelic effects of the drug. Thus 2-bromo-LSD, which is very weak or inactive in its psychedelic effects, is still about as potent as LSD at the dopamine receptors.

Direct evidence for the interaction of LSD with the dopamine receptor derives from studies demonstrating binding of ^{3}H-LSD to dopamine receptors in the caudate nucleus. In areas of the brain that possess negligible levels of dopamine, such as the hippocampus, ^{3}H-LSD binding appears to involve predominantly the serotonin receptor [1]. By contrast, in the caudate, besides labeling serotonin receptors, about 15 to 20 per cent of ^{3}H-LSD binding is inhibited with high affinity by dopamine [3]. By adding nonradioactive serotonin to reduce the binding of ^{3}H-LSD to serotonin receptors, one can study interactions of ^{3}H-LSD with dopamine receptors in caudate membranes relatively selectively. ^{3}H-Dihydroergokryptine, which binds to serotonin and α-noradrenergic receptors in the brain, can also be demonstrated to label dopamine receptors in the caudate to a certain extent [30].

NEUROLEPTIC EFFECTS ON DOPAMINE RECEPTORS AND PHARMACOLOGICAL ACTIONS

The affinities of neuroleptics of diverse structures for dopamine receptors labeled with ^{3}H-haloperidol correlates closely with pharmacological actions of these drugs in animals and man [10,11,25] (Figure 2). It is quite striking that one can predict clinical potencies from *in vitro* effects of drugs, since average clinical doses vary in different patients. Moreover clinical potencies are determined by variable absorption, metabolism and penetration of drugs into the brain. Presumably over a wide range of drugs these factors tend to equalize. The close correlation between clinical potencies and affinity for dopamine receptors is by no means fortuitous.

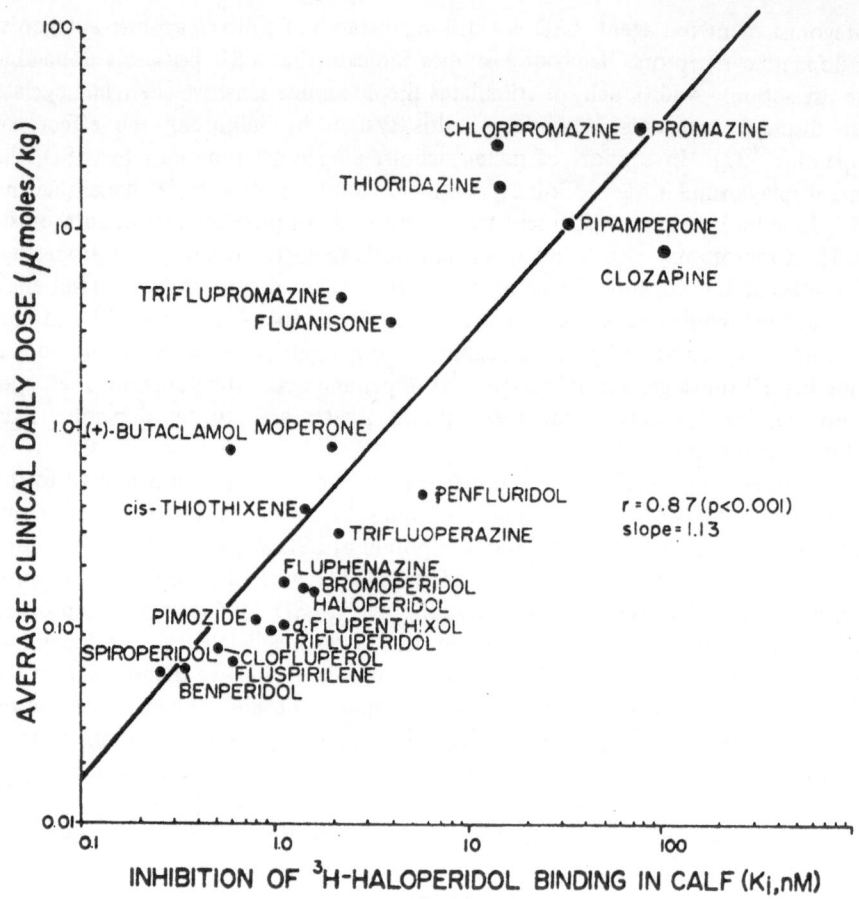

FIGURE 2. Correlation of clinical potencies and affinities of neuroleptics for dopamine receptors labeled with ^{3}H-haloperidol. Adapted from reference 25.

Thus, potencies of neuroleptics in competing for α-noradrenergic and other neurotransmitter receptors do not correlate at all with clinical potency [21].

Because the clinical activity of drugs is difficult to quantify, it would be desirable to relate binding properties of neuroleptics with pharmacological actions in animals. Apomorphine induces a sterotyped sniffing, licking, and gnawing in rodents which derives from stimulation of postsynaptic dopamine receptors in the corpus striatum and/or nucleus accumbens. The ability of neuroleptics to block apomorphine induced stereotyped behavior predicts the clinical potencies of drugs and has been used by the pharmaceutical industry as a screening test. By causing a direct release of dopamine, amphetamine elicits similar stereotyped behavior, and blockade of these amphetamine effects also predicts clinical activities of neuroleptics. Interestingly, in higher animals such as cats, monkeys and chimpanzees, the stereotyped behavior provoked by amphetamine tends to resemble such behavior observed in human amphetamine addicts on high doses close to the time of onset of amphetamine psychosis. Since amphetamine psychosis provides a fairly impressive model of some forms of schizophrenia, one can construe that stereotyped behavior in rodents as an animal drug model of a human drug model of schizophrenia. In any event, potencies of neuroleptics in competing for ^{3}H-haloperidol binding correlate quite closely with their potencies in blocking apomorphine and amphetamine induced stereotyped behavior [10,11],

Neuroleptics are also among the most effective antiemetics known. Presumably they act by blocking dopamine receptors in the chemoreceptor trigger zone in the brainstem. The ability of neuroleptics to prevent apomorphine induced emesis also correlates closely with their affinity for ^{3}H-haloperidol binding sites [10,11].

Affinities of neuroleptics for dopamine receptors labeled by ^{3}H-apomorphine and ^{3}H-dopamine do not correlate well with pharmacological activity in animals and man [10]. Butyrophenones account for most of the discrepancy in the correlation. Thus, these most potent neuroleptics in terms of pharmacological activity *in vivo*, are relatively weak both in competing for ^{3}H-agonist binding sites and in inhibiting the dopamine sensitive adenylate cyclase. Perhaps ^{3}H-dopamine labels the cyclase which is a distinct macromolecule from the dopamine receptors labeled by ^{3}H-butyrophenones.

The impressive correlation between the clinical antischizophrenic actions of neuroleptics and their blockade of dopamine receptors labeled with ^{3}H-haloperidol or ^{3}H-spiroperidol is more striking than has been observed for any other biochemical effect of the drugs. It therefore seems likely that this action is intimately associated with the antischizophrenic effects of the drugs. Whether or not dopamine synapses are the locus of schizophrenic pathophysiology remains a subject for further study.

There are of course other links between dopamine and schizophrenia. Besides eliciting an acute paranoid psychosis in nonschizophrenics, amphetamine in small doses dramatically exacerbates patients' schizophrenic symptoms and L-dopa produces similar effects. Thus one can titrate schizophrenic symptoms by manipulating, with drugs, synaptic levels of dopamine. This still does not directly indicate that the dopamine receptor is the site of schizophrenic disturbances. One would have to demonstrate a specific abnormality in some parameter related to dopamine in the brains of schizophrenics and to show that the abnormality was not related to drug treatment or other artifactual sources.

VARIATIONS IN EXTRAPYRAMIDAL SIDE EFFECTS DERIVE
FROM EFFECTS ON MUSCARINIC RECEPTORS

If neuroleptics elicit both extrapyramidal side effects and their therapeutic actions by blocking dopamine receptors and if the properties of the dopamine receptors involved in both actions are similar, then at therapeutic doses, all neuroleptics should elicit similar incidences of extrapyramidal side effects. Yet, neuroleptics vary widely in these actions. Butyrophenones and piperazine phenothiazines, such as fluphenazine (Prolixin; Permitil) produce the highest incidence of extrapyramidal side effects. The piperidine phenothiazines, thioridazine (Mellaril), has a substantially lesser tendency to produce these effects, while clozapine produces few if any extrapyramidal side effects. Pimozide, which is structurally related to the butyrophenones, tends to produce relatively few extrapyramidal side effects. What might account for these discrepancies?

One possibility is that dopamine receptors in whatever parts of the brain are involved in schizophrenic symptoms differ in drug specificity from striatal receptors. Two good candidates for the site of antischizophrenic effects of neuroleptic agents are the nucleus accumbens and olfactory tubercle, both parts of the limbic system, well known to be involved in emotional regulation. We could find no difference among these regions in the relative potencies of neuroleptics of high, medium, and low tendencies to provoke extrapyramidal side effects [10]. A recently discovered dopamine pathway has terminals in the frontal and cingulate cerebral cortex which could represent sites for antischizophrenic effects. Thus far, dopamine receptors in these areas have not yet been characterized.

We considered an alternative explanation for the differential tendencies of various neuroleptics to provoke extrapyramidal side effects based on their known anticholinergical actions. All neuroleptics produce some atropine-like side effects such as dry mouth, blurry vision and difficulty in urinating. Prior to the advent of L-dopa, atropine and synthetic relatives were the major treatment modality for Parkinson's disease. We speculated that atropine-like properties of some neuroleptics might tend to alleviate the extrapyramidal symptoms provoked by the dopamine receptor blockade elicited by the same drugs. Indeed, we [26] and others [19] found an inverse correlation between potency of these agents in competing for muscarinical cholinergical receptors labeled with a potent radioactive muscarinical antagonist and their incidence of extrapyramidal side effects.

SEDATIVE AND HYPOTENSIVE EFFECTS OF NEUROLEPTICS RELATE
TO AFFINITIES FOR α-NORADRENERGIC RECEPTORS

Sedative and hypotensive actions of neuroleptics represent serious side effects interfering with drug use. We wondered whether these effects might be attributed to blockade of α-noradrenergic receptors. The ability to label α-receptors in our laboratory provided an opportunity to test this hypothesis [17]. One does not merely wish to know affinities for α-receptors *in vitro*. Since doses, blood levels, and presumably brain levels of the drugs vary widely, one really wants to know the extent to which various drugs will occupy α-receptors in the brain when the different drugs are employed at equivalent therapeutic doses. One drug might be a much more potent α-blocker than a second drug, but if the first is an extremely potent dopamine receptor blocker, it will probably be employed in very low doses and might attain so much lower brain levels than the second drug that in clinical practice it would in fact cause less central α-receptor blockade. One can evaluate the relative α-receptor effects of drugs at therapeutic doses simply by measuring the ratio of affinities for α-noradrenergic and dopamine receptors, since affinities for dopamine receptors predict clinical doses. Drugs with low ratios of potencies as α-antagonists to potencies as dopamine antagonists would be anticipated to elicit a substantial amount of α-blockade at blood and brain levels of the drugs required for dopamine receptor blockade. By contrast, drugs with high ratios will be employed clinically at the very low doses required to secure dopamine receptor blockade and so, in general, would be less likely to elicit side effects associated with α-receptor blockade.

Butyrophenones such as haloperidol and spiroperidol and piperazine phenothiazines such as fluphenazine and trifluoperazine have a relatively low propensity to elicit hypotension and sedation and display ratios of potencies for α-noradrenergic to dopamine receptors greater than ten (Table 1) [21]. By contrast, promazine and clozapine, the neuroleptics with the greatest tendency to cause orthostatic hypotension and sedation, have ratios less than 0.2. The butyrophenone droperidol is one of the most sedating of the butyrophenones (and indeed is utilized as an adjunct to anesthesia) and is the most potent of the butyrophenones in competing for α-receptor binding.

Thus the profile of neuroleptic drugs as inhibitors of dopamine, muscarinic cholinergic and α-noradrenergic receptor binding determined *in vitro* can provide a valuable approximation of both therapeutic and side effects. Besides increasing our understanding of how these drugs act, these simple binding techniques may afford a useful screen for developing new agents with fewer side effects.

TARDIVE DYSKINESIA: APPARENT DOPAMINE RECEPTOR SUPERSENSITIVITY RELATED TO AN AUGMENTED NUMBER OF DOPAMINE RECEPTOR BINDING SITES

Chronic treatment with neuroleptics elicits motor abnormalities in both man and animals. Tardive dyskinesia with abnormal movements of facial muscles and extremities is a major and sometimes virtually irreversible consequence of long term neuroleptic drug use. Lowering the dose or terminating drug treatment frequently worsens the symptoms, while increasing the dose may alleviate them.

TABLE 1. Neuroleptic Effects on α-Adrenergic and Dopamine Receptors

	K_i, ^{3}H-WB-4101, nM (α-adrenergic)	K_i, ^{3}H-haloperidol, nM (dopamine)	Ratio K_i WB-4101 K_i haloperidol	Hypotension sedation
Phenothiazines				
Alkylamino				
Triflupromazine (Vesprin)	3.4	2.1	1.6	++
Chlorpromazine (Thorazine)	5.2	10.2	0.5	++++
Promazine (Sparine)	11	72	0.15	+++
Piperidine				
Thioridazine (Mellaril)	5.4	15	0.34	++++
Piperazine				
Fluphenazine (Prolixin)	9.9	0.9	11	+
Trifluoperazine (Stelazine)	46	2.1	22	+

Dibenzodiazepine			
Clozapine	17	120	0.14	++++
Thioxanthenes			
cis-Thiothixene (Navane)	6.6	1.5	4.4	+
trans-Thiothixene	150	145	1.0	+
α-Flupenthixol	10	1.0	10	+
β-Flupenthixol	36	48	0.8	+
Butyrophenones			
Droperidol	0.7	1.0	0.7	+++
Haloperidol	12	1.4	8.4	+
Spiroperidol	18	0.25	73	+
Classic α-antagonists			
Phentolamine	3.6	—	—	—
Phenoxybenzamine	4.0	—	—	—

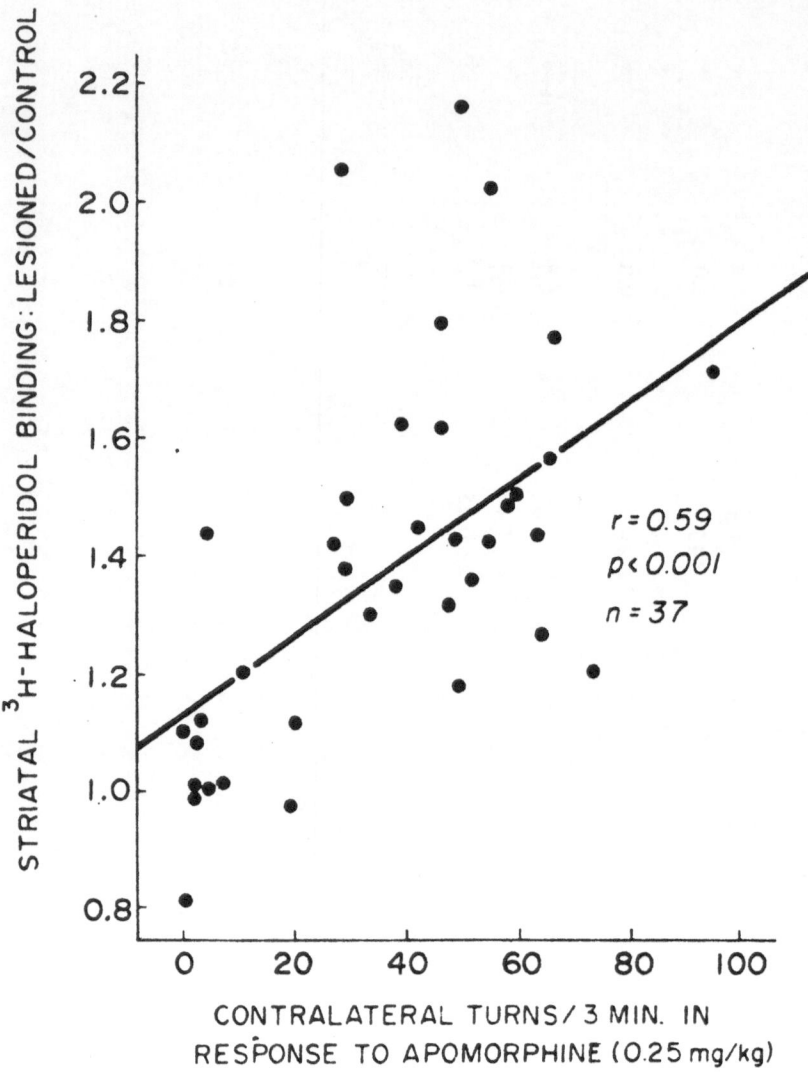

FIGURE 3. Increased ^{3}H-haloperidol binding in the lesioned striatum and behavioral supersensitivity to apomorphine following unilateral nigrostriatal 6-hydroxydopamine lesion. Rotational behavior to apomorphine (0.25 mg/kg s.c.) was measured between two and seven months after unilateral injection of 6-hydroxydopamine into the substantia nigra. Striatal ^{3}H-haloperidol binding was assayed in the lesioned and control striatum of each rat separately between one and ten weeks later and is expressed as the percentage increase in specific cpm ^{3}H-haloperidol bound in the lesioned/control striatum at four concentrations of ^{3}H-haloperidol (0.4–4.0 nM).

Such observations suggest that tardive dyskinesia represents supersensitivity of dopamine receptors which "fight back" after having been blocked for a prolonged period. Indeed, the excessive abnormal movements of tardive dyskinesia are reminiscent of the side effects of high doses of L-dopa.

One animal model in which behavioral supersensitivity of dopamine receptors is well established involves the "denervation supersensitivity" produced by 6-hydroxydopamine induced lesions of the nigrostriatal dopamine pathway. The rotation elicited by apomorphine in animals with unilateral nigrostriatal lesions is a useful behavioral measure of supersensitivity at striatal dopamine receptors. What might be the biochemical mechanism underlying the behavioral supersensitivity? We found an augmentation in the number of ^{3}H-haloperidol binding sites in these animals. The extent of the increase correlates closely with behavioral supersensitivity as measured by apomorphine induced rotation [12] (Figure 3).

After chronic administration of neuroleptics, animals become hypersensitive to the ability of apomorphine to elicit stereotyped behavior. Using this animal model of tardive dyskinesia, we attempted to determine whether the behavioral supersensitivity of dopamine receptors was also associated with a change in binding sites. Treating rats for three weeks with haloperidol (or fluphenazine) elicited at 20 per cent augmentation in ^{3}H-haloperidol binding [5]. By contrast similar treatment with a higher dose of promethazine, which lacks antischizophrenic activity, failed to enhance receptor binding. Reserpine, which functionally reduces stimulation of dopamine receptors by depleting the brain of dopamine, also augmented ^{3}H-haloperidol binding, while amphetamine produced no changes. Enhanced binding was due to an increase in the number of binding sites with no change in their affinity.

A SIMPLE SENSITIVE AND SPECIFIC RADIORECEPTOR
ASSAY FOR BLOOD NEUROLEPTICS

The failure of many patients to respond to substantial doses of neuroleptics appears to be related at least in part to failure to attain adequate blood levels. Clearly the physician wants to administer high enough doses to reach therapeutic levels even in patients who absorb the drug poorly or are extremely rapid metabolizers. On the other hand, one wants to utilize the lowest effective dose of neuroleptics, because the likelihood of eliciting tardive dyskinesia increases as a function of increasing dose. For these reasons it would be desirable to monitor patient blood levels of neuroleptics routinely. Until now, this has not been feasible, because available techniques are either too complicated for routine clinical application or do not measure all the neuroleptics in everyday use.

Since neuroleptics compete with ^{3}H-haloperidol or ^{3}H-spiroperidol at substantially lower concentrations than occur in the blood of patients at therapeutic doses, one can measure neuroleptic levels in plasma, serum or red cells by quantifying their ability to reduce ^{3}H-butyrophenone binding [14]. No extraction procedure is necessary as small volumes of plasma, serum or red cell extracts do not interfere substantially with receptor binding. Under these conditions neuroleptics which are bound to blood components dissociate during the assay procedure.

This radioreceptor assay offers potential for routine application to all patients receiving the drugs. The procedure is sensitive. Since the clinical potencies of neuroleptics are well correlated with their affinity for dopamine receptors, the sensitivity of the assay is greater for drugs which are employed therapeutically at lower doses.

The assay is selective for neuroleptics. Only dopamine receptor antagonists compete for ^{3}H-butyrophenone binding. Levels of circulating catecholamines are much too low to pose any problems. In screens of many drugs at various chemical classes, no drugs other than neuroleptics compete significantly except for ergots. Ergots which do bind to dopamine receptors are not employed clinically in the therapy of schizophrenics. Thus this assay can be utilized in patients receiving a variety of drugs besides neuroleptics. In patients treated with two or more different neuroleptics, one cannot measure the absolute concentration of each drug.

The radioreceptor assay detects metabolites which compete for dopamine receptor binding and thus may be therapeutically active. A substantial portion of the therapeutic activity of chlorpromazine may be attributed to its metabolite 7-hydroxychlorpromazine which has almost as much affinity for dopamine receptors as chlorpromazine itself [14]. Since blood levels of 7-hydroxychlorpromazine vary among patients widely and are often higher than those of chlorpromazine, measuring chlorpromazine alone is unlikely to provide the most effective indicator of the patients' levels of antischizophrenic drug activity.

The radioreceptor assay is simple to perform. As many as a hundred assays can be conducted in a morning.

CONCLUSIONS

In summary, studies of the dopamine receptor have enhanced our understanding of the therapeutic actions of neuroleptic drugs. Evidence that these agents exert their therapeutic actions by blocking dopamine receptors is as strong as the evidence for the mechanism of action of almost any drugs in clinical medicine. Effects of neuroleptics on muscarinical cholinergical and α-noradrenergic receptors can predict the relative propensities of the drugs to elicit extrapyramidal and sedative-hypotensive side effects, respectively. Tardive dyskinesia has been shown to be related to an augmentation in the number of dopamine receptors, which may elicit behavioral supersensitivity. Finally, dopamine receptor binding has provided an approach to a simple and specific radioreceptor assay for neuroleptics which may enhance the therapy of schizophrenic patients.

Neurotransmitters are to psychiatry what hormones are to endocrinologists; they are the body chemicals which account for therapeutic effects of medications and possibly for disease states as well. Elucidating interactions of psychotropic drugs with neurotransmitters, particularly with receptor sites, has clarified the actions of drugs in psychiatry. Perhaps these approaches will in the not-too-distant future shed light on the pathophysiology of psychiatric disturbances themselves.

REFERENCES

1. Bennett, J. P., Jr. and Snyder, S. H. Stereospecific binding of d-lysergic acid diethylamide (LSD) to brain membranes: Relationship to serotonin receptors. *Brain Res. 94*:523-544, 1975.

2. Burt, D. R., Enna, S., Creese, I., and Snyder, S. H. Dopamine receptor binding in the corpus striatum of mammalian brain. *Proc. Nat. Acad. Sci. (USA) 172*: 4655-4659, 1975.

3. Burt, D. R., Creese, I., and Snyder, S. H. Binding interactions of LSD and related agents with dopamine receptors in the brain. *Mol. Pharmacol. 12*: 631-638, 1976.

4. Burt, D. R., Creese, I., and Snyder, S. H. Characteristics of [^{3}H] haloperidol and [^{3}H] dopamine binding associated with dopamine receptors in calf brain membranes. *Mol. Pharmacol. 12*:800-812, 1976.

5. Burt, D. R., Creese, I., and Snyder, S. H. Antischizophrenic drugs: Chronic treatment elevates dopamine receptor binding in brain. *Science 196*:326-328, 1977.

6. Carlsson, A. and Lindqvist, J. Effect of chlorpromazine and haloperidol on formation of 3-methóxytyramine and normetanephrine in mouse brain. *Acta Pharmacol. Toxicol. 20*:140-144, 1963.

7. Caron, M. G., Raymond, V., Lefkowitz, R. J., and Labrie, F. Identification of dopaminergic receptors in anterior pituitary: Correlation with the dopaminergic control of prolactin release. *Fed. Proc. 36*:278, 1977.

8. Creese, I., Burt, D. R., and Snyder, S. H. Dopamine receptor binding. Differentiation of agonist and antagonist states with [^{3}H] dopamine and [^{3}H]-haloperidol. *Life Sci. 17*:993-1002, 1975.

9. Creese, I., Burt, D. R., and Snyder, S. H. The dopamine receptor: Differential binding of d-LSD and related agents to agonist and antagonist states. *Life Sci. 17*:1715-1720, 1975.

10. Creese, I., Burt, D. R., and Snyder, S. H. Dopamine receptor binding predicts clinical and pharmacological potencies of antischizophrenic drugs. *Science 192*:481-483, 1976.

11. Creese, I., Burt, D. R., and Snyder, S. H. Dopamine receptors and average clinical doses. *Science 194*:546, 1976.

12. Creese, I., Burt, D. R., and Snyder, S. H. Dopamine receptor binding enhancement accompanies lesion-induced behavioral supersensitivity. *Science 197*: 596-598, 1977.

13. Creese, I., Schneider, R., and Snyder, D. H. ^{3}H-Spiroperidol labels dopamine receptors in pituitary and brain. *Eur. J. Pharmacol. 46*:377-381, 1977.

14. Creese, I. and Snyder, S. H. A novel, simple and sensitive radioreceptor assay for antischizophrenic drugs in blood. *Nature 270*:180-182, 1977.

15. Davis, P. W. and Brody, T. M. Inhibition of Na^+K^+ activated adenosine triphosphatase activity in rat brain by substituted phenothiazines. *Biochem. Pharmacol. 15*:703-710, 1966.

16. Deniker, P. Introduction of neuroleptic chemotherapy into psychiatry. In: *Discoveries in biological psychiatry*, edited by F. J. Ayd, Jr. and B. Blackwell. Philadelphia: Lippincott, 1970, pp. 155-164.

17. Greenberg, D. A., U'Prichard, D. C., and Snyder, S. H. α-Noradrenergic receptor binding in mammalian brain, differential labeling of agonist and antagonist states. *Life Sci. 19*:69-76, 1976.

18. Kebabian, J. W., Petzold, G. L., and Greengard, P. Dopamine-sensitive adenylate cyclase in caudate nucleus of rat brain and its similarity to the "dopamine receptor." *Proc. Nat. Acad. Sci. (USA) 69*:2145-2149, 1972.

19. Miller, R. J. and Hiley, C. R. Anti-muscarinic properties of neuroleptics and drug-induced Parkinsonism. *Nature (London) 248*:596-597, 1974.

20. Miller, R. J., Horn, A. S., and Iversen, L. L. The action of neuroleptic drugs on dopamine-stimulated adenosine cyclic 3',5'-monophosphate production in rat neostriatum and limbic forebrain. *Mol. Pharmacol. 10*:759-766, 1974.

21. Peroutka, S. J., U'Prichard, D. C., Greenberg, D. A., and Snyder, S. H. Neuroleptic drug interactions with norepinephrine α-receptor binding sites in rat brain. *Neuropharmacol. 16*:549-556, 1977.

22. Pert, C. B. and Snyder, S. H. Opiate receptor binding of agonists and antagonists affected differentially by sodium. *Mol. Pharmacol. 10*:868-879, 1974.

23. Seeman, P. Erythrocyte membrane stabilization by local anesthetics and tranquilizers. *Biochem. Pharmacol. 15*:1753-1766, 1966.

24. Seeman, P., Chau-Wong, M., Tedesco, J., and Wong, K. Brain receptors for antipsychotic drugs and dopamine: Direct binding assays. *Proc. Nat. Acad. Sci. (USA) 72*:4376-4380, 1975.

25. Seeman, P., Lee, T., Chau-Wong, M., and Wong, K. Antipsychotic drug doses and neuroleptic/dopamine receptors. *Nature 261*:717-719, 1976.

26. Snyder, S. H., Banerjee, S. P., Yamamura, H. I., and Greenberg, D. Drugs, neurotransmitters and schizophrenia. *Science 184*:1243-1253, 1974.

27. Snyder, S. H., U'Prichard, D. C., and Greenberg, D. A. Neurotransmitter receptor binding in the brain. In: *Psychopharmacology: A generation of progress*, edited by M. A. Lipton, A. DiMaccio, and K. F. Killam. New York: Raven Press, 1978, pp. 361-370.

28. Snyder, S. H., Burt, D. R., and Creese, I. The dopamine receptor of mammalian brain: Direct demonstration of binding to agonist and antagonist states. *Neuroscience Symposia 1*:28-49, 1976.

29. Spirtes, M. A. and Guth, P. S. Effects of chlorpromazine on biological membranes. I. Chlorpromazine-induced changes in liver mitochondria. *Biochem. Pharmacol. 12*:37-46, 1963.

30. Tittler, M., Weinreich, P., and Seeman, P. New detection of brain dopamine receptors with [^{3}H] dihydroergokryptine. *Proc. Nat. Acad. Sci. (USA) 74*: 3750–3753, 1977.

31. Von Hungen, K., Roberts, S., and Hill, D. F. Interactions between lysergic acid diethylamide and dopamine-sensitive adenlyate cyclase systems in rat brain. *Brain Res. 94*:57–66, 1975.

The Fate of Catecholamines and Its Impact in Psychopharmacology

Julius Axelrod, Ph.D.

The past 25 years have seen remarkable changes in the growth of the
neurosciences. These developments
have had far-reaching influences in
psychiatry, pharmacology, endocrinol-
ogy, and cell biology. Central to this
growth has been the increase in our
knowledge of neurotransmitters, the
chemical messengers of nerves. The
concept of chemical neurotransmission
was first conceived by a graduate student at Cambridge University, T. R.
Elliot, about 75 years ago. He made
the astute observation that a newly discovered constituent of the adrenal medulla,
the catecholamine adrenaline, has the same effect on a number of organs as that
produced by stimulation of the sympathetic nerves. Elliot then proposed that
sympathetic nerves produce their effects by releasing a chemical substance similar
to adrenaline. It took many years to provide experimental proof for chemical
neurotransmission. In 1921, Otto Loewi, an Austrian pharmacologist, provided the
proof of chemical neurotransmission by a beautiful and compelling experiment.
Loewi placed two frog hearts in a vessel with a physiological solution bathing both
hearts. Upon stimulation of the vagus nerve of the first heart, the rate of the second
was reduced. This indicated that stimulation of a nerve in the heart released a
chemical substance which slowed the beat of the second heart. The vagus nerve
transmitter was soon identified as acetylcholine by Henry Dale. At about the same

Research in the Schizophrenic Disorders: The Stanley R. Dean Award Lectures, vol. 2, edited
by R. Cancro and S. R. Dean. Copyright © 1985 by Spectrum Publications.

time Walter Cannon showed that sympathetic nerves could release an adrenaline-like substance which he called sympathin. Cannon also found that sympathin is released in stress, fear, and anxiety. In 1946, Ulf von Euler isolated the sympathetic nerve transmitter and identified it as noradrenaline.

I was intrigued by a paper by Barger and Dale published in 1910 in which they showed that compounds related in structure to the catecholamine adrenaline can mimic the action of the sympathetic nervous systems. These compounds were named "sympathomimetic amines." Many sympathomimetic amines such as amphetamine, ephedrine and mescaline affect behavior. To get some insight into the action of these amines, I began studies on the fate of amphetamine and ephedrine in the body. I found that amphetamine and ephedrine were metabolized by a variety of chemical changes involving hydroxylation and demethylation. This work also led to the discovery of enzymes that transform these drugs that were later to be shown to metabolize most drugs.

The introduction of antidepressants and antipsychotic drugs in the 1950s had a profound effect in psychiatry as well as pharmacology. At about the same time the neurotransmitters, noradrenaline and serotonin, were found to be present in the brain, it was observed that reserpine, a drug used to treat mental illness, depleted these amines in the brain. Pharmacological studies showed that LSD antagonized the pharmacological action of serotonin. It was also apparent that the hallucinogens, mescaline and LSD, resembled the neurotransmitters noradrenaline and serotonin in chemical structure. Monoamine oxidase inhibitors, compounds that blocked the metabolism of noradrenaline and serotonin, were introduced as antidepressant agents. All of these events set the stage for an intense research effort to show the relationship between neurotransmitters and mental illness.

FATE OF NORADRENALINE IN SYMPATHETIC NERVES

In 1957, at a seminar of the Laboratory of Clinical Science, Seymour Kety described an intriguing observation by Hoffer and Osmond that adrenochrome, an oxidation product of adrenaline, produced hallucinations. They proposed that schizophrenia might be due to an abnormal metabolism of adrenaline. At that time little was known about the metabolism of adrenaline in the body. Because of my research in the metabolism of compounds related to adrenaline, such as ephedrine and amphetamine, I initiated studies on the fate of catecholamines in the body. I spent three frustrating months searching for an enzyme that converted adrenaline to adrenochrome. Just at that time I read an abstract that reported that patients with an adrenaline-forming tumor, pheochromocytoma, excreted large amounts of 3-methoxy-4-hydroxymandelic acid (VMA). This immediately gave me an important clue to a possible route for catecholamine metabolism. Since VMA is closely related in structure to adrenaline, a likely route for the metabolism could be via methylation of the 3-hydroxy group. If this were so, then an intermediate

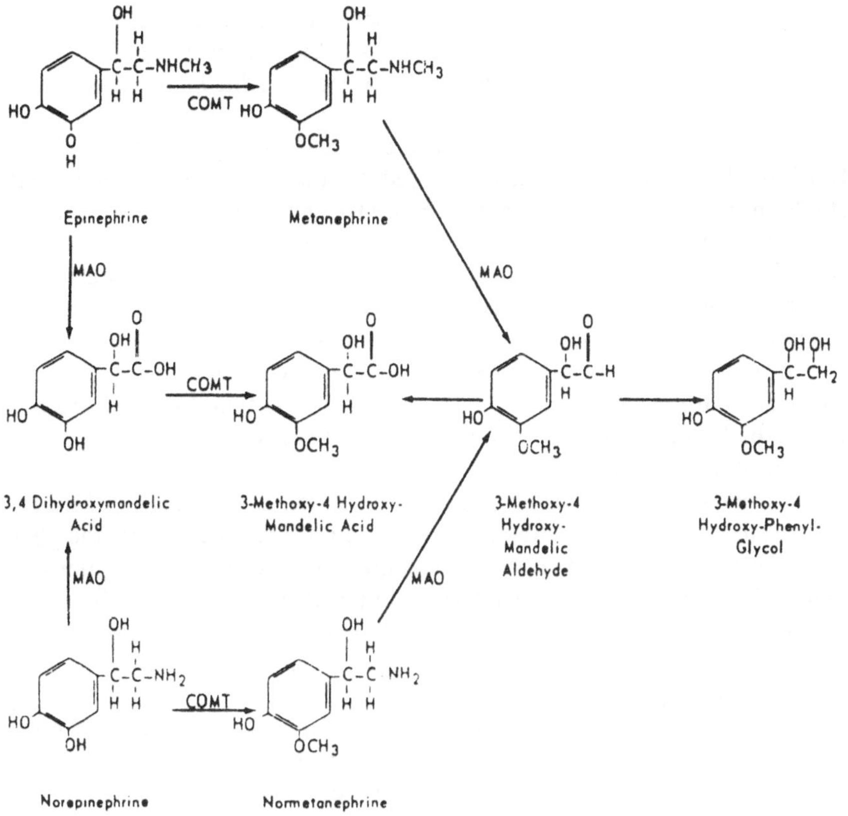

FIGURE 1. Route of metabolism of adrenaline (epinephrine) and noradrenaline (norephinephrine).

product should be 3-methoxyadrenaline. In a series of experiments I found that 3-methoxyadrenaline which we later named metanephrine was a normal constituent in nerves and tissues. After the administration of adrenaline (epinephrine) or noradrenaline (norepinephrine), there were large increases in the excretion of O-methylated metabolites of these catecholamines. Using the methyl donor S-adenosylmethionine, the enzyme that O-methylates catecholamines, catechol-O-methyltransferase (COMT) was discovered. As a result of these experiments the metabolic route of catecholamines was elucidated (Figure 1). Catecholamines were found to be metabolized by two enzymes, one involving COMT and the other MAO. In the course of this work several new metabolites of catecholamines were identified, normetanephrine, metanephrine, 3-hydroxyphenylglychol (MHPG) and methoxytyramine. MPHG was reported to be a major metabolite of noradrenaline in the brain and was subsequently found to be decreased in certain affective disorders. When the metabolic fate of adrenaline was elucidated, a study comparing its metabolism in normal and schizophrenic subjects was made possible. No difference in the metabolism of adrenaline in normal or schizophrenic subjects was found.

As a consequence of the discovery of COMT, we synthesized radioactive [^{3}H] methyl-S-adenosylmethionine. This led to the discovery of several other methyltransferase reactions which metabolized many biologically active biogenic amines: phenylethanolamine N-methyltransferase, the enzyme that forms adrenaline; histamine N-methyltransferase, an enzyme that inactivates histamine; indoleamine N-methyltransferase, an enzyme that can make endogenous hallucinogens; and protein carboxylmethylase, an enzyme that methylates the carboxy group of proteins. Recently, we described an enzyme that methylates phospholipids in membranes. This enzyme was found to be critical in the transduction of receptor-mediated biological signals through membranes. This discovery has led to a change in the direction of my research in the past three years.

A necessary property of neurotransmitter action is the capacity of tissues to terminate its action rapidly. About 20 years ago it was believed that the physiological effects of catecholamine neurotransmitters were ended by the action of the enzyme MAO. However, after complete inhibition of the enzyme, the physiological effects of catecholamines still persisted. Inhibition of COMT also failed to terminate the effects of catecholamine administration, indicating that there were other mechanisms for rapidly ending the actions of catecholamines. A clue for a rapid inactivating mechanism for catecholamines came with the use of radioactive noradrenaline. When [^{3}H] noradrenaline was injected into cats, it persisted in tissues rich in sympathetic nerves long after its physiological effects disappeared. This suggested that the catecholamine was taken up in sympathetic nerves and held there. To prove this, we denervated the superior cervical ganglia of a cat on one side only and injected [^{3}H] noradrenaline. The injected radioactive catecholamine was taken up on the innervated but not the denervated side. These experiments indicated that noradrenaline is inactivated by rapid uptake into sympathetic nerves and stored there in a physiologically inactive form. This finding provided a stimulus for further work on the mechanism of action of adrenergic and psychoactive drugs. In subsequent work using [^{3}H] noradrenaline radioautography and electron microscopy, we showed that the catecholamine neurotransmitters were taken up by sympathetic nerves and stored in dense core vesicles. When sympathetic nerves that had previously taken up [^{3}H] noradrenaline were stimulated the radioactive transmitter was released.

Studies using electron microscopy and histology clarified the cell structure of the sympathetic neurons. It consists of a cell body where the enzymes that make catecholamines are localized, a long axon, and highly branched nerve terminals. The nerve terminals contain thousands of swellings or varicosities. The enzymes that synthesize the neurotransmitters are translocated from the cell body to the varicosities in the nerve terminals by a process of axoplasmic transport. It is at the varicosities that the neurotransmitter is liberated. Catecholamine neurotransmitters are released by exocytosis, a process in which the storage vesicle fuses with the outer membrane of the varicosity and forms an opening through which not only the neurotransmitter is discharged into the synaptic cleft, but also other contents of the vesicle, including the enzyme dopamine-β-hydroxylase, are released. The neurotransmitter then binds to a specific recognition site on the surface of the effector cells in close juxtaposition to the nerve terminal. When the neurotransmitter interacts with the receptor, it sets off a series of complex chemical and physical changes in the membrane, permitting the cell to carry out its specific function. Most of the neurotransmitter is then rapidly inactivated by reuptake into the sympathetic neurons. Some of the catecholamines are metabolized by COMT and by MAO in the effector cell, and some are discharged into the circulation where they are physically removed and ultimately metabolized in the liver. A model for the fate of catecholamine neurotransmitters is depicted in Figure 2.

The initial work on the fate of catecholamines was obtained using peripheral tissues. The brain contains dopaminergic, noradrenergic and adrenergic neurons and it was of interest to study the metabolism and physiological disposition of catecholamines in the brain. Because catecholamines do not cross the blood brain barrier, information regarding the storage, release, and metabolism of these amines was obtained by intraventricular or intracisternal injections of [^{3}H] noradrenaline into the rat brain. After such injections of [^{3}H] noradrenaline, it was selectively taken up into nerve terminals and stored in dense core vesicles. Upon release, the noradrenaline was metabolized by O-methylation and deamination. Like the peripheral nerve, the catecholamines in the brain can be inactivated through reuptake by the nerve terminals.

EFFECT OF PSYCHOACTIVE DRUGS

Once the mechanism whereby catecholamines are released and inactivated was described, it was possible to understand the action of many drugs. Using radioactive noradrenaline we were able to show the effect of amphetamine and other sympathomimetic amines. Amphetamine was found to have two actions: it released catecholamines from their storage sites and blocked the reuptake of these amines (Table 1). This then produces an excessive amount of catecholamines to interact with their respective receptors. The psychotic action of amphetamine is considered to be due, to a great extent, to its ability to release catecholamines and to activation of postsynaptic receptors, mainly dopaminergic. Blocking the dopamine receptor by antipsychotic drugs such as chlorpromazine overcomes the behavioral effects of amphetamine. Cocaine, another psychoactive drug, acts by blocking the inactivation of noradrenaline by reuptake and causes an excessive release of this amine in the brain (Table 1). We have already observed that antidepressant drugs block the inactivation of catecholamines by reuptake in peripheral

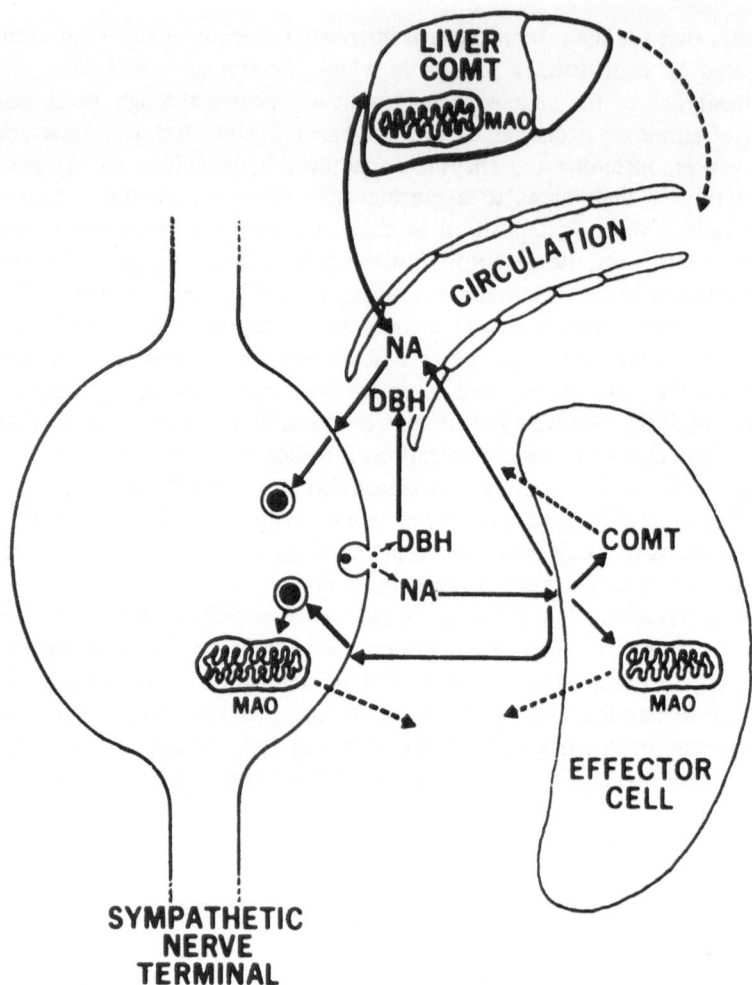

FIGURE 2. Fate of noradrenaline at the sympathetic nerve terminal.

nerves. A relationship between the clinical effectiveness of antidepressant drugs and their ability to block reuptake in brain noradrenergic nerves has been discovered.

Reserpine depletes catecholamines from brain aminergic neurons and sometimes causes clinical depression in patients who receive excessive amounts of the drug (Table 1). This observation, plus the fact that monoamine oxidase inhibitors could also be used to relieve affective disorders, led to the proposal of the catecholamine hypothesis of depression. This hypothesis states that mental depression is caused by a decreased availability of catecholamines in the brain. Affective disorders can be relieved by drugs that increase the amount of catecholamines

TABLE 1. Effect of Psychoactive Drugs on Catecholamines

Drug	Effect on catecholamines	Behavioral effect
Reserpine	Depletes	Causes depression
Amphetamine	Releases and blocks reuptake in nerves	Paranoid psychosis
Cocaine	Blocks reuptake, causes release in brain	Psychosis
Tricyclic antidepressants	Blocks reuptake	Relieves depression
Monoamine oxidase inhibitors	Blocks metabolism	Relieves depression
Antipsychotics	Blocks dopamine receptors	Relieves psychosis

available to these postsynaptic receptors. Although this hypothesis has been subjected to critical comments, it stimulated a considerable amount of productive research in biological psychiatry.

BIOSYNTHESIS OF CATECHOLAMINES AND THEIR REGULATION

Catecholamines are synthesized in the nerve tissue and adrenal medulla by four enzymes: tyrosine hydroxylase (TH), dopa decarboxylase (DDC), dopamine-β-hydroxylase (DBH) and phenylethanolamine N-methyltransferase (PNMT) as follows: tyrosine $\underrightarrow{TH}$ dopa $\underrightarrow{DDC}$ dopamine $\underrightarrow{DBH}$ noradrenaline $\underrightarrow{PNMT}$ adrenaline. Although catecholamines are in a state of flux, continuously being synthesized, released and metabolized, they maintain a constant level in nerves. This is due to a variety of adaptive mechanisms that mainly control the biosynthetic enzymes. They are rapid and slow regulating mechanisms for the control of catecholamine synthesis. Rapid regulation involves the rate-limiting enzyme in the biosynthesis, tyrosine hydroxylase, through a negative feedback mechanism. Low levels of catecholamines in the neurons, resulting from increased nerve activity, changes the kinetic properties of TH so that it more rapidly converts tyrosine to dopa. High levels of catecholamine inhibit this enzyme. The molecular mechanism for this regulation is due to phosphorylation and dephosphorylation of TH which changes the affinity of the enzyme for the substrate tyrosine and the cofactor tetrahydrobiopterin.

A slower adaptive process for catecholamine biosynthesis is caused by prolonged activity of sympathetic nerves and the adrenal medulla. Increasing activity of the adrenosympathetic system by drugs, stress or nerve stimulation causes a slow

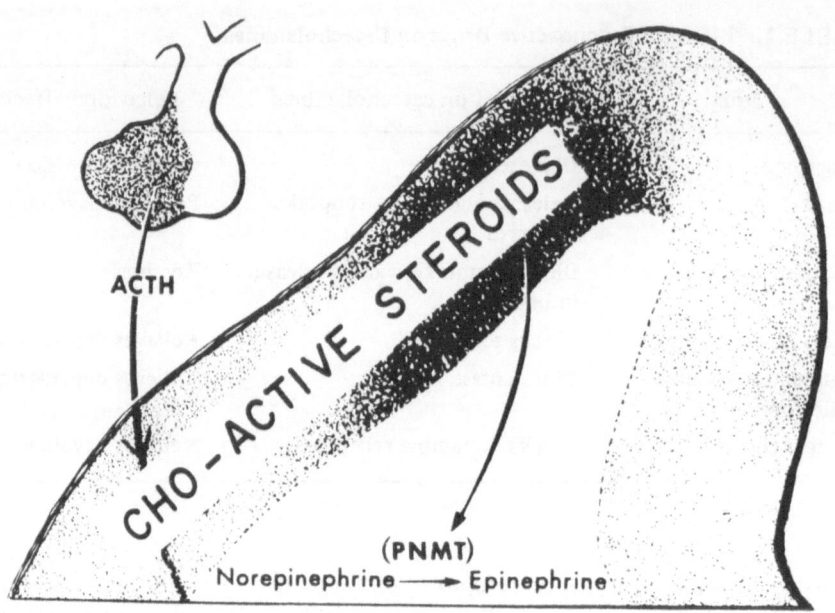

FIGURE 3.   Regulation of the adrenaline-forming enzyme PNMT by ACTH
and glucocorticoids.

but steady increase in the amount of TH, DBH, PNMT in sympathetic ganglion, cell bodies in brain, and adrenal gland. The release of these biosynthetic enzymes can be interrupted by a preganglionic denervation indicating that the induction of these enzymes is a transsynaptic event.

There is a relationship between catecholamine biosynthesis and the endocrine system. The synthesis of adrenaline from noradrenaline by PNMT in the adrenal medulla is regulated by ACTH and glucocorticoids (Figure 3). We have shown that the removal of the pituitary results in a profound reduction of PNMT activity in the adrenal gland of the rat. This activity is restored by injecting ACTH or gluco-corticoids. Thus, high levels of glucocorticoids arising from the adrenal cortex are necessary to maintain the adrenaline-forming enzyme, PNMT.

STRESS AND CATECHOLAMINES

The role of the adrenosympathetic system in stress was recognized by Walter Cannon many years ago. Experimental stress in animals caused by immobilization, cold, electroshock, insulin, etc., causes a steady increase in tyrosine hydroxylase, dopamine hydroxylase and PNMT in the adrenal gland and nerves. This increase is due to a transsynaptic induction of these enzymes. Psychosocial stress in mice also elevates the activity of the catecholamine biosynthetic enzyme.

During the past few years, extremely sensitive radioenzymatic methods for the accurate measurement of noradrenaline and adrenaline in plasma of animals and humans were developed. This was made possible by the use of the enzymes COMT and PMNT which transfer a highly radioactive [^{3}H] methyl group from S-adenosyl-methionine to catecholamines. Recently there have been many studies examining the effect of mild and severe stress on the adrenal medulla and/or sympathetic nerves by measuring the plasma adrenaline and noradrenaline levels. Mild stress such as mental arithmetic, public speaking, and smoking causes an elevation of adrenaline and noradrenaline. Severe stress such as cold, hyperglycemic hemorrhage, and surgery results in large increases in plasma catecholamine levels.

CLINICAL IMPLICATIONS OF CATECHOLAMINES

The discoveries concerning catecholamines in the past decade have had a marked influence on our understanding and treatment of cardiovascular, neurological, and mental diseases. One of the more important developments in medicine was the introduction of effective drugs for the treatment of cardiovascular diseases such as hypertension. Many of the drugs used for the treatment of hypertension (reserpine, guanethidine, α-methyl dopa) affect the storage, release, or metabolism of catecholamines. The involvement of the sympathetic nervous system in cardiovascular diseases was suspected for many years. In a study using rats in which hypertension was induced experimentally by deoxycorticosterone and NaCl, we found that the turnover of catecholamines was markedly increased in the sympathetic nerves of the heart. A direct relationship between the degree of increase in catecholamine turnover and the extent of elevation of blood pressure was found. Removal of salt from the diet and the administration of natruretic drugs reduced blood pressure in hypertensive rats and brought the catecholamine turnover in the heart back to normal values. After further work it became apparent that hypertension was caused by disturbances of adrenergic nerves in the central nervous system. Using immunological and microdissection techniques, it was found that there was an increased activity of these adrenaline-containing nerves in the brainstem in experimental hypertension. The administration of drugs that block PNMT, the adrenaline-forming enzyme, reduced the blood pressure in rats with experimental hypertension.

The introduction of dopa for the treatment of Parkinsonian disease by Hornykeiwicz and Cotzias was a direct outgrowth of basic catecholamine research. It was observed that the depletion of the catecholamines dopamine and noradrenaline in the brain, by reserpine, causes Parkinson-like symptoms in rats. To increase the catecholamine levels in the brain, dopa was administered. Catecholamines do not cross the blood brain barrier but dopa does. Dopa is converted to dopamine in the brain and also overcomes the Parkinson-like symptoms in re-serpine-treated rats. This prompted an examination of the dopamine concentration in brains of patients who died of Parkinson's disease. A marked deficiency in this amine was found in the corpus striatum of these subjects. Dopa was then admin-istered to patients with the disease with remarkable clinical effectiveness.

The possible role of catecholamines in affective disorders was noted above. For many years, dopamine was suspected to be involved in schizophrenia. Dopa-mine receptors were found to be present in the brain. It was soon found that anti-psychotic drugs blocked these receptors (Table 1). Further work then showed that antipsychotic drugs bind to dopamine receptors in the brain with an affinity related to their potency as therapeutic agents for schizophrenia. Disturbances in catechol-aminergic nerves were also found in neurological diseases such as dystonia and dysautonomia.

SYMPATHETIC NERVES, THE PINEAL GLAND AND CIRCADIAN RHYTHMS

My interest in the pineal gland stemmed from the discovery of melatonin (5-methoxy-N-acetylserotonin) in the gland by A. Lerner some 20 years ago. Melatonin is related in structure to the neurotransmitter serotonin and it has an O-methyl group. Because of the chemical structure of melatonin and the observa-tion that the pineal, an organ of unknown function, is innervated by noradrenaline-containing nerves, we were prompted to study how this compound was synthesized and regulated. We soon found the enzyme that makes melatonin and elucidated the biosynthetic pathway from tryptophan as follows: tryptophan → serotonin → N-acetylserotonin → melatonin. The melatonin-forming enzyme hydroxyindole-O-methyltransferase (HIOMT) was found to be almost exclusively localized in the pineal gland.

Circadian rhythms of serotonin, N-acetylserotonin and melatonin were found in the pineal gland. Serotonin levels were highest during the daytime and N-acetyl-serotonin and melatonin were highest at night. This prompted a series of experi-ments in our laboratories and elsewhere to find out how these 24-hour cycles were generated. Denervating the sympathetic nerves leading to the pineal abolished these rhythms. Preganglionic denervation and lesions in the suprachiasmatic nucleus in the hypothalamus also abolished the serotonin, N-acetylserotonin and melatonin rhythms. All of these observations indicated that the 24-hour rhytm of indole-amine in the pineal arises from the suprachiasmatic nucleus. This nucleus sends its diurnal message via the sympathetic nerves to the pineal gland. Sympathetic nerves appeared to release more noradrenaline at night than during the daytime.

The rhythmic release of noradrenaline at night results in an increase of N-acetylserotonin in the pineal. This provided an opportunity to study adrenergic receptor response and its regulation. The pineal cell was found to have β-adren-ergic receptors, which upon stimulation with noradrenaline released from nerves

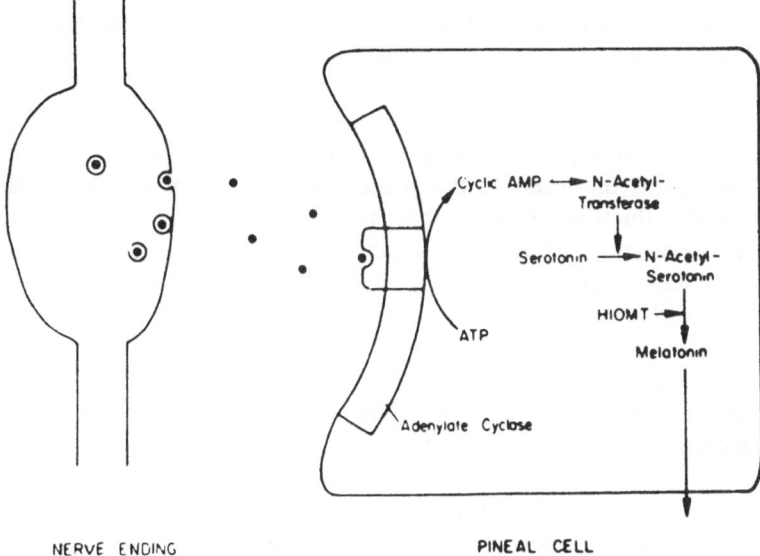

FIGURE 4. Action of noradrenaline (0) in the synthesis of melatonin
in the pineal cells.

generated cyclic AMP from adenylate cyclase, which then caused a marked increase
of N-acetylserotonin, the precursor to melatonin (Figure 4). When rats were left in
constant light there was a diminished release of noradrenaline. This resulted in an
increase in the responsiveness of the β-adrenergic receptors. With the increased
nerve-firing of the pineal at night, β-adrenergic receptors became desensitized.
Changes in sensitivity of β-adrenergic receptors, as well as release of noradrenaline
produced by daylight and darkness, provided an amplification and dampening
system whereby small changes in the release of catecholamines produced large
increases and decreases in the melatonin levels in the pineal.

Recently, accurate methods for the measurement of melatonin in human
plasma have been devised. Humans, like animals, appear to have circadian rhythms
of melatonin which are regulated by changes in sensitivity of β-adrenergic receptors.
Plasma melatonin levels in man should provide insight regarding the responsiveness
of β-adrenergic receptors in normal and disease states.

LIPIDS AND THE TRANSMISSION OF CATECHOLAMINE
NEUROTRANSMITTER SIGNALS

In the past three years the research of my laboratory has been directed
towards the biochemical mechanisms whereby catecholamines and other neuro-

transmitters transfer their messages through membranes. The transduction of neurotransmitter signals through the cell membranes permits cells to carry out their specific function. Cell membranes consist of a lipid bilayer in which neurotransmitter receptor molecules are imbedded. Receptor molecules face the outside of the membrane while adenylate cyclase, an enzyme that generates the second messager, cyclic AMP, faces the inside of the membrane. Cyclic AMP is a key compound which, through the transfer of the phosphate group, turns cells on or off. We have found two enzymes in cell membranes that methylate phosphatidylethanolamine to form phosphatidylcholine. In the process of methylation these lipids are translocated from the inside to the outside of the membrane and also increase the fluidity of the membrane.

The interaction of catecholamines with β-adrenergic receptors stimulates phospholipid methylation and increases membrane fluidity. This then enhances the lateral mobility of the receptor on the cell surface and offers a greater chance to couple with adenylate cyclase facing the inside of the membrane. This then generates more cyclic AMP which then activates the cell to carry out its function. Our work on phospholipid methylation has expanded and led to other exciting areas of research.

REFERENCES

1. Axelrod, J. Noradrenaline: Fate and control of its biosynthesis. *Les Prix Nobel*. Imprimerieal Royal P. A. Norstedt & Soner, Stockholm, 1971, pp. 189–208; *Science 173*:598–606, 1971.
2. Axelrod, J. Neurotransmitters. *Scientific American 230*:59–71, 1974.
3. Axelrod, J. The pineal gland: A neurochemical transducer. *Science 184*:1341–1348, 1974.
4. Axelrod, J. Biochemical and pharmacological approaches in the study of sympathetic nerves. In: *The neurosciences: Paths of discovery*, edited by F. G. Worden, J. P. Swazey, and G. Adelman. Cambridge, MA: The MIT Press, 1975, pp. 191–208.
5. Glowinski, J. and Axelrod, J. Effects of drugs on the disposition of H^3-norepinephrine in the rat brain. *Pharmacol. Rev. 18*:775–785, 1966.
6. Hirata, F. and Axelrod, J. Phospholipid methylation and biological signal transmission. *Science 209*:1082–1090, 1980.

The Dopamine Hypothesis: Variations on a Theme

Richard Jed Wyatt, M.D.

The dopamine hypothesis of schizophrenia, like many neurochemical notions of psychiatric disorders, has evolved largely out of beliefs and increasing knowledge of how the medications we use to treat psychiatric disorders work. In the dopamine hypothesis' simplest form, it is deduced that because antipsychotic or neuroleptic medications block dopamine function [12,64], the schizophrenic patient has a functional excess of brain dopamine. Generally, it is thought that this functional excess of dopamine is located in the limbic system. During recent years this hypothesis has been bolstered by some evidence of postsynaptic excess—namely, an increased number of brain dopamine receptors. A characterization of what this basic alteration may produce, however, is not usually attempted.

Here, I examine the role of dopamine in the normal brain and review the data supporting the current dopamine hypothesis of schizophrenia. Also, I describe an alternative hypothesis suggesting that some schizophrenic patients have a dopaminergic system that is uninvolved in the process, or has diminished function.

DOPAMINE STRUCTURE AND METABOLISM

Dopamine is the simplest of catecholamines; catecholamines have as their nucleus dihydroxyphenylethylamine. The phenylethylamines are important in their

Research in the Schizophrenic Disorders: The Stanley R. Dean Award Lectures, vol. 2, edited by R. Cancro and S. R. Dean. Copyright © 1985 by Spectrum Publications.

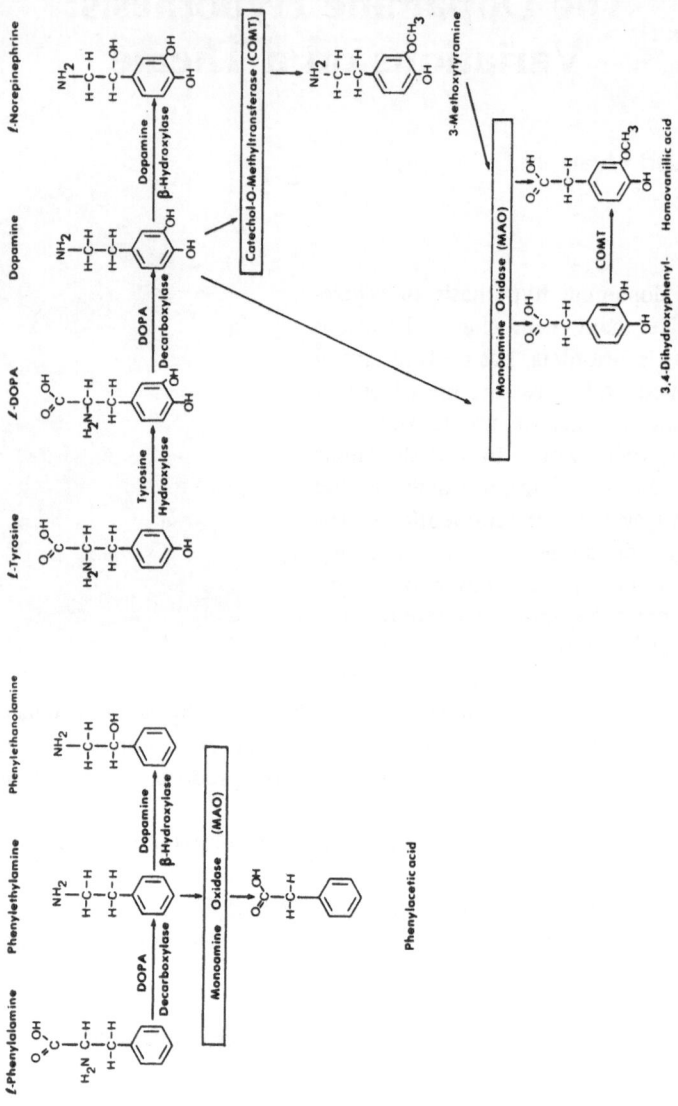

FIGURE 1. Dopamine metabolism. Figure demonstrates similarities between dopamine metabolism and phenylethylamine metabolism.

own right because they are endogenous substances, as well as the backbone of numerous drugs such as amphetamine [75]. As seen in Figure 1, the metabolism of the catecholamines and phenylethylamines is very similar.

Dopamine is synthesized from tyrosine, first with addition of a second hydroxyl group by tyrosine hydroxylase. L-dopa, the product, is then de-carboxylated to dopamine by L-aromatic amino acid decarboxylase. Within the neuron, dopamine can be converted to the other catecholamines by dopamine-beta-hydroxylase (DBH) or metabolized to an inactive product which, in turn, is removed from the central nervous system. In the latter instance, the final product can be 3-methoxytryramine, dihydroxyphenylacetic acid (DOPAC) or homovanillic acid (HVA). Homovanillic acid, however, is the most abundant end product.

DOPAMINE IN THE NORMAL BRAIN (Figure 2)

The monoamine pathways in the human brain, for the most part, have not been well described. Our knowledge about them comes mainly from histofluores-cent studies in animals, particularly the rat [16,28]. Implementation of the histo-fluorescence technique is one of many revolutions that has taken place in the neurosciences during the last 30 years. Using the tendency of the monoamines to fluoresce, the neuroanatomist can see the cell bodies and is able to trace the mono-amine-containing fibers throughout the brain. For the first time, neuroscientists know that neuronal pathways not only have a beginning and an end, but are asso-ciated with a specific neurochemical.

Monoamine neuron cells bodies are located in small, relatively compact nuclei. In particular, dopamine neuron cell bodies are found primarily in the sub-stantia nigra within the mesencephalon. These neurons and their outflow fibers form the nigro striatal tract.

Slightly more anterior and medial are other dopamine cell bodies that pro-ject to the limbic system and the forebrain. These cells and their tracts are called the mesolimbic and mesocortico systems. More anterior yet, are dopaminergic neurons of the arcuate nucleus that form the tuberoinfundibular tract providing dopaminergic innervation to median eminence and pituitary gland. Recently [33], it has been demonstrated that some dopamine-containing neurons project to the spinal cord. Finally, still within the brain, are short dopaminergic neurons within the retina.

The arcuate and the retinal dopaminergic neurons are particularly important for studies of schizophrenia since they can both be studied by relatively noninvasive techniques. Dopamine in the tuberoinfundibular tract, in part, controls prolactin and growth hormone secretion and has been extensively studied in schizophrenia. Retinal dopaminergic function, which can be examined with the electroretinogram, has not been extensively studied in schizophrenia. [48]

A very important characteristic of dopamine neurons is their extensive branching. In 1966, Andén and colleagues [2] estimated that a single dopaminergic

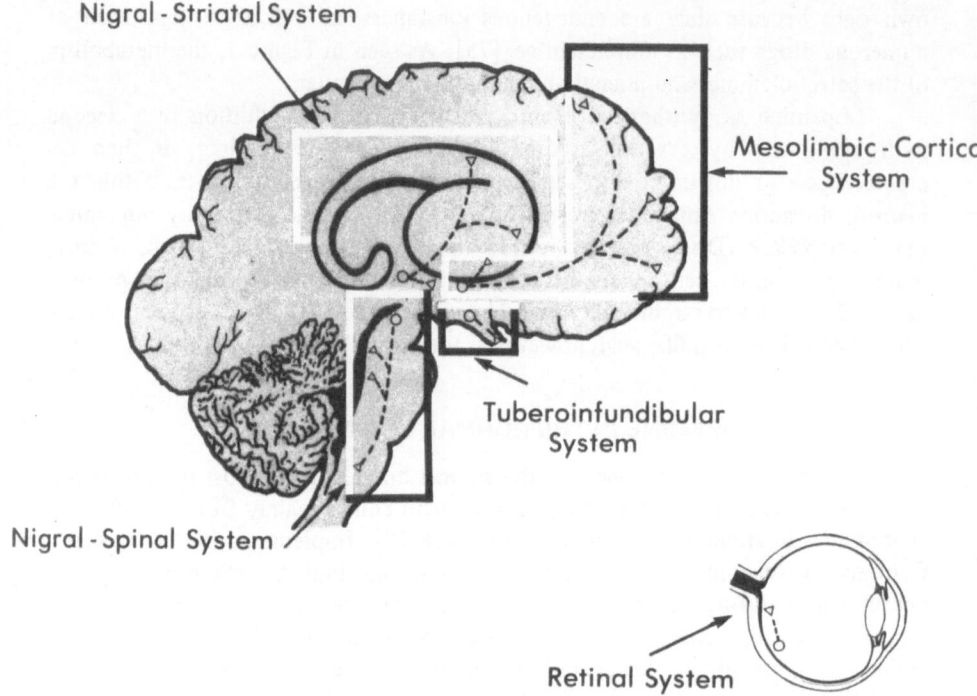

FIGURE 2. Dopamine pathways in the brain.

neuron in the rat substantia nigra gives rise to one half million synapses within the caudate and putamen. Within each varicosity are hundreds to several thousand dopamine-containing vessels capable of releasing dopamine onto the dopamine receptors. This wide distribution of influence, coming from only a few neurons buried deep in the brain, suggests that dopamine has a relatively nonspecific but important function.

DOPAMINE RECEPTORS

If dopamine is in the brain to modulate tone of particular brain functions, there must be receiving antennae or receptors to pick up its message. In fact, there appear to be more than one type of receptor. Given, for the most part, that the way these receptors have been studied, has been by determining the drug binding characteristics, they may more properly be called binding sites. The term receptor implies a clearer demonstration of physiological function than has been shown by

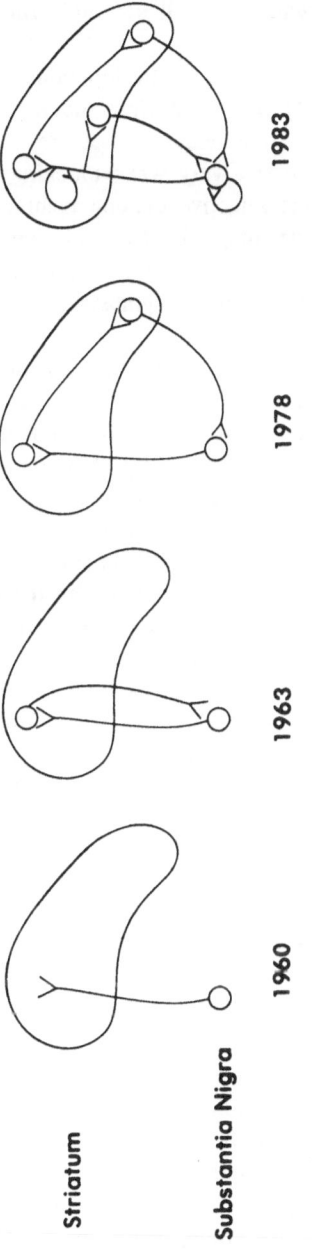

Striatum

Substantia Nigra

1960   1963   1978   1983

Nigral - Striatal System

FIGURE 3. Diagram of changing view over time of nigral-striatal system, with increasing complex of feedback loop from striatum.

most studies. Nevertheless, for convenience I will use the terms binding site and receptor interchangeably.

The type of binding site that is coupled with dopamine-sensitive adenylate cyclase has been designated D-1 [34]. Phenothiazines block the ability of dopamine to stimulate dopamine-sensitive adenylate cyclase. In contrast, the more potent butyrophenone neuroleptics have a very weak blocking affect. A second type of receptor, D-2, is labeled using selective binding techniques with butyrophenones as well as the other dopamine antagonists (or neuroleptics) in close correlation with their relative clinical potencies. There may be a third type of receptor. If it exists, the D-3 receptor may bind dopamine agonists, such as dopamine itself and apomorphine.

Much effort has gone into determining where on the neuron or other brain elements these receptors are located. In the striatum, for instance, there are receptors located on the axons projecting from other brain regions, such as the cortex. There are receptors on neurons intrinsic to the striatum and a receptor on the dopamine terminal that autoregulates the dopamine neurons' firing rate.

Finally, within the striatum itself, dopamine dendrites release dopamine that impinges directly on the dopamine neuron and, presumably, there is a dopamine receptor located there. Unfortunately, the location of the receptor must be inferred because of our current inability to actually see the receptor [13].

Another approach to defining dopamine receptors has been to separate them from other brain tissues by cellular fractionation techniques [26]. These studies have demonstrated that there is an astroglial adenylate cyclase that can be enriched by the cellular fractionation in parallel to enrichment of a neuroleptic receptor. These studies indicate, further, that adenylate cyclase is inhibited in a manner that correlates with neuroleptic clinical potency. Astroglial adenylate cyclase is present in the caudate but not in the neocortex.

THE HISTORICAL ROLE OF DOPAMINE IN SCHIZOPHRENIA

Amphetamine and Amphetamine-Like Stimulants

In 1938, Young and Scoville [81] noted that amphetamine in nonschizophrenic patients could produce a paranoid-like psychosis that to many observers was virtually identical to paranoid schizophrenia. This observation has been substantiated many times, both in clinical practice and in experimental situations [71]. Amphetamine is an important part of the dopamine hypothesis because amphetamine is known to release brain dopamine, producing a functional excess of dopamine at the receptor site [57].

While it is clear that amphetamine is capable of producing a paranoid psychosis, true paranoid schizophrenia is a relatively uncommon syndrome occurring somewhat later in life than other forms of schizophrenia [62]. Most young schizophrenic patients with paranoid features have a substantial nonparanoid disturbance as well. Also, although amphetamine does cause dopamine release at low amphetamine doses, the high, sustained doses required to produce a paranoid psychosis suggest that another process is required for the production of psychosis [76].

Neuroleptic Blockade of Dopamine Receptors

Although the neuroleptics became available for clinical use in the 1950s, it was not until 1963 that Carlsson and Lindqvist [12] ingeniously demonstrated that dopamine might be involved in the mechanism of action of these agents. Carlsson and Lindqvist administered chlorpromazine, haloperidol (antipsychotics of two different classes—phenothiazine and butyrophenone), promethazine (a phenothiazine without antipsychotic effects), and phenoxygenazamine (an alpha-noradrenergic blocking agent) to mice. They then measured brain norepinephrine and the norepinephrine metabolite, normetanephrine, dopamine and its metabolite, 3-methoxytyramine. As can be seen from Table 1, the change found unique to the antipsychotics was their ability to raise the dopamine metabolite 3-methoxytyramine. The investigators concluded that the antipsychotics block dopamine receptors and that the increase in 3-methoxytyramine was due to a feedback loop that in turn produced increased dopamine turnover.

Dopamine-Related Enzymes

A number of attempts have been made to measure the enzymes that form and degrade dopamine in schizophrenic patients and compare them with controls (see Figure 1).

Enzymes of Dopamine Formation

Wyatt et al. [73] and Crow et al. [15] found that tyrosine hydroxylase activity was normal in several different areas of the brains of schizophrenic patients. While one group of investigators [61] found that erythrocyte decarboxylation of 14-C-labeled-dopa was greater in schizophrenics with thought disorder and hallucinations compared with schizophrenic patients in remission and normal controls, Wyatt et al. [73] found that direct measurement of dopa decarboxylase in the brains of autopsied schizophrenic patients was not different than controls.

TABLE 1 Study of Carlsson and Lindqvist [12] Showing Antipsychotics, as Opposed to Nonantipsychotics, to Be Associated with an Elevation of the Mouse Brain Metabolite of Dopamine, 3-Methoxytyrosine

Drug	Catecholamines		Catecholamine metabolites	
	Norepinephrine	Dopamine	Normetanephrine	3-Methoxytyramine
Antipsychotics				
Chlorpromazine	100	116	325	224
Haloperidol	91	93	200	229
Nonantipsychotics				
Promethazine	141	131 ·	100	73
Phenoxybenzamine	104	105	200	88

Values are expressed as a percent of predrug baseline.

Enzymes of Dopamine Catabolism

Because of their interest in norepinephrine and its possible relationship to anhedonia, Wise, Baden, and Stein [70] predicted and found a deficit of dopamine-beta-hydroxylase in the brains of autopsied schizophrenics compared with controls. Wyatt et al. [79] and Cross et al. (1978) found no such deficit. Wyatt et al. [73] did find a negative correlation between dopamine-beta-hydroxylase activity and morgue-to-autopsy time as well as neuroleptic dose. It was felt that this relationship might explain at least some of the deficit found by Wise, Baden, and Stein [70].

In addition to measuring dopamine-beta-hydroxylase in the brain, a number of investigators have measured it in both plasma and cerebrospinal fluid. While some investigators [7,24] have found dopamine-beta-hydroxylase in plasma to be low in schizophrenic patients, the majority of studies [19] have not found a decrease. Similarly, Lerner et al. [41] reported no significant differences in cerebrospinal fluid dopamine-beta-hydroxylase between schizophrenic and psychiatric patients with other disorders.

Normally, dopamine-beta-hydroxylase activity is measured with substances that inhibit an endogenous dopamine-beta-hydroxylase inactivator. When Yu et al. [82] measured plasma dopamine-beta-hydroxylase activity in the usual manner, they found no differences between schizophrenic patients and controls. When, however, they omitted the substances that inhibit the dopamine-beta-hydroxylase inactivator, they found that schizophrenic patients had a significantly lower enzyme activity than did controls. Nevertheless, when Wyatt et al. [73] measured dopamine-beta-hydroxylase inhibitory activity in the postmortem brains of schizophrenic patients, they found no difference between experimental groups and controls.

The story of monoamine oxidase activity in schizophrenia is long and complicated. In 1941, Birkhauser [5] first described decreased monoamine oxidase in brains of autopsied schizophrenic patients. While numerous subsequent studies have not found decreased monoamine oxidase activity in the brains from autopsied schizophrenic patients, decreased monoamine oxidase activity has been found in the platelet, lymphocyte, and muscle. In 1982, DeLisi et al. [17] reviewed all the studies of peripheral monoamine oxidase activity published through early 1981. They concluded that there had been 38 independent studies of monoamine oxidase in platelets and six in lymphocytes, skeletal muscle or skin fibroblasts. Of these studies, 31 found a significant decrease in enzyme activity between schizophrenic patients and controls. Of the 13 reporting no significant decrease six had a trend towards a decrease.

These numbers leave little doubt that there is a group of schizophrenic patients that have decreased peripheral monoamine oxidase activity. The relatively small decreases (the greatest reported mean decrease from normal is 68 per cent) [43] that are present, however, make it seem unlikely that the alteration is of direct physiological importance. Furthermore, although early reports [77] indicated that neuroleptics had little or no effect on monoamine oxidase activity, a number of subsequent studies have indicated that platelet monoamine oxidase activity is significantly decreased by neuroleptics. For example, DeLisi et al. [18] demonstrated that there was a 23 per cent decline in platelet monoamine oxidase activity over a four-month period when schizophrenic patients who had been off medications were given neuroleptics.

Dopamine Concentrations

Postmortem studies of brain dopamine concentrations in schizophrenic patients have produced mixed results. Owen et al. [46], using a radiochemical, found increased dopamine in the caudate but not in the nucleus accumbens. Mackay et al. [42], using a radioenzymatic assay found an increase in both brain areas, particularly in early onset schizophrenics (between age 15 to 24). On the other hand, using more sensitive and specific techniques Kleinman et al. [38] found no alteration in dopamine concentrations in the nucleus accumbens.

Dopamine Metabolities

Studies of dopamine metabolite concentrations in the brain, cerebrospinal fluid and urine of schizophrenic patients have also produced mixed results. On the whole, the results are not impressive regarding either an increase or decrease of dopamine-related substances [25]. One recent study from Moscow [1] indicates that schizophrenic patients off drugs had only nine per cent of the dihydroxyphenylacetic acid in their urine as did normals. Dopamine and homovanillic acid concentrations were normal in these patients.

Dopamine Receptors

Studies measuring dopamine receptors in postmortem brain using labeled spiperone and other dopamine antagonists [40] have found increased numbers of binding sites in the basal ganglia and nucleus accumbens as well as an increased dissociation constant (decreased affinity). Similar findings have been observed by a number of investigators [38]. Because binding site number [10,53] has been demonstrated to increase in animals given neuroleptics for relatively short periods (one to three weeks), it is important to determine if the changes found in postmortem schizophrenic brains is due to previous neuroleptic administration. Owen et al. [46] and Seeman [40] noted an increased number of spiperone binding sites in a small group of schizophrenic patients who had not received neuroleptic medication for some time prior to death. The increase in binding sites, however, was much smaller than in patients known to have received neuroleptics.

More recently, Mackay et al. [42] indicated that the results from receptor binding studies in postmortem schizophrenic brains was probably due to the presence of neuroleptics since patients who had not been on neuroleptics the month before death did not have the receptor changes. Those patients on neuroleptics had increased receptor number.

Morphological Changes and Dopamine

For well over 50 years researchers [27] have described large ventricles in the brains of schizophrenic patients [68]. In fact, in Jacobi and Winkler's study [27], 18 of 19 chronic schizophrenic patients were reported to have hydrocephalus. With the introduction of computed tomography (CT) into psychiatric research [32] large cerebral ventricles have been found in 15 controlled studies [67] with only three exceptions [4,30,44]. While the issue of whether or not the large ventricles are secondary to some aspect of being schizophrenic, such as medication treatment or being in the hospital, has not been resolved [59], the evidence that is available suggests that ventricular enlargement occurred before these factors could become important. For example, there are now three studies [45,52,66] that have found that some first-break schizophrenic patients have large ventricles.

Crow [14] has suggested that schizophrenia be divided into two syndromes. Type I, which has no cellular loss or large ventricles is associated with the symptoms of hallucinations, delusions and thought disorder (positive symptoms). Type I schizophrenia tends to be acute, have a good response to neuroleptics and no intellectual impairment, be reversible and related to an increased number of dopamine receptors.

Type II schizophrenia has cellular loss, larged ventricles, affective flattening, poverty of speech, negative symptoms and loss of drive. The patients tend to be more chronic, have a poorer response to medications, have intellectual impairment and irreversible outcome. According to Crow, the two syndromes represent two or more diseases that at times are temporarily overlapping and at other times are quite separate. Thus, a patient may either start as a Type II, such as simple schizophrenia, or move from Type I to Type II. A patient who at one time has a diagnosis of schizophreniform psychosis but who has progressed to Bleuerian hebephrenia would be an example of movement from Type I to Type II. A patient may also stay a Type I, as in paranoid schizophrenia. It should be noted that the Type I–II distinction is not substantially different from previous subgroupings, such as the process-reactive designation. While Type I patients are thought to have a functional excess of dopamine occurring through increased dopamine receptors (increased dopamine receptor number may be secondary to neuroleptic treatment), it is possible that some patients with defect-state schizophrenia (and morphological abnormalities) may behave as if they have a faulty dopaminergic system or too little dopamine in some parts of the brain. The addition of the correlations with structural abnormalities gives this distinction a fresh appearance, and perhaps serves to increase scientific leverage.

Negative Symptoms

Clinicians have struggled with the problem of how to divide schizophrenic patients into meaningful subgroups using behavioral or phenomenological descriptors. With few exceptions these attempts have met with relatively meager success. One approach that makes intuitive sense and is used by both clinicians and researchers is to divide patients into categories of positive (delusions, hallucinations, formal thought disorder, and bizarre behavior) and negative or defect-state symptoms (affective flattening, alogia, avolition, impaired attention and impoverished thinking). Hughlings–Jackson (1931) made this division concerning a diagnosis of insanity during the last century. Nineteenth century insanity, at least as Hughlings–Jackson used the term, referred to clear neurological conditions such as the post-epileptic period.

A number of writers have observed that schizophrenia evolves over many years and that patients change with time. Certainly some patient's symptoms, which early in the course of the illness call out for attention, decrease with time. For example, Bridge et al. [9] summarized several studies that showed delusions and hallucinations as more prominent during the early years of schizophrenia than subsequently. Pfohl and Winokur [47] retrospectively followed 52 hebephrenic/catatonic schizophrenic patients who required long-term hospitalization. Hallucinations and delusions tended to have an early onset and disappear over time. Flat affect seemed to develop late and persisted.

In 1969, Cancro [11] suggested that disorders of object relations, affect, and thought disorder were part of the same process because they were highly correlated in schizophrenia. Strauss et al. [56] discussed positive and negative symptoms, as well as disorders of personal relationships, as being associated with prognosis. Their conclusion was that positive symptoms had relatively little prognostic value while negative symptoms generally predicted a poor prognosis.

More recently, Andreason et al. [3] have further defined and correlated these symptoms with ventricular size. As Johnstone et al. [32] showed in their initial study, Andreasen et al. found that patients with large ventricles tended to have more negative symptoms while patients with normal size ventricles tended to have positive symptoms. Affective flattening and formal thought disorder were prominent about equally in both groups. On a similar note Takahashi et al. [58] found a relationship between brain atrophy and "non-productive symptoms."

EVIDENCE THAT FOR SOME SCHIZOPHRENIC PATIENTS THE DOPAMINE SYSTEM IS EITHER UNINVOLVED WITH THE PROCESS, OR IS DEFICIENT

In this section, I review some evidence not systematically collected, some schizophrenic behavior is consistent with normal of hypoactivity of the dopamine system. As will be seen, the animal research and much of the human data present arguments by analogy. The animal work, involving destroying the dopamine system, produces a behavioral syndrome somewhat analogous to the multirational syndrome of certain schizophrenic patients. The studies showing that some schizophrenic patients do not have a correlation of behavior with prolactin secretion, do not respond to apomorphine, or do not respond to neuroleptic medications are consistent with either of the above possibilities. The findings that neuroleptics do not decrease blinks in some schizophrenic patients and that cerebrospinal fluid homovanillic acid is decreased are more consistent with the dopamine system being hypoactive.

Animal Studies

Neuroleptic Treated Rats

Wise [69], studying rats that were given neuroleptics (and consequently had decreased functional dopamine) found that there is, ". . . a selective attenuation of multivational arousal which is (a) critical for goal-directed behavior, (b) normally induced by reinforcers and associated environmental stimuli, and (c) normally accompanied by the subjective experience of pleasure . . ." (p. 39). Wise likens this antidopaminergic or neuroleptic-induced behavior to human anhedonia.

Dopamine Depleted Rats

Rats given the neurotoxin, 6-hydroxydopamine (6-OHDA), lose dopamine containing neurons. When the lesion is made unilaterally and dopamine-containing neurons are lost on only one side of the brain, rats lose their ability to orient toward sensory stimuli on the side of the body opposite the lesion. This condition is called sensory neglect and is attributed to a failure in the sensory motor integrating mechanism [54]. Recently, sensory neglect has been corrected by grafting embryo substantia nigra to the lateral part of the striatum [21].

Mesolimbic Cortical Dopamine [23]

In rats, destruction of the mesolimbic cortical dopamine projection system, sparing the striatal dopamine system, produces deficits in exploratory behavior to a novel object. Apomorphine, but not the norepinephrine agonist clonidine, restores the exploratory behavior. The apomorphine restoration of exploration is blocked by the dopamine antagonist pimozide. Fink and Smith intrepret their data to mean that mesolimbic dopamine neurons provide arousal mechanism for integration of normal behavior.

Human Studies

Parkinson's Disease

While Parkinson's disease is primarily a disorder of motor function, there also can be a constellation of defect-state symptoms. Progressive mild intellectual impairment may develop. There may be apathy or depression, lack of spontaneous speech, and problems with concept formation [29]. Of considerable interest is that Parkinson's disease is at least in part due to a brain dopamine deficit and is corrected by agents that increase brain dopamine, such as L-dopa. On the other hand, too vigorous treatment with L-dopa or other dopamine-like agents produces a psychosis with positive symptoms.

Economo's Encephalitis or Encephalitis Lethargica

Economo described a striking hypokinetic state that he called "psychic torpor" [22]. Economo found that such patients remained by themselves for hours without suffering boredom. In a 1921 report (The Association for Research in Nervous and Mental Diseases), Hohman described the stuporous reactions of encephalitis lethargica that were often mistaken for schizophrenia, the major difference being the schizophrenic patients' ability to use their voluntary nervous system.

In that same report, Grossman described the preference of some patients with encephalitis lethargica to be left alone. As late sequelae, the patients became restless, unable to concentrate, and had little interest in their surroundings. The Diagnostic and Statistical Manual of Mental Disorders II (1968) described some patients who survived the disease as having, ". . . apparent indifference to persons and events ordinarily of emotional significance, such as the death of a family member . . ." (p. 27). Encephalitis lethargica is thought to have led to subsequent loss of dopamine neurons and Parkinson's disease. It is unclear how much of the symptoms described above were due to dopaminergic involvement, rather than alterations of other brain structures.

Akinesia

Akinesia can be a side effect of neuroleptic (dopamine blocking) treatment. While it is usually thought to consist of decreased gestures, shortened stride and rigid posture; it can also involve non-motoric behavior [35]. There may be apathy, lack of goal-directedness, and lack of spontaneous speech. These symptoms can be corrected by lowering the neuroleptic dose or by using antiparkinsonian medications.

Prolactin Secretion

Prolactin secretion is in part under tuberoinfundibular dopamine control. Increased dopamine activity decreases prolactin release while decreased dopaminergic function increases prolactin release. While a number of investigators have examined resting prolactin concentration as a window into the brain and found no abnormalities, a study by Kleinman et al. [38] recently divided a sample of unmedcated chronic schizophrenic patients into those with normal and large ventricles. The resting prolactin concentrations were found to be inversely correlated with the thought disorders syndrome on the Brief Psychiatric Rating Scale (BPRS). This inverse correlation of the prolactin and thought disorder rating is consistent with the conventional dopamine hypothesis. The patients with large ventricles, however, had no significant correlation between their resting serum prolactin concentrations and thought disorder. The absence of this correlation in the large ventricle patients makes it seem less likely that there is a hyperdopaminergic syndrome in these patients.

Apomorphine Response

In small doses the dopamine agonist, apomorphine, should act presynaptically and decrease dopamine neuronal firing. Thus, it is not suprising to find reports of apomorphine making schizophrenic patients better, worse or unchanged [25]. In one preliminary study [31], a small dose of apomorphine was administered to ten chronic schizophrenic patients. The ten patients also had CT scans. None of the six patients with abnormal CT scans became worse, while three of the patients with normal scans became worse. While it is not possible to determine which of these responses is normal, the Jeste et al. [31] study does suggest that increasing dopamine activity aggravates psychotic activity only in those patients with normal ventricles.

Clinical Response to Neuroleptics

There have been a small number of studies in which response to neuroleptics has been correlated with ventricular size. In the first such study, Weinberger et al. [65] found ten chronic schizophrenic patients with large ventricles and ten similar patients with normal size ventricles. Both groups of patients were admitted to the study because they had previously failed to respond to conventional treatments. Brief Psychiatric Ratings Scales (BPRS) were filled out by nurses at the end of a four week drug-free period and at the end of an eight week neuroleptic treatment period. The ten patients with normal size ventricles had almost 28 units of improvement on the BPRS, while the patients with large ventricles had no change.

In a follow-up to their 1980 study, Weinberger et al. [67] studied 33 patients. Seven of these patients could not be maintained off medication because of behavior regression. These seven patients had smaller ventricles than those patients able to complete the drug-free period. One interpretation of these results might be that neuroleptics are more efficacious in treating normal ventricle patients than large ventricle patients. Large ventricle patients appear not to be responsive to dopamine blocking drugs. Schultz et al. [52] studied 12 schizophrenic or schizophreniform patients. There was a near significant correlation between the ventricular size and BPRS change score for neuroleptic treatment. The normal ventricular size group had 17 units of BPRS improvement while large ventricle patients had no change on neuroleptics (p = .05). Smith et al. [56], found patients with smaller ventricles responded better on the BPRS than larger ventricle patients to three weeks of fixed dose neuroleptic treatment.

Blink Response to Neuroleptics

Spontaneous blink rate is at least in part regulated by brain dopamine. Blink rate increases as dopamine function increases and decreases when dopamine function is diminished [37]. In monkeys, neuroleptics decrease spontaneous blinking. Neuroleptics also decrease blinking in schizophrenic patients, but not in schizophrenic patients with large ventricles [37]. The latter suggests that schizophrenic patients with large ventricles do not respond in a normal fashion to dopamine blockade.

Cerebrospinal Fluid Homovanillic Acid

As discussed previously, most studies have found no differences in dopamine metabolites in schizophrenic patients. Recently, however, van Kammen et al. [63] studied 16 schizophrenic patients with abnormal CT scans and 12 with normal scans. At a time when the patients had been off medications, cerebrospinal fluid homovanillic acid was measured, homovanillic acid was 215 nmol/ml in the abnormal scan patients while it was 135 in patients with normal scans. In an earlier study, Bowers [8] found good prognosis schizophrenic patients had a mean homovanillic acid concentration that was 142 ng/ml compared with 95 ng/ml for poor prognosis patients.

CONCLUSIONS

The evidence that there is an increase in functional dopamine activity in schizophrenia is largely indirect, supported by analogy (amphetamine) or deduction (neuroleptics' mechanism of action). Where there is direct evidence, such as from autopsied material (increased dopamine concentration or increased receptor number), the findings have not been widely replicated (increased dopamine) or appear to be secondary to neuroleptic medication (increased receptors). If there is a functional excess of dopamine in some schizophrenic patients it could be nonspecific, or secondary to medications, stress, or a more basic disease process.

One of the major components of the dopamine hypothesis is the correspondence between the milligram potency of the neuroleptics in schizophrenia and their ability to block dopamine receptors [72]. It is striking, however, that this same relationship holds for affective disorder as well as other neuropsychiatric illnesses where neuroleptics may be used [80]. This suggests that neuroleptic-dopamine blockade is downstream from the primary deficit and that the dopamine blocking function of neuroleptics prevents the expression of symptoms rather than altering a fundamental process related to schizophrenia. One possibility is that in some schizophrenic or psychotic patients, there is an initial process that stimulates an increase in dopamine and that that process decreases with time.

Now that findings, spanning over 50 years, show that some schizophrenic patients have morphological alterations in their brains, researchers can focus on the cause of these alterations. Is it primary or secondary to being ill? Is it a failure of the brain to fully develop, a metabolic deterioration, an immunological or infectious attack? Or, is there something about being psychotic that produces the changes? While the results of studies indicate that the findings are nonspecific that is, they occur in other neuropsychiatric disorders) there is no current indication that they are secondary to schizophrenia.

Animal and human data suggest that there may be a subgroup of patients with a dopaminergic system that either is not involved in the psychosis, or for which there is diminished function. If further studies confirm what is now only a suggestion, whether the abnormality is primary or secondary will still be at issue. Did whatever produced the morphological abnormalities also destroy dopaminergic fibers, particularly those fibers passing near the ventricular system? Were these fibers first stimulated and ultimately injured or destroyed, producing negative symptoms? Whatever the answers to these questions might be, there does not appear to be a generalized dopamine deficiency. While schizophrenic patients may have some adventitious movements, they clearly do not have Parkinson's disease. The dopamine deficit, if it exists, may occur in parts of the dopaminergic system that does not deal with movement, such as the mesolimbic and mesocortical pathways.

The original dopamine hypothesis was that there is a functional excess of dopamine in the brains of schizophrenic patients. If there is a subset of patients in whom a variation of the dopamine hypothesis described here holds, patient research will require attempts to divide populations of patients into subgroups. Perhaps dividing patients into those with and without morphological abnormalities will be of help. Bleuler [6] said in his famous book, "Under the term dementia praecox or schizophrenia we thus subsume a group of diseases. . . . They have many common symptoms and similar prognosis. Nevertheless, their clinical pictures may be extremely varied. This concept may be of temporary value only inasmuch as it may later have to be reduced (in the same sense as the discovering of bacteriology necessitated the subdivision of the pneumonias in terms of various etiological agents)." The concept of schizophrenia as a single entity may have now outlived its usefulness.

And, as the subgrouping of schizophrenic patients is refined, so too may be the role of dopamine.

REFERENCES

1. Ahoxuma, I. P. and Kogan, B. M. Some characteristics of dopamine metabolism in schizophrenia. *Zhurnal Neuropatol i Psikh S.S. Korsakova* (Russian), *81*:1332–1347, 1981.

2. Andén, Fuxe, K., Hamberger, B., and Hokfelt, T. A quantitative study on the nigro-neostriatal dopamine neuron system in the rat. *Acta Physiol. Scand.* *67*:306–312, 1966.

3. Andreasen, N. C., Olsen, S. A., Dennert, J. W., and Smith M. R. Ventricular enlargement in schizophrenia: Relationship to positive and negative symptoms. *Am. J. Psych. 132*:297–302, 1982.

4. Benes, F., Sunderland, P., Jones, B. D., LeMay, M., Cohen, B. M., and Lipinski, J. F. Normal ventricles in schizophrenia. *Br. J. Psych. 141*:90–93, 1982.

5. Birkhauser, V. H. Cholinesterase and monoamine oxidase in the central nervous system. *Schweizerishe Medizinishe Wochenschrift 71*:750–752, 1941.

6. Bleuler, E. Dementia Praecox or the Group of Schizophrenias, 1911. Translated by J. Zinkin. New York: International University Press, 1950.

7. Böök, J. A., Wetterberg, L., and Modrzewska, K. Schizophrenia in North Swedish geographical isolate, 1900–1977. Epidemiology, genetics and biochemistry. *Clin. Genetics 14*:373–394, 1978.

8. Bowers, M. B., Jr. Central dopamine turnover in schizophrenic syndromes. *Arch. Gen. Psych. 31*:50–54, 1974.

9. Bridge, T. P., Cannon, H. E., and Wyatt, R. J. Burned-out schizophrenia: Evidence for age effects on schizophrenic symptomatology. *J. Gerontology 33*:835–839, 1978.

10. Burt, D. R., Creese, I., and Snyder, S. H. Antischizophrenic drugs: Chronic treatment elevates dopamine receptor binding sites. *Science 196*:326–328, 1977.

11. Cancro, R. Prospective prediction of hospital stay in schizophrenia. *Arch. Gen. Psych. 20*:541–546, 1969.

12. Carlsson, A. and Lindqvist, M. Effect of chlorpromazine and haloperidol on formation of 3-methoxytyramine and normetanephrine in mouse brain. *Acta Pharmacol. Toxicol. 20*:140–144, 1963.

13. Church, A. C., Kleinman, J. E., and Wyatt, R. J. Immunocytochemical localization of D-2 binding sites in rodent forebrain. Abstracts, Society for Neuroscience, 1982, p. 717.

14. Crow, T. J. Positive and negative schizophrenic symptoms and the role of dopamine. *Br. J. Psych. 139*:251-254, 1981.
15. Crow, T. J., Baker, H. F., Cross, A. J., Joseph, M. H., Lofthouse, R., Clover, V., and Killpack, W. S. Monoamine mechanism in schizophrenia. Postmortem neurochemical findings. *Br. J. Psych. 134*:249-256, 1979.
16. Dahlström, A., and Fuxe, K. Evidence for the existence of monoamine-containing neurons in the central nervous system. I. Demonstration of monoamines in the cell bodies of brain stem neurons. *Acta Physiol. Scand.*, Suppl. 232, *62*:1-55, 1964.
17. DeLisi, L. E., Wise, C. D., Bridge, T. P., Phelps, B. H., Potkin, S. G., and Wyatt, R. J. Monoamine oxidase and schizophrenia. In: *Biological Markers in Psychiatry and Neurology*, edited by E. Usdin and I. Hanin. New York: Pergamon Press, 1982, pp. 79-96.
18. DeLisi, L. E., Wise, C. D., Bridge, T. P., Potkin, S. G., and Wyatt, R. J. A probable effect of neuroleptic medication on platelet monoamine oxidase activity. *Psych. Res. 4*:95-107, 1981.
19. DeLisi, L. E., Wise, C. D., Potkin, S. G., Zalcman, S. G., Phelps, B. H., Lovenberg, W., and Wyatt, R. J. Dopamine-beta-hydroxylase, monoamine oxidase and schizophrenia. *Biol. Psych. 15*:510-519, 1980.
20. Diagnostic and Statistical Manual of Mental Disorders. Second Edition. Prepared by the Committee on Nomenclature and Statistics of the American Psychiatric Association. Washington, D.C.: American Psychiatric Association, 1968.
21. Dunnett, S. B., Björklund, A., Stenevi, V., and Iversen, S. D. Grafts of embryonic substantia nigra reinnervating the ventrolateral striatum ameliorate sensorimotor impairments and akinesia in rats with 6-OHDA lesions of the nigrostriatal pathways. *Brain Res 229*:209-217, 1981.
22. Economo, C. F. *Encephalitis lethargica: Its sequelae and treatment*, edited and translated by K. O. Neuman. London: Oxford University Press, 1931.
23. Fink, S. J. and Smith, G. P. Mesolimbicortical dopamine terminal fields are necessary for normal locomotor and investigatory exploration in rats. *Brain Res. 199*:359-384, 1980.
24. Fujita, K., Ito, T., Maruta, K., Terodavia, R., Bepper, H., Nakagami, Y., Kato, Y., Nagatsu, T., and Kato, T. Serum dopamine-beta—hydroxylase in schizophrenic patients. *J. Neurochem. 30*:1569-1572, 1978.
25. Haracz, J. L. The dopamine hypothesis: An overview of studies with schizophrenic patients. *Schizo. Bull. 8*:438-469, 1982.
26. Henn, F. A. Dopamine: A role in psychosis of schizophrenia. In: *Schizophrenia as a Brain Disease*, edited by F. A. Henn and H. A. Nasrallah. New York: Oxford University Press, 1982, pp. 176-195.
27. Jacobi, W. and Winkler, H. Enephalographische studien au chronisch schizophrenen. *Arch. Psychiat. Nervenkr. 81*:299-332, 1981.
28. Jacobowitz, D. M. Monoaminergic pathways in the central nervous system. In: *Psychopharmacology: A Generation of Progress*, edited by M. A. Lipton, A. DiMascio and K. F. Killam. New York: Raven Press, 1978, pp. 119-129.
29. Javoy-Agid, F. and Agid, Y. Is the mesocortical dopamine system involved in Parkinson's disease? *Neurol. 30*:1326-1330, 1980.

30. Jernigan, T. L., Zatz, L. M., Moses, J. A., and Berger, P. A. Computerized measures of cerebral atrophy in schizophrenics and normal controls. *Arch. Gen. Psych. 39*:765–770, 1982.

31. Jeste, D. V., Zalcman, S., Weinberger, D. R., Bigelow, L. B., Kleinman, J. E., Rogol, A., Gillin, J. C., and Wyatt, R. J. Apomorphine response and subtyping of schizophrenia. *Prog. Neuropsychopharmacol* (in press).

32. Johnstone, E. C., Crow, T. J., Frith, C. D., Husband, J., and Kreel, L. Cerebral ventricular size and cognitive impairment in chronic schizophrenia. *Lancet* 924–926, 1976.

33. Karoum, F., Commissiong, J. C., Neff, N. H., and Wyatt, R. J. Regional differences in catecholamine formation and metabolism in the rat spinal cord. *Brain Res. 212*:361–366, 1981.

34. Kebabian, J. W. and Calne, D. B. Multiple receptors for dopamine. *Nature 727*: 92–96, 1977.

35. Klein, D. F., Gittleman, R. and Quitkin, A. (Eds.) Diagnosis and drug treatment of psychiatric disorders: Adults and Children. Baltimore:Williams and Wilkins, 1980.

36. Kleinman, J. E., Bridge, T. P., Karoum, F., Speciale, S., Staub, R., Zalcman, S., Gillin, J. C., and Wyatt, R. J. Biochemical abnormalities in post-mortem brain. In: *Perspectives in schizophrenia research*, edited by C. Baxter and T. Melnechuk. New York: Elsevier, 1980, pp. 227–236.

37. Kleinman, J. E., Karson, C. N., Weinberger, D. R., Freed, W. J., Berman, K. F., and Wyatt, R. J. Eye-blinking in chronic schizophrenic patients grouped by ventricular brain ratios. *Am. J. Psych.* (in press).

38. Kleinman, J. E., Weinberger, D. R., Rogol, A. D., Bigelow, L. B., Klein, S. T., and Wyatt, R. J. Relationships between plasma prolactin concentrations and psychopathology in chronic schizophrenia. *Arch. Gen. Psych. 39*:655–657, 1982.

39. Lee, T., and Seeman, P. Elevation of brain neuroleptics/dopamine receptors in schizophrenia. *Am. J. Psych. 137*:191–197, 1980.

40. Lee, T., Seeman, P., Tourtelotle, W. W., Farley, I. J., and Hornykeiwicz, O. Binding of ^{3}-H-neuroleptic and ^{3}H-apomorphine in schizophrenic brains. *Nature 274*:897–900, 1978.

41. Lerner, P., Goodwin, F. K., van Kammen, D. P., Post, R. M., Major, L. F., Ballenger, J. C., and Lovenberg, W. Dopamine-beta–hydroxylase in cerebrospinal fluid of psychiatric patients. *Biol. Psych. 13*:685–694, 1978.

42. Mackay, A. V. P., Iversen, L. L., Rossor, M., Spokes, E., Arregui, I., Creese, I., and Synder, S. H. Increased brain dopamine and dopamine receptors in schizophrenia. *Arch. Gen. Psych. 39*:991–997, 1982.

43. Meltzer, H. Y. and Stahl, S. M. Platelet monoamine oxidase activity and substrate preferences in schizophrenic patients. *Res. Comm. Chem. Pharmacol. 7*:419–431, 1974.

44. Mundt, C. H., Radii, W., and Gluch, E. Computertomographishce untersuchungen der Liquorraume an chronisch schizophrenen patienten. *Nervenarzt 51*:743–748, 1980.

45. Nybäck, H., Wisel, F. A., Berggren, B. M., Hindmarsh, T. Computed tomography of the brain in patients with acute psychosis and in healthy volunteers. *Acta Psychiat. Scand. 65*:403–414, 1982.

46. Owen, F., Crow, T. J., Powlter, M., Cross, A. J., Longden, A., and Riley, G. J. Increased dopamine receptor sensitivity in schizophrenia. *Lancet 2*:223–225, 1978.

47. Pfohl, B. and Winokur, G. The evolution of symptoms in institutionalized hebephrenic/catatonic schizophrenics. *Br. J. Psych. 141*:567–572, 1982.

48. Raese, J. D., King, R. J., Barnes, D., Berger, P. A., Barchas, J. D., Maror, M., and Hock, P. Retinal oscillatory potentials in schizophrenia; Implications for assessment of dopamine transmission in man. *Psychopharmacol. Bull. 18*:72–78, 1982.

49. Report on the papers and discussions of the investigation on acute epidemic encephalitis (lethargic encephalitis). Association for Research in Nervous and Mental Diseases, New York: 1920. Paul B. Hoeber, 1921.

50. Rosenblatt, J. E., Shore, D., Neckers, L. M., Perlow, M. J., Freed, W. F., and Wyatt, R. J. Effects of chronic haloperidol on caudate ^{3}H-spiroperidol binding in lesioned rats. *Eur. J. Pharmacol. 60*:387–388, 1979.

51. Schultz, S. C., Sinicrope, P., Kishore, P., and Friedel, R. O. Treatment response and ventricular brain enlargement in young schizophrenic patients. *Abstracts of the annual meeting of the American College of Neuropsychopharmacology*, 1982, p. 58.

52. Schultz, S. C., Sinicrope, P., Koller, M., Kishore, P., and Friedel, R. O. Treatment response and ventricular brain ratio in young schizophrenic patients. *Abstracts of the Society for Biological Psychiatry* 1982, p. 75.

53. Schwartz, J. C., Baudry, M., Martes, M. P., Costentin, J. and Protais, P. Increased *in vivo* binding of ^{3}H-pimozide in mouse striatum following repeated administration of haloperidol. *Life Sci. 23*:1785–1970, 1978.

54. Siegfried, B. and Bures, J. Conditioning compensates the neglect due to unilateral 6-OHDA lesions of substantia nigra in rats. *Brain Res. 167*:139–155, 1979.

55. Smith, R. C., Largen, J. Shvartsburd, A., and Calderon, M. CAT scans and clinical response to neuroleptic medication in schizophrenia. *Abstracts of the Annual Meeting of the American College of Neuropsychopharmacology*, 1982, p. 58.

56. Strauss, J. S., Carpenter, W. F. and Bartko, J. J. The diagnosis and understanding of schizophrenia. Part III: Speculations on the processes that underlie schizophrenia symptoms and signs. *Schizophrenia Bull. 11*:61–79, 1974.

57. Synder, S. H. Catecholamines in the brain as mediators of amphetamine psychosis. *Arch. Gen. Psych. 27*:169–179, 1972.

58. Takahashi, R., Inabi, Y., Inanaga, K., Kato, N., Kumashiro, H., Nishimkura, T., Okuma, T., Otsuki, S., Sakai, T., Sato, T., and Shimazono, Y. CT scanning and the investigation of schizophrenia. In: *Third world congress of biological psychiatry*, edited by B. Jansson, C. Perris, and C. Struwe. Amsterdam: Elsevier (in press).

59. Tanaka, T., Hazama, H., Kawakara, R., and Kobayusulu, K. Computerized tomography of the brain in schizophrenic patients. *Acta Psychiat. Scand. 63*:191–197, 1981.

60. Taylor, J. (Ed.). *Selected Writings of John Hughlings-Jackson*. London, Hodder and Stoughton, 1932.

61. Tran, N., Laplante, M., and Lebel, E. Decarboxylation of radioactive DOPA by erythrocytes in schizophrenia. *Br. J. Psych. 118*:465–466, 1971.

62. Tsuang, M. T. and Winokur, G. Criteria for subtyping schizophrenia: Clinical differentiation of hebephrenia and paranoid schizophrenia. *Arch. Gen. Psych. 31*:43–47, 1974.

63. van Kammen, D. P., Mann, L. S., Sternberg, D. E., Scheinin, M., Marder, S. G., Rieder, R. O., and Linnoila, M. Spinal fluid dopamine-beta-hydroxylase activity and homovanillic acid levels in schizophrenic patients with brain atrophy. *Science* (in press).

64. van Rossum, J. M. The significance of dopamine-receptor blockade for the mechanism of action of neuroleptic drugs. *Arch. Int. Pharmacodyn. 160*:492–494, 1966.

65. Weinberger, D. R., Bigelow, L. B., Kleinman, J. E., Klein, S. T., Rosenblatt, J. E., and Wyatt, R. J. Cerebral ventricular enlargement in chronic schizophrenia: Association with poor response to treatment. *Arch. Gen. Psych. 37*:11–14, 1980.

66. Weinberger, D. R., DeLisi, L. E., Perman, G. P., Targum, S., and Wyatt, R. J. Computed tomography in schizophreniform disorder and other acute psychiatric disorders. *Arch. Gen. Psych. 39*:778–783, 1981.

67. Weinberger, D. R., Karson, C. N., Bigelow, L. B., and Wyatt, R. J. Cerebral ventricular size and response to neuroleptic treatment. *Abstracts of the Annual meeting of the American College of Neuropsychopharmacology*, 1982, p. 57.

68. Weinberger, D. R., Wagner, R. L., and Wyatt, R. J. Neuropathological studies of schizophrenia: A selective review. *Schizo. Bull.* (in press).

69. Weinberger, D. R. and Wyatt, R. J. Cerebral morphology in schizophrenia. In: *Schizophrenia as a brain disease*, edited by F. Henn and H. A. Nasrallah. New York: Oxford University Press, 1982, pp. 148–175.

70. Wise, R. A. Neuroleptics and operant behavior: The anhedonia hypothesis. *Behav. Brain. Sci. 5*:39–87, 1982.

71. Wise, C. D., Baden, M. M., and Stein. L. Post-mortem measurement of enzymes in human brain: Evidence of a central noradrenergic deficit in schizophrenia. *J. Psychiat. Res. 11*:185–198, 1974.

72. Woodrow, K. M., Reifman, A. and Wyatt, R. J. Amphetamine psychosis: A model for paranoid schizophrenia? In: *Neuropharmacology and behavior*, edited by B. Haber and M. H. Aprison. New York: Plenum Press, 1978, pp. 1–22.

73. Wyatt, R. J. Biochemistry and schizophrenia (Part IV). The neuroleptics–Their mechanism of action: A review of the biochemical literature. *Psychopharmacol. Bull. 12*:5–50, 1976.

74. Wyatt, R. J., Erdelyi, E, Schwartz, M., Hermann, H., and Barchas, J. D. Difficulties in comparing catecholamine-related enzymes from the brains of schizophrenics and controls. *Biol. Psych. 13*:317–334, 1978.

75. Wyatt, R. J. and Gillin, J. C. The transmethylation hypothesis: A quarter of a century later. *Psych. Ann. 6*:33–49, 1976.

76. Wyatt, R. J., Gillin, J. C., and Stoff, D. M., Moja, E. A., and Tinklenberg, J. R. β-phenylethylamine (PEA) and the neuropsychiatric disturbances. In: *Neuroregulators and psychiatric disorders*, edited by E. Usdin, J. Barchas, and D. Hamburg. New York: Oxford University Press, 1977, pp. 31–45.

77. Wyatt, R. J., Moja, E. A., Karoum, F., Stoff, D. M., Potkin, S. G., and Kleinman, J. E. Phenylethylamine, dopamine and norepinephrine in schizophrenia. In: *Apomorphine and other dopaminomimetics*, edited by G. U. Corsini and G. L. Gessa. New York: Raven Press, 1981, pp. 39–44.

78. Wyatt, R. J. and Murphy, D. L. Low platelet monoamine oxidase activity and schizophrenia. *Schizo. Bull.* 2:77–89, 1976.

79. Wyatt, R. J., Potkin, S. G., Kleinman, J. E., Weinberger, D. R., Luchins, D. L., and Jeste, D. V. The schizophrenia syndrome: Examples of biological tools for subclassification. *J. Nerv. Ment. Dis.* 169:100–112, 1981.

80. Wyatt, R. J., Schwartz, M. A., Erdelyi, E., and Barchas, J. D. Dopamine-β-hydroxylase activity in brains of chronic schizophrenic patients. *Science 187*: 368–369, 1975.

81. Wyatt, R.J. and Torgow, J. S. A comparison of equivalent clinical potencies of neuroleptics as used to treat schizophrenic and affective disorders. *J. Psych. Res. 13*:91–98, 1976.

82. Young, D. and Scoville, W. B. Paranoid psychosis in narcolepsy and the possible danger of benzedrine treatment. *Med. Clin. N. Am. 22*:637–646, 1938.

83. Yu, P. H., O'Sullivan, K. S., Keegan, D., and Boulton, A. A. Dopamine-beta-hydroxylase and its apparent endogenous inhibitory activity in the plasma of some psychiatric patients. *Psych. Res. 3*:205–210, 1980.

The Contributions of Franz J. Kallmann to the Genetics of Schizophrenia

John D. Rainer, M.D.

FRANZ J. KALLMAN

The work of Franz Kallmann was so closely fused with his own life experiences, his scientific and medical ideals, his delights and his enthusiasms, that it would be impossible to do for him what he would have done tonight—to place in his own perspective almost 40 years of investigation on two continents. I shall try to describe that portion of his research dealing with schizophrenia, recalling perhaps some data or emphases that have been relatively neglected. Before that, however, a few words about Kallmann as a person might supply some hints about why he worked and how he worked, what he was driving at, and by what interests and energies he was moved.

Franz Kallmann was a many sided person. He was a scientist in the broadest sense with a fertile imagination, a thorough knowledge of subject matter and method, a scanning interest in all of human activity, and the constant ability to frame richly suggestive hypotheses and to formulate careful research plans for their investigation. At the same time, he was always a good physician, a knower of men, a student of human fortitude and weakness, a family counselor, and a clinical psychiatrist in the noblest tradition.

Research in the Schizophrenic Disorders: The Stanley R. Dean Award Lectures, vol. 2, edited by R. Cancro and S. R. Dean. Copyright © 1985 by Spectrum Publications.

[NOTE: Biographical sketch on Dr. Kallmann is on page 292.]

Some years ago, Kallmann prepared an autobiographical chapter, for a volume which was never published, entitled "That Rare Specimen—a Psychiatrist Concerned With Genetics." This chapter obviously just antedated the general rapprochement between clinical psychiatry and genetics sparked by the refinement of laboratory techniques as well as the detailed study of individual differences among infants and children. Kallmann's most general contribution throughout the years was to catalize that rapprochement between psychiatry and genetics, a task made uniquely possible for him by his intimate professional affiliation with national and world leaders in both of these disciplines.

In that autobiographical chapter, Kallman said: "It will be found that a close relationship exists between the two disciplines, psychiatry and genetics. What is a good psychiatrist? In my opinion he has to be a good physician and, above all, a medical man who takes a humanitarian interest in the psychological and emotional problems of his patients. What is a good human geneticist? While his essential prerequisite is a scientific attitude or scientific detachment, in the final analysis scientific data are drawn from people and are intended to be applied to people. Here, to my mind, is the link between the work of the psychiatrist and that of the human geneticist."

Kallmann went on to describe his early life in Silesia with many revealing personal details. Suffice it to say here that he was the son of a hard working and socially minded physician and surgeon, beloved in his town. He received a classical education. Memorizing the Iliad in Greek, defending his dissertation in Latin were part of his preparation and foreshadowed the keen linguistic ability so evident in his own writings, all the more remarkable for being couched in an adopted tongue. Kallman received his medical degree at the University of Breslau in 1919 and embarked upon psychiatric training there, showing an early interest in the forensic field. After his marriage and some busy years in private practice, he decided in 1925 that he wanted to join the clinic of the eminent Professor Karl Bonhoeffer in Berlin where the highest standards of clinical psychiatry were fostered. Bonhoeffer, apparently a demanding but broad-minded teacher, saw to it that Kallman enrolled in a special training course at the newly founded Berlin Psychoanalytic Institute where he studied under Rado and Alexander. Kallmann also studied neurophathology under Creuzfeldt and began a lifelong interest in the study of schizophrenia. His first approach was anatomical. Creuzfeldt had held the theory that the disorder was caused by a disproportional growth of two phylogenetically different parts of the brain and had developed an intricate technique of dissection. At the time, therefore, Kallmann was dividing his days between clinical service and work in histopathology. At visiting hours, he recalled, "the head nurse knew she could reach me in the laboratory when relatives of patients wanted to talk to a physician. Swayed by some pragmatic notions of her own, she often announced a group of waiting visitors with the statement, 'Doctor, the familial taint is here.' It was just because that nurse was so domineering that I undertook my first family study to 'disprove' the inheritance of schizophrenia" [76, p. 18]. Although, according to Kallmann's reminiscences, Bonhoeffer thought the desire to blend psychiatry with genetics was somewhat unusual, he did not turn down his request to apply for a fellowship at the Research Institute in Munich, later known as the Max Planck Institute, which was then headed by Ernst Rüdin. It was at that time that Kallmann organized the first family study based on the incidence of schizophrenia among the children and grandchildren as well as the brothers and sisters of a sample of patients suffering from that disorder. At first assisted only by his wife, he gradually enlisted the aid of a group of loyal nurses in the Berlin hospitals. Commuting between Berlin and Munich, he developed the methods of statistical procedure and field investigation, the careful diagnostic appraisal, and the psychiatric insight which marked his first and subsequent studies. By 1934, he was becoming increasingly anxious to leave Germany, where he found the political abuse of genetics for racial purposes, including sterilization laws for the mentally ill, not at all consistent either with the Hippocratic responsibility of the physician or with the genetic data on schizophrenia which he was beginning to collect. In 1936, Kallmann came to the United States and with the generous support and assistance of many leaders in American psychiatry, neurology, and medicine—including Dr. Nolan D. C. Lewis, Dr. David Levy, and Dr. Franz Boaz—he joined the staff of the New York State Psychiatric Institute and the faculty of Columbia University. He brought with him to this country the manuscript of his first master volume, *The Genetics of Schizophrenia* [12]. With some assistance, he translated the book into English and it was published in New York in 1938. I would like now to look into that study in some detail.

Having failed to disprove his head nurse, Kallmann introduced this volume with a discussion of the need for scientific exploration and a disclaimer that the discovery of genetic predisposition to schizophrenia signifies therapeutic nihilism. "We need hardly mention," he said, "that even a successful clinical program of psychosomatic therapy in the field of schizophrenia would (not) minimize the practical value of modern hospital care based on efficient methods of symptomatic treatment, complete psychological understanding of psychopathological mechanisms, and a well planned system of improved, mentally hygienic living conditions" (p. 3). Since schizophrenia had the tendency to manifest itself relatively late in life and since its clinical forms and severity varied so much, Kallmann felt even at that time that a eugenic program would be difficult and undependable. On the other hand he insisted that a *sine qua non* for any intelligent thinking on the subject was to determine not only the nature and genetic pattern of the mental abnormalities transmitted to the descendants, direct and collateral, of schizoprenic patients but also to determine something about the relative fertility of such patients and their families. Since it was not and is still not possible to identify the genotypical structure in each family member, Kallmann relied on the only genetic method available, the empirical determination of the hereditary prognosis of families of affected individuals, the contingency method devised by Rüdin.

The material for the 1938 study consisted of 1,087 index cases or probands, making up the total number of schizophrenic case histories available in the archives of the Herzberge hospital in Berlin for the years 1893 to 1902. Since they represented a random group untouched by any selective process, these patients could indeed be designated as "probands" according to the criteria of the contingency method. It remained only to follow the indicated relatives of each of these cases regardless of the difficulties involved. Kallmann describes, with so large a series of probands, how hard it was to trace family members; nevertheless this was done. Kallmann explicitly states that unlike some previous studies, the cases which he followed did not represent only those who were most accessible or those who had surviving children or other relatives; such a group would be overweighted with paranoid schizophrenics who are usually the oldest and have had the most children before their first hospitalization.

The diagnosis of schizophrenia was made according to consistent standards and excluded all schizophrenics developing after the age of 40, thus avoiding "climacteric and presenile psychoses frequently having a schizoform character" (p. 11). All records were personally reviewed and a diagnosis was made without reference to the earlier diagnosis or to any notes on hereditary conditions in the family of the patient. Cases with specific neurological symptoms, particularly those associated with alcohol or syphilis, were excluded. After four years of work the total number of persons in the entire material amounted to 13,851, made up of the 1,087 probands, their husbands, wives, and parents, their direct descendants, their siblings, and their nephews and nieces. A little more than half of the probands had no children and hence no direct descendants.

The cases were divided along medical and symptomatological lines into hebephrenic, catatonic, and paranoid groups, in which there was a degree of disintegration of personality during the course of the disease; and the simple group which ran a mild course without any considerable deterioration, recovering without gross defect. In describing the hebephrenic and catatonic groups, Kallmann laid stress on "a relatively early and definite check in the development of personality; on emotional inadequacy with affective disharmony and change of the primary psychic activity; on disorders of association, and on a progressive deterioration of personality clearly associated with the process of schizophrenia" (p. 21). For hebephrenia he regarded affective dementia as a particular characteristic, and he did not consider the presence of chronic or occasional hallucinations and unorganized delusions as a ground for excluding hebephrenia. In catatonia the decisive point was psychomotor dementia—"cases in which motor disturbances were prominent, namely the agitated, stuporous, and cataleptic forms" (p. 21). Paranoid cases, as well as simple cases, were marked by the absence of definite dementia, the former determined by the predominance of paranoid delusions, but without domination by hallucinations and with onset before the age of 40. The simple group, finally, was restricted to patients who, since their discharge and for at least 30 years, had remained "permanently free from gross psychotic disturbances. They must have continued tolerably useful in their social and occupational relations and nevertheless, have shown the

TABLE 1. Distribution of the Probands According to Sex and Form of Schizophrenia

	Female		Male		
By groups	Absolute	Percent	Absolute	Percent	Total
H	300	60.9	193	39.1	493
C	143	56.5	110	43.5	253
P	93	62.0	57	38.0	150
S	92	60.9	59	39.1	151
D	19	47.5	21	52.5	40
Total number	647	59.5	440	40.5	1087

From Kallmann, *The Genetics of Schizophrenia*, New York: J. J. Augustin, 1938, p. 23.

definite stamp of the earlier schizophrenic process" (p. 23). The decisive diagnostic criteria remained "a more or less marked bent in their life-curve, in the presence of some slight symptoms of schizophrenic deterioration" (p. 23). The total group (see Table 1) consisted of 493 hebephrenic cases, 253 catatonic, 150 paranoid, and 151 simple, plus 40 doubtful cases which were omitted from further statistical analysis. Females predominated, ranging from 57 per cent of the total in the catatonic group to 62 per cent in the paranoid, and the females tended to be from one to five years older than the males. Kallmann attributed this sex distribution to his decision to omit automatically all persons with a history of alcoholism or syphilis, most of whom were males. Discussing at length the social background, occupation, and medical history of the ancestors of his probands, he concludes: "for the present we can prove schizophrenia only in a bare ten per cent of the immediate ancestry of schizophrenic patients, but the total number of ancestors obviously tainted with the hereditary predisposition to schizophrenia and belonging to the heredity circle of schizophrenia comes, in the direct line, to at least one third of the parents, uncles and aunts" (pp. 42 to 43). Ruling out earlier vague attempts to consider every form of psychopathic abnormality or organic nervous disease in the ancestors of schizophrenics as manifestations of a hereditary taint, he considered these as the minimum figures for diagnosed schizophrenia. He found them at this point to be consistent with a recessive mode of heredity in which there would be no reason to expect to find schizophrenia in the parents of an index case, and in which most affected individuals would be born to carrier or heterozygotic parents. Further evidence for the recessivity of schizophrenia was supplied later in the book.

Before turning to the important data on the frequency of schizophrenia in descendants of his schizophrenic index cases, Kallmann discussed at length the

TABLE 2. Differential Fertility of Probands in Comparison with the
General Population

	Celibacy rate	Birth rate per marriage	Child mortality 0–4 years
Schizophrenic probands (1892–1902)			
HC group	60.1%	3.0	45.6%
PS group	23.9%	4.6	42.5%
General population of Berlin[a]			
1880	–	4.04	–
1890–1900	29%	–	34.0%
General population of Prussia (1896–1901) (Woytinski)			
Rural inhabitants	–	5.3	–
Urban factory workers	–	4.4	–
General population of Germany (1884–1900)	–	4.4–4.8	–

From Kallmann, *The Genetics of Schizophrenia*, New York: J. J. Augustin, 1938, p. 66.
[a] Annual Statistical report of the city of Berlin, Vol. 88, Pages 21 and 89.

fertility and mortality of the probands and their descendants. His chief findings in
that section (see Table 2) were that the mortality of the children of all index cases
exceeded the corresponding mortality for the general population by about 10 per
cent. As far as fertility was concerned, while the paranoid and the simple schizophrenic patients married about as often and had about as many children as the general population, in the hebephrenic and catatonic groups only half as many of the
patients married, and the birth rate was about three quarters that of the normal;
one third of the children born to the latter groups of schizophrenic patients were
born after the onset of the disease, while the paranoid cases had most of their
children before onset.

Foreshadowing his future concern with responsible genetic counseling, Kallmann said at this point "we need hardly mention that a general and legal compulsory sterilization simply cannot be considered as the eugenic method which would
be applicable to the heterozygotic taint carriers predisposed to schizophrenia. We
must reject it if only because this procedure seems unsuitable on human, medical,

and methodological grounds even for the hereditarily diseased patients themselves. We are rather inclined to believe that in countries with high ethical standards and moral disciplines, the liberty of the individual to determine his own fate within the limits of his natural sovereign rights and the voluntary submission of every citizen to public measures adopted for the perpetuation of common happiness and security belong to the finest and most indestructible ideals of mankind. Accordingly, the methods of education in biology, official bureaus of eugenic guidance and marriage counsel, mandatory health certificates before marriage and, if necessary, the legal prohibition of marriage seem preferable to us both on personal and scientific grounds" (pp. 68 to 69). At the same time, although there seemed at that time to be selective factors lowering the marriage and birth rate—not only among the patients themselves but as Kallmann found, among their brothers and sisters—this limitation of reproduction was in no sense adequate to eliminate spontaneously the predisposition to schizophrenia. We shall see that one of Kallmann's last studies was a reevaluation of the marriage and fertility patterns of schizophrenics in New York State.

Among the challenging implications of the case material at this point was Kallmann's first observation that there was a steep rise in mortality from tuberculosis in comparison with the average population, not only for the schizophrenic patients, but for their children as well (see Table 3). The mortality in the children occurred mostly in the second decade of life and was almost five times that of the general population. The increase was also considerable in brothers and sisters of the index cases. The only other cause of death in which the index cases as well as their siblings showed an increase was suicide. Kallmann was interested in these figures both as indicating possible lethal factors tending to limit the number of schizophrenic persons in the population, and also as representing possible biological correlations, especially in the case of tuberculosis.

The key chapter in *The Genetics of Schizophrenia*, Chapter 4, describes the frequency of schizophrenia in the direct and indirect descendants of the probands. The chapter is introduced by some methodological and diagnostic statements. The rationale for such elaborate studies is given by the statement, "since genetics cannot estimate the frequency of disease in the descendants simply by theoretical figures of hereditary prognosis, nor disclose it by cross-breeding of pure strains, we are confined at present to obtaining the probability of disease for the various descent groups of schizophrenics by the accurate interpretation of experience" (p. 99). Kallmann describes the Weinberg Abridged Method to correct for variations in the age distribution; namely for the fact that among the relatives of the index cases, there are some who will be too young to have developed the disease clinically, so that a pure prevalence figure would represent too low an estimate of those who would develop schizophrenia if they lived through the age of 45. Briefly described, the interval from the 15th to the 44th year was considered as the danger period for the manifestation of schizophrenia. In determining the total number of relatives in a given category, one counted therefore only half of the persons who were between

TABLE 3. Total Mortality from Suicide and Tuberculosis in the Various Descent Groups

	Suicides		Mortality from tuberculosis, based on deaths between 10–59 years
	Based on all births	Based on deaths between 10–59 years	
Children of probands	1.3	11.6	58.5
Grandchildren	0.3	8.3	60.0
Siblings	1.8	8.0	33.0
Nephews and nieces	0.8	6.6	38.7
Panse's general population of Berlin	0.4	2.4	14.3

From Kallmann, *The Genetics of Schizophrenia*, New York: J. J. Augustin, 1938, p. 91.

the ages of 15 and 44. All persons who were 45 or older were reckoned in full, while those who were below the age of 15 at the time of ascertainment or at the time they dropped out of sight, were omitted entirely. A detailed discussion of the abridged method of Weinberg and a comparison with other methods designed to accomplish the same correction are given in this chapter. Regarding diagnosis, Kallmann pointed out that in the chapter to follow all children and grandchildren of index cases classified as schizophrenic could be found described clinically, one by one, in a series of accurate vignettes, so that his diagnostic tenets could be checked at any time and could be employed similarly by other investigators for further genetic studies. Kallmann stated at this point, "the accuracy of psychiatric diagnosis depends much more on whether psychopathologic attributes are correctly interpreted and evaluated in proper technical terms than on whether the right final diagnosis is deduced from the sum of the identified symptoms. The best diagnostic system is worthless if it is incorrectly applied and, in the last analysis, the accurate recognition and appraisal of schizophrenic manifestations must depend on the subjective delicacy of touch and personal ability of the individual investigator" (pp. 101–102). He went on to discuss the diagnosis of schizoid personality, a category included in all of his statistics, although he was unable to decide whether "an eccentric borderline case is a homozygotic carrier of the predisposition to schizophrenia with inhibited manifestation, or the most definite type of a germinally affected taint-carrier (heterozygote), or perhaps only a symptomatic schizoid type without direct connection to the heredity-circle of schizophrenia." In respect to the proof of diagnosis, "psychiatry and genetics are in about the same position as

TABLE 4. Comparative Taint Survey in the Various Descent Groups
of the Proband Families

	The present study estimated according to the abridged method			
		Expectation of schizophrenia		
			Clinical subgroups	
	Total number of adults under observation	Total material	Nuclear group	Peripheral group
Children	1000	16.4%	20.9%	10.4%
Grandchildren	543	4.3%	5.1%	2.9%
Great grandchildren	29	–	–	–
Siblings	2581	11.5%	12.9%	8.9%
Half sisters and half brothers	101	7.6%	7.6%	–
Nephews and nieces	1654	3.9%	4.7%	3.4%
Normal average population		0.85%		

Adapted from Kallmann, *The Genetics of Schizophrenia*, New York: J. J. Augustin, 1938, p. 145.

internal medicine was in the prebacteriological era when there was the difficulty of deciding whether a suppurative coating on the tonsils was diphteritic or nonspecific, or whether an infectious fever was caused by influenza bacilli or other types of bacteria. For want of other criteria, it is necessary for the psychiatrist to utilize experiences of extensive clinical observation, studying the various psychopathological manifestations most carefully, and limiting the sources of error as far as possible" (p. 104).

The statistical findings in this chapter are fairly well known in general, but their detailed breakdown is worth repeating. For the *children* of the index cases, the following expectancy figures for the four clinical subgroups were derived (see Table 4): hebephrenic 20.7 per cent, catatonic 21.6 per cent, paranoid 10.4 per cent, simple 11.6 per cent. The close approximation between the hebephrenic and catatonic expectancy and between the paranoid and simple expectancy led Kallman, at this point, to bracket the two former groups together as the nuclear group and the latter groups together as the peripheral group. The expectancy of schizophrenia in the offspring of a patient depends, of course, upon the diagnosis of the

other parent. In those cases in which the other parent had some psychiatric manifestation, the expectancy was 21 per cent overall, while for those in whom the
other parent was normal, it was 11.9 per cent. For illegitimate children it was highest of all, 26.3 per cent. A significant finding noted at this point was that there
were no significant risk differences between children of male and female probands,
nor between children born before and after the disease onset of the proband. The
overall expectancy for children of schizophrenics was 16.4 per cent, about 19 times
as high as the schizophrenic expectancy in the general population, taken on the
basis of a number of studies quoted in the text as 0.85 percent.

In the case of grandchildren the expectancy figures were 4.3 per cent for
schizophrenia, still about five times greater than the general population, although
if one omitted those grandchildren who had a schizophrenic parent as well as the
proband grandparent, the rate dropped to 1.3 per cent.

In his discussion of siblings, Kallmann preceded the figures with his statement
that theoretically the frequency ought to be greater in the siblings than in the
children if one is dealing with a recessive hereditary trait; with about a one per
cent normal average frequency in the population, he calculated that the frequency
of heterozygotes would be about 18 per cent if there were complete manifestation
and the marriage rate of heterozygotes were normal. Actually these last two conditions were not met, since marriage rate, fertility, and disease manifestation varied
according to the individual disease form and in the different generations. The overall figures established by the survey were 11.5 per cent for the sibs; 7.6 per cent
for the half-sibs; 19 per cent for the sibs of those cases with one schizophrenic
parent. Turning to the four clinical subdivisions, the schizophrenia rate in the
sibs was 12.6 per cent for hebephrenic; 13.4 per cent for catatonic; 8.9 per cent for
paranoid; and 9.5 per cent for simple. It is of interest that the figures for schizoid
personality in the sibs and for that matter in the children of index cases did not
vary significantly among the four diagnostic groups.

The rate for nephews and nieces of schizophrenic patients was 3.9 per cent.
However, these figures actually ranged from 1.8 per cent for those nephews and
nieces who had phenotypically normal parents to 21.4 per cent for those with one
schizophrenic parent and 50.0 per cent for those with two schizophrenic parents.
These figures, Kallmann said, "form the final link in the chain of our differential
taint study on the various cross-breeding proportions among the descendants of
schizophrenics, terminating with a rather conclusive proof of the recessive hereditary transmission in schizophrenia" (p. 133).

Chapter 4 concludes with a noteworthy theoretical discussion of the nature
of the hereditary predisposition factor in schizophrenia. Kallmann gave further
substantiation for the fact that the schizophrenia rate in sibs is indeed higher in the
nuclear group than in the peripheral group and ascribed this to a biological cause
rather than to a social pattern having to do with selection of marriage partners, age
of manifestation, and the like. Nevertheless, he found that the four different
disease forms do not form different hereditary predispositions; the children of

hebephrenic or paranoid patients are not always in turn hebephrenic or paranoid, and two or more schizophrenics in the same series of children do not manifest the same form of psychosis. As a matter of fact, he pointed out that only about half of the schizophrenias in the children and grandchildren correspond to the disease form of the probands, while, even in the same sibship, the number of similar psychoses is only 73 per cent. He concluded, therefore, that the individual form of schizophrenia is determined not solely by the special nature of the hereditary predisposition, but by a series of other factors. These are not racial, he said in this 1938 book, quoting with tongue in cheek a German writer who said only Nordic people develop pure types of schizophrenia; rather he went on to indicate his belief that the phenotypic manifestation depended on varying hereditary constitutional characteristics of the individual, including the "inner environment," with the "phenotypical and somatogenous components of the individual constitution" (p. 152). A discussion followed of the recessive nature of the basic inherited factor versus the theory of dominance with incomplete penetrance and some of the dihybrid theories that were offered. Kallmann supported the former theory and mentioned here for the first time some twin figures, those of Luxenburger and Rosanoff, in which the concordance for schizophrenia in identical twins was about 70 per cent. "The inhibition of manifestation," Kallmann said, "in more than a quarter of identical twins with a homologous predisposition to schizophrenia, is to be explained only by the assumption that these cases lack exciting dispositional factors" (p. 153). There were only 41 twin pairs among the descendants in the present investigation; 19 sets died in childhood, and in 12 others only one twin survived the 20th year. Of the remaining ten pairs, seven were of the same sex, but of these only one pair showed any psychiatric abnormalities—the rather famous identical twin sisters, Kaete and Lisa, raised since infancy by two different uncles and concordant for catatonic schizophrenia—and there were two opposite sex discordant pairs.

Since another source of presumed homozygotes was the offspring of two schizophrenic parents, Kallmann investigated the schizophrenia rate among this group and in his total material found it to be 68.1 per cent, again short of the expected 100 per cent. Kallmann's detailed formulation explaining the variations in manifestation by assuming constitutional factors as determining elements in the development of the individual disease form is worth further description. "The constitution of an individual is viewed," he said, "as the phenotypical resultant of the impact of all environmental influences on the sum of all genetic strains." "Hebephrenic and catatonic psychoses," he speculated, "tend to occur when there is in certain organs a particular inferiority that is based on a specific hereditary predisposition and is constantly exposed to the influence of dispositional physical processes running their course in every body and frequently excited by unfavorable components of the environment." In those severe forms however, this occurrence might be confined, he said, "to individuals whose physical constitution does not provide, especially during the critical stages of generative endocrine change, for repressive factors adequately resistant to the urge toward manifestation of the

schizophrenic predisposition. Correspondingly, the paranoid and simple schizophrenias might occur only after the original restraining mechanism has ceased to operate,
either through gradual exhaustion of the physiological forces before and during the
climacteric and presenility or in simple schizophrenic cases, under the impact of
sudden somatic processes markedly reducing the resistance of the total organism
for a period of time. . . . To express these concepts of the phenotypical development of schizophrenia most simply, we may say that the mode of manifestation
tends to be variable and depends in each instance on the result of the individual interplay among genetic, constitutional, and environmental factors" (pp. 161-163).

The following chapter described in detail all descendant cases as well as some
of the other interesting families and pedigrees, including twins; and the one after
that discussed the absence of an increase in other psychopathological abnormalities
in the schizophrenic families—namely, epilepsy, syphilis, feeble-mindedness, and
psychopathy. Kallmann then turned to the study of the genetic relation between
schizophrenia and tuberculosis and concluded with his thoughts about the constitutional resistance factors in both conditions. Indeed he found the mortality from
tuberculosis considerably increased in all the catagories of relatives of schizophrenics, greater in the nuclear group than in the peripheral group, and greatest in those
categories of relatives with the highest expectancy of schizophrenia. He suggested
that the anatomical substratum for the low resistance in both of these conditions
was an "inherited tissue insufficiency rather than a similarity of physical structure
or some toxic endocrine organic change per se" (p. 256) and that the most plausible
explanation at the time was that an impaired reactive ability or weakness of the
reticuloendothelial system was the common constitutional basis. Kallmann quoted
in this chapter the experimental findings of Meyer who found by the use of the
Congo red method that the capacity for storage in schizophrenics was diminished
and by use of the cantharidin blister technique of Kauffmann, that there was an
inadequate reactive capacity of the cells of the reticuloendothelial system—possibly
correlated with a deficiency in the production of antibodies. At that time, the
methods employed for testing the efficiency of the reticuloendothelial apparatus
were crude and Kallmann regarded the theory as only a working hypothesis, one to
which he returned many years later.

The book we are describing concludes with a discussion of clinical and
eugenic implications. Kallmann reiterates that sterilization of schizophrenic patients
would eliminate only from two to three per cent of the total number of schizophrenics and that the results would in no way justify the means. Clinically, he
describes the interaction between heredity and environment as follows: "It is too
often forgotten that only predispositions are inherited and never attributes as such,
and that the phenotypical manifestation of the trait in the individual depends on
the sum of all the given environmental conditions" (p. 264). The inhibition of
schizophrenia in over 30 per cent of the offspring of two schizophrenic parents corresponding with the twin figures that others had obtained indicated to him the importance of other somatic or dispositional factors. Eugenic measures by systematic

enlightenment of individual families did not interfere, he said, "with the clinical problem of psychiatry to prevent and combat the manifestation of the schizophrenic trait in the tainted individual born in spite of prophylactic measures. To achieve the therapeutic optimum, all psychiatric efforts must be concentrated on identifying the somatic processes responsible for the phenotypical development of the hereditary predisposition ... (intervening) to improve gradually the remedial success already achieved in a minority of schizophrenics by nonspecific mutation treatment (insulin, sulfosin) or other methods ... in order to reach the goal of modern medicine: simultaneous prevention and healing" (pp. 268, 272).

I have gone into a great deal of detail in describing *The Genetics of Schizophrenia* since, in the first place, the book has long been out of print and is not easily available, and, in the second place, it contains many leads to future investigations and many formulations, as well as detailed data, which are worth reviewing in discussing Franz Kallmann's investigations and career. Moving on, in one of his first papers written in the United States, [13] Kallmann said he was convinced that systematic instruction of marriage partners would be enough to effect an adequate reduction of postpsychosis fertility rate and that early marriage should not be recommended to young schizophrenic females as a therapeutic procedure. He recommended in 1938 [11] the availability and use of contraception for those couples who could understand the nature of the problem.

By 1941 Kallmann had already begun to study manic-depressive disease [20] and for our purpose, we may indicate some of the differential diagnostic criteria he set down at that time. Manic-depressive disease was marked for him by alternating episodes of depression and elation which do not show real hallucinations, dissociation, delusions, or a definite tendency to mental deterioration. He referred to his ongoing twin study for indeed, upon coming to New York State, he decided to obtain a larger group of individuals presumably predisposed to schizophrenia by searching for the identical twins of schizophrenic patients. He wrote at that time that the main purpose of twin studies was not to show concordance but to study those pairs that differ in onset or production, in order to prevent or to heal the condition, and that curability and inheritability were not incompatible.

An interesting paper in 1941 [21] written together with Drs. Barrera, Hoch, and Kelley on the role of mental deficiency in the incidence of schizophrenia reported no correlation between these two except by coincidence. In 1942 [22], Kallmann described further his concepts of the heredoconstitiutional mechanisms of predisposition and resistance (see Figure 1). He considered a scheme in which the schizoid personality was either a heterozygote with little resistance or a homozygote with strong resistance; the schizophrenic, a homozygote with little resistance; and the mild schizophrenic, a homozygote with intermediate resistance; and he discussed the possibility of a normal clinical appearance in a homozygote with strong resistance or a heterozygote with intermediate resistance. The resistance, in any case, he considered as a multifactorial constitutional mechanism. He pointed out that in the twins which he was currently studying, the

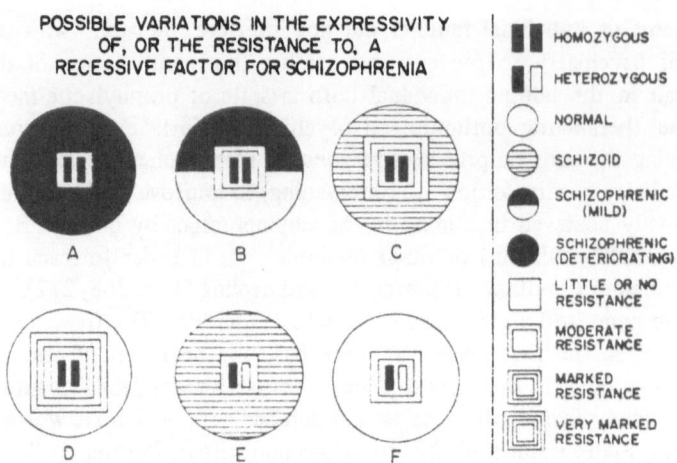

POSSIBLE VARIATIONS IN THE EXPRESSIVITY
OF, OR THE RESISTANCE TO, A
RECESSIVE FACTOR FOR SCHIZOPHRENIA

FIGURE 1. From Kallmann, and Barrera, *Amer. J. Psychiat.*, 1942, 98, 546.

nonschizophrenic twin was always the physically stronger, taller, and heavier, and far more resistant to infections and other ailments than the twin who did develop schizophrenia. A number of case studies with pictures were presented at that time, and, in connection with Sheldon, the patients were somatotyped and schizophrenia found to be correlated with the asthenic type, resistance with the athletic. During those years, Kallmann was engaged concurrently in a large-scale twin study of tuberculosis per se and in a prototypic family study reported in 1943 with Reisner [25], he found 87 per cent concordance for tuberculosis in monozygotic twins; 26 per cent in dizygotic twins ranging from 20 per cent in opposite-sex pairs to 30 per cent in same-sex pairs, 26 per cent in full siblings, 12.9 per cent in half-siblings, 7.1 per cent in marriage partners—all compared with 1.37 per cent in the general population. In this paper he described his perfected twin family method, soon to be reported in connection with the schizophrenia study.

In 1946, there appeared what is perhaps Kallmann's best known publication [30], a paper on the genetic theory of schizophrenia, read the previous year at the annual meeting of the American Psychiatric Association. The report summarized the result of 10 years' intensive survey of twin index cases. By describing them as "collected from the resident populations and new admissions of all mental hospitals under the supervision of the New York State Department of Mental Hygiene" (p. 311). Kallman did not do justice to the many hours and months of personal journeying through the hinterlands accompanied always by his wife and variously by distinguished psychiatrists like Slater, Strömgren, Hurst, and Svendsen. In the 794 twin index cases, there was an excess of females over males of almost 20 per cent. About 70 per cent were unmarried, and 68 per cent of the total number belonged to the nuclear groups.

TABLE 5. Schizophrenia Rates for Relatives of Schizophrenic Twin Index Cases

| | Number | Rates[a] | |
		Uncorrected	Corrected
Parents	1191	9.1	9.2
Spouse	254	2.0	2.1
Step siblings	85	1.4	1.8
Half siblings	134	4.5	7.0
Full siblings	2741	10.2	14.3
Two-egg cotwins	517	10.3	14.7
One-egg cotwins	174	69.0	85.8

Adapted from Kallmann, The genetic theory of schizophrenia. *Am. J. Psych.*, 1946, 103, p. 313.

[a] Uncorrected = all schizophrenia and all persons over age 15. Corrected = definite schizophrenia and ½ persons age 15 to 44 + all over 44.

The actual prevalency rates for schizophrenia uncorrected for age in the various twin index cases were as follows: parents, 9.1 per cent; husbands and wives, 2.0 per cent; step-siblings, 1.4 per cent; half-siblings, 4.5 per cent; full siblings, 10.2 per cent; dizygotic co-twins, 10.3 per cent; monozygotic co-twins, 69.0 per cent (see Table 5). The figures which are more generally known are those which have been corrected by the abridged Weinberg method: parents, 9.2 per cent; husbands and wives, 2.1 per cent; step-sibs, 1.8 per cent, half-sibs, 7.0 per cent; full sibs, 14.3 per cent; dizygotic co-twins, 14.7 per cent; monozygotic co-twins, 85.8 per cent. It is of interest to note that a reexamination of the records of the co-twins done in later years showed that out of 103 monozygotic co-twins diagnosed as schizophrenic, all but two had themselves been in a mental hospital; likewise all but four of 47 affected dizygotic co-twins.

Kallmann explained the difference between the 68 per cent risk previously obtained for the offspring of two schizophrenic parents and the 85.8 per cent risk for monozygotic co-twins by the fact that the parents had already been distinguished by having had the chance to marry and produce offspring, and were probably marked by better genetic resistance factors than the nonselected series of twins. Emphasized again was the importance of the fact that the morbidity rate in monozygotic co-twins, as in the offspring of two schizophrenic parents, was less than 100 per cent. From a biological standpoint, Kallmann said, the finding classifies schizophrenia as both preventable and potentially curable. The implication was that the main schizophrenic genotype is not fully expressed either in the absence of any particular factor of a precipitating nature or in the presence of strong constitutional defense mechanisms which, in turn, are partially determined by heredity. Kallmann took pains to point out that the risk figures "do not mean however that heredity is effective in only 70 to 85 per cent of schizophrenic cases or that it is essential merely to the extent of 70 to 85 per cent in any one case" (p. 315).

As one might put it today, the measures represent the interaction between genetic and nongenetic factors in determining variability within a given population, and not a division into two mutually exclusive categories. Most striking of all was the observation that "over 85 per cent of the siblings and dizygotic co-twins did not develop schizophrenia, although about ten per cent of them had a schizophrenic parent, all of them had a schizophrenic brother or sister, and a large proportion shared the same environment with these schizophrenics before and after birth." There was no correlation between schizophrenia in the co-twin and instrumental delivery or premature birth. The concordance in dizygotic twins seemed to be independent of their sex (see Table 6). The total concordance was 14.3 per cent for male pairs, 14.9 per cent for female pairs; for opposite sex pairs, it was 10.5 per cent where the male was the index case, 10.2 per cent where the female was the index case; for same sex pairs it was 17.4 per cent for male pairs, and 17.7 per cent for female pairs. In the monozygotic pairs, there was some difference between separated and nonseparated twins, bearing in mind that in this series separated twins were those who were not living together for five years prior to disease onset in the index twin. The rates were 77.6 per cent for separated pairs, 91.5 per cent for nonseparated pairs, the overall figure having been given as 85.8 per cent. Only in 17.6 per cent of the monozygotic twin pairs was the age of onset simultaneous. In 52.9 per cent it ranged from one month to four years, and in 29.4 per cent the difference in age of onset was over four years. Regarding the clinical course of the disease, there were important similarities in the monozygotic pairs and great dissimilarities in the dizygotic pairs, pointing again to the multifactorial constitutional resistance factors. For the first time an excess of consanguineous matings was found among the parents of the index cases, with about 5.7 per cent of the twin index pairs originating from consanguineous matings of nonschizophrenic parents. Kallman's general conclusion to this paper was as follows: "Psychiatrically it should be evident that the genetic theory of schizophrenia, as it may be formulated on the basis of experiment-like observations with the twin family method, does not confute any psychological concepts of a descriptive or analytical nature, if these concepts are adequately defined and applied. There is no genetic reason why the manifestations of a schizophrenic psychosis should not be described in terms of narcissistic regression or of varying biological changes such as defective homeostasis or general immaturity in the metabolic responses to stimuli. Genetically, it is also perfectly legitimate to interpret schizophrenic reactions as the expression of either faulty habit formations or progressive maladaptation to disrupted family relations. The genetic theory explains only why these various phenomena occur in a particular member of a particular family at a particular time" (p. 320).

TABLE 6. Variations in the Schizophrenia Rates of Siblings and Twin Partners According to Sex and the Similarity or Dissimilarity in Environment

	Siblings of twin index cases			Dizygotic co-twins			Monozygotic co-twins		
	Male	Female	Total number	Male	Female	Total number	Separated	Nonseparated	Total number
Same sex	15.9	16.3	16.1	17.4	17.7	17.6	77.6	91.5	85.8
Opposite sex	12.5	12.0	12.3	10.5	10.2	10.3	–	–	–
Total number	14.0	14.5	14.3	14.3	14.9	14.7	77.6	91.5	85.8

Adapted from Kallmann, The genetic theory of schizophrenia. *Am. J. Psych*, 1946, 103, p. 316.

These principles were developed further in a paper written in 1948 [40], in which Kallmann discussed such concepts as the ability to be normal, behavioral plasticity, genes and fate, and the genetic approach to therapy. In this paper and a paper published in the American Psychopathological series on failures in psychiatric treatment [45], he pointed out the close correlation between sudden weight loss and onset of schizophrenic symptoms in some of the twins whom he followed. In the latter paper he also discussed some of the unconscious objections to heredity that he found among his fellow psychiatrists and reiterated that there was no fixed relation between a gene itself and the character caused by it, that "therapeutic action against an inherited disorder is possible either by rendering the underlying main gene less penetrant or by changing its expression through carefully directed management of vital environmental factors or through methodical stimulation of secondary modifiers" (p. 136), and that the need for studies of dissimilar twins was very great.

Those were the years in which the postwar ascendancy of interest in psychodynamics coincided with the low point of attention to genetic factors in psychiatry, and indeed they were difficult years for Kallmann as a pioneer in this area. Nevertheless while continuing to collect schizophrenic twins, he was engaged in a number of other fruitful activities including the study of over 2000 senescent twin pairs, the founding of the American Society of Human Genetics, and participation in an increasing number of national and international meetings.

At one of those conferences, the First International Congress of Psychiatry in Paris in 1950, Kallmann summarized not only the twin material in schizophrenia, but the smaller series of manic-depressive twins that he had meanwhile collected. In his report to that congress [56] he addressed himself to the problem of nosology, deriving from the twin data on the one hand a biological disparateness between schizophrenia and manic-depressive psychosis and on the other, a partial association between schizophrenia and involutional psychosis. While he came to consider that manic-depressive psychosis was genetically homogeneous and conditioned by a dominant gene with incomplete penetrance, he felt that "the diagnostic category of involutional psychosis is either less homogeneous clinically or more complex pathogenetically than are those of schizophrenia and manic-depressive psychosis. . . . There is some reason to believe that a few involutional cases might actually be late-developing and attenuated processes of schizophrenia precipitated only by the impact of the involutional period of life, which threatens loss of security" (p. 13). He had found no twin pairs with both schizophrenia and manic-depressive psychoses, no increase in manic-depressive psychosis in the families of schizophrenic twins or vice versa, but some increase in involutional psychosis in the families of schizophrenics. Recapitulating some of his diagnostic criteria, he wrote at that time, "Except for paranoid syndromes occurring in previously nonpsychotic persons during the involutional and senile periods of life, all the 'mixed' psychoses causing a disintegrative bend in personality development were regarded as basically schizophrenic processes and were generally classified as such unless clearly symptomatic or associated with gross organic pathology. In accordance with this latitude in diagnosing a schizophrenic psychosis, the classification of schizophrenia was not reserved solely for the episodic and deteriorating forms of hebephrenic, catatonic and paranoid coloring but it was extended to include the simple, atypical, and 'schizo-affective' varieties showing only a very slow tendency to deterioration, the acute confusional states precipitated by extreme stress, and the 'panneurotic' borderline cases without delusions or hallucinations described by Hoch and Polatin as the pseudoneurotic type of schizophrenia. . . . Our diagnosis of schizophrenia rested on the constellative evaluation of basic personality changes observed in association with a whole group of possible psychopathological mechanisms, rather than on the presence of any particular type of symptomatology. As a general principle greater diagnostic importance was attached to the demonstrable effect of a 'bending' curve of personality development than to any surface similarities to pathognomonic textbook descriptions, especially when this bend was found in conjunction with such malignant features as xenophobic pananxiety, loss of the capacity for free associations, inability to maintain contact with reality, (autistic and dereistic attitudes toward life), or a compulsive tendency to omnipotential thought generalization" (pp. 8-9). In that Paris address, Kallmann said that "in future twin studies . . . emphasis should be placed on clarifying the intricate interplay of gene-specific biochemical dysfunctions, general constitutional (adaptational) modifiers and precipitating outside factors arising from the effect of certain basic imperfections in the structure of modern human societies" (pp. 22-23). With now 953 schizophrenic index cases the corrected expectancy rate for parents was 9.3 per cent, for half-sibs 7.1 per cent, for full sibs 14.2 per cent, for dizygotic co-twins 14.5 per cent, and for monozygotic co-twins 86.2 per cent.

This major summarizing paper was followed within two years by the Thomas W. Salmon Lectures at the New York Academy of Medicine which were published in 1954 as a monograph [76]. In this book, Kallmann reviewed his own data and those of other investigators in schizophrenia in the framework of a volume discussing at length general concepts of heredity in mental disorder and genetics in mental health planning. Kallmann dealt for the first time there with the problems of childhood or preadolescent schizophrenia. He noted that the excess of males in this younger group as opposed to the excess of females in the older schizophrenic patients was as yet unexplained, and he mentioned that he was conducting an investigation of preadolescent twin index cases in which the majority of the concordant twin pairs were of the male sex and the concordance rates of both zygosity groups apparently lower than for the adult forms of schizophrenia.

The complete report on preadolescent schizophrenia was published in 1956 by Kallmann and Roth [96] and described 52 sets of twins, 17 monozygotic and 35 dizygotic. By that time Kallmann was inclined to explain the excess of males in terms of an increased biological vulnerability. He stated that "the diagnostic criteria were generally on the conservative side . . . very young children who presented a picture of psychosis with mental deficiency, perhaps simulating a severe intellectual defect as a result of a very early schizophrenic process, were not included in the sample. . . . All diagnoses were made by one investigator and strictly on the basis of the clinical history of the child, without prior knowledge of the family background. A distinct change in the behavior of a child who previously seemed to develop normally was regarded as a crucial diagnostic feature. The most frequently observed symptoms were diminished interest in the environment, blunted or distorted affect, peculiar conduct especially in motor activity, diffuse anxiety with phobias and vague somatic complaints, bizarre thinking with a tendency toward exaggerated fantasies, and hallucinations." There appeared to be no etiological significance in the birth order of the index cases, and the ages of their parents at the time of their birth were within the normal range. The expectancy rates for schizophrenia in the co-twins of the preadolescent cases were: dizygotic twins 22.9 per cent, monozygotic twins 88.2 per cent. These were uncorrected figures; since this study dealt with schizophrenics under the age of 15, it was not thought advisable to correct age differences by the abridged Weinberg method. If one considered only preadolescent schizophrenia in the co-twins, the rates were 17.1 per cent for dizygotic twins, 70.6 per cent for monozygotic twins. In parents the schizophrenia rate was 8.8 per cent, in sibs 9.0 per cent. (For the study of sibs, the 52 twins were supplemented by 50 single born preadolescent schizophrenics.) In the course of this study, two observations were made which pointed the way to some of Kallmann's later investigations. First a small rise in the schizophrenia rate of the parents of the preadolescent cases was found over that noted in previous studies. It was suggested that this increase might indicate that the present population of schizophrenics had a slightly better chance of becoming parents then was true in previous samples. Four years later Kallmann was to begin a study of changing mating and fertility trends in schizophrenia. In the second place, the quality of the parental homes of the preadolescent schizophrenic cases was investigated and a close relationship found between inadequate homes and severe maladjustment, between unstable parents and schizophrenia in the children. While these findings were compatible with both genetic and environmental theories, it was noted that two thirds of the normal twins and other brothers and sisters came from the same inadequate homes.

Returning to the adult form of schizophrenia, Kallmann wrote [79] that the vulnerability to schizophrenia might be described psychodynamically as an integrative pleasure deficiency leading to adaptive incompetence, a formulation for which he gives credit to Sandor Rado. Indeed Rado, Kallmann's former teacher and then the Director of the Columbia University Psychoanalytic Clinic, had formulated his theory of the "schizotype" based on his pupil's work and marked by the particular psychodynamic constellations of the pleasure deficiency and a proprioceptive defect.

In the 1950s, keeping pace with certain current trends in genetics, Kallmann was discounting the importance of the difference between dominant and recessive inheritance, preferring the dynamic approach of the Goldschmidt school which tried to understand phenomena in terms of "gene-specific molecular processes and developmental systems with all their interaction, embryonic regulation, and integration. In this frame of reference, the concept of Mendelian heredity (with or without simple segregation ratios) becomes more or less synonomous with 'chromosomal heredity.' Pertinent environmental factors which mold, and formative elements which secure behaviorable malleability on the human level are viewed as 'end products of the same evolutionary process'" [122, p. 341].

In the American Handbook of Psychiatry [121] he wrote: "For several reasons it still seems appropriate to interpret the main adaptive defect in a potentially schizophrenic person as the result of a basically recessive unit factor, although it is quite irrelevant with respect to the overall genetic theory whether the given main gene is classified as primarily dominant or recessive, or fails altogether to show simple segregation ratios. The modern concept of human heredity does not hinge 'upon the counting of simple Mendelian ratios' and it seems that most pathological conditions in man are neither fully dominant or completely recessive but only relatively so, depending 'upon the facility with which the trait is detectable in the heterozygote;'" and "the question of how to designate a mode of transmission that seems distinguished by a specific unit factor for the entire syndrome, plus a number of modifying genes responsible for the variable clinical expressions of the main genotype, is more or less a matter of semantics" (p. 191).

The late years of the 1950's and the early 1960's brought with them the well known advances in genetics including the study of human chromosomes, the increased understanding of the molecular biology of heredity, and the current search for knowledge about control and feedback mechanisms. In his writings between 1960 and 1965, Kallmann incorporated these modern concepts of genetics into his theoretical framework and taught them to psychiatrists. He conceived of the interaction of biochemical, molecular, subcellular, and cellular mechanisms with threshold factors to produce the individual behavioral phenotype and the family and social expressions thereof, as indicated in a diagram adapted from one of his more recent publications [138] (see Figure 2). It was appropriate that for the 25th anniversary of the department of medical genetics a symposium on expanding goals in genetics in psychiatry [145] included discussion not only of the genetics of disordered behavior, but of the most recent progress in basic cytological and biochemical genetics from various parts of the world.

In the last five years, Kallmann devoted most of his attention to two consuming projects, a marriage and fertility study of schizophrenic patients and a psychiatric and genetic study of the deaf. The former study was formulated in accord with growing concern for the problems of genetic counseling, a field in which Kallmann had much to say. He regarded genetic counseling as a form of psychotherapy, as a dialogue between the physician and the family in which not only their specific questions regarding genetic risk, but also their wishes, fears, and hopes could be discussed and they could be helped to come to a reasonable conclusion. In the case of schizophrenia it was important within that framework to know whether more schizophrenic patients were marrying than had been the case

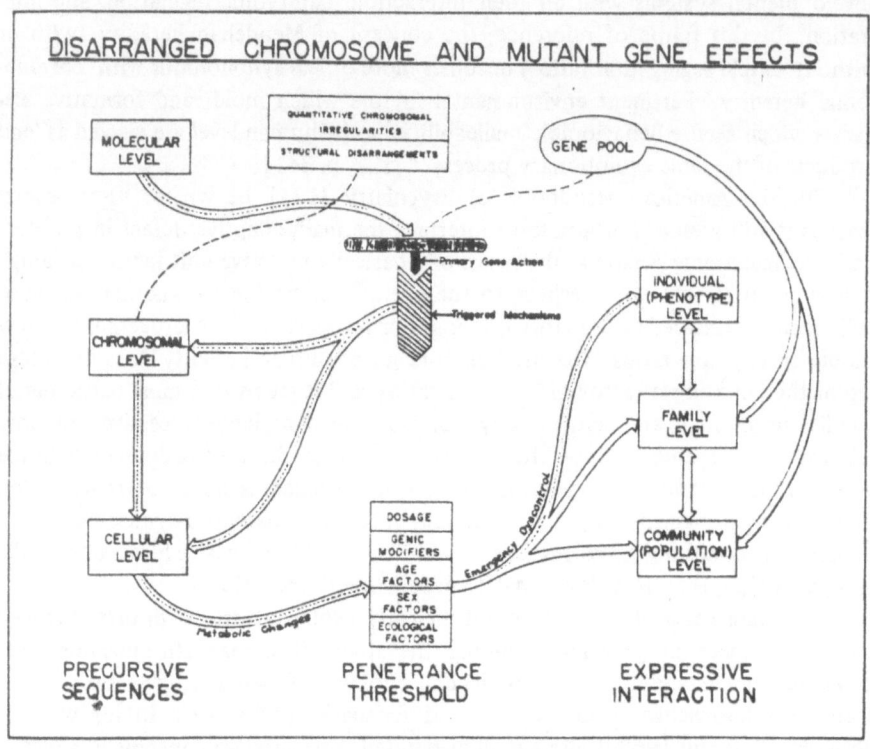

FIGURE 2. Adapted from Kallmann, in *Comparative Epidemiology of the Mental Disorders*, Grune & Stratton, New York, 1961, p. 236.

at the time of the earlier study and whether they were having more children. Preliminary reports [176] by Erlenmeyer-Kimling, Rainer, and Kallmann on this study (which is still in progress) indeed indicated an increase in the marriage rate, in the divorce rate, and in the reproductivity rate among schizophrenic patients which is higher than that for the general population (see Table 7). Comparing a group of schizophrenic patients admitted in the mid-1930s with a group in the mid-1950's after modern treatment methods were instituted, the authors noted not only this change in the marriage and fertility pattern but equally significant, an increase in the fertility rate in brothers and sisters of schizophrenic hospital patients. These findings were taken to indicate that any selective disadvantage which schizophrenia may have had in terms of its evolutionary history is now in the process of disappearing. In one of the last papers that Kallmann wrote [168] he described the developmental history of children born of two schizophrenic parents. Many of these children were found to be displaced from their homes for various periods of

TABLE 7. Reproductive Rates of Schizophrenics in Two Survey Periods and Comparison with General Population

Survey period	Number of children per 100 persons			General population females	Reproductivity of schizophrenic females expressed as percent of general population
	Schizophrenics				
	Both sexes	Males	Females		
1934–36	60	50	70	120	58.3
1954–56	90	60	130	150	86.7

Adapted from data of Erlenmeyer-Kimling, Rainer, and Kallmann. Current reproductive trends in schizophrenia: A psychiatric-genetic survey. In P. H. Hock & J. Zubin (eds.) *Psychopathology of Schizophrenia*, New York: Grune & Stratton, 1966.

time during their formative years. Kallmann said that since any study, genetic or otherwise, indicated a poorer prognosis for these children, it was clearly indicated that "until the biological disciplines exploring one or the other facet of the complex etiology of schizophrenia come up with a reliable method for the prevention or treatment of this disorder, more attention should be given to the attainable goals of marriage and parenthood counseling in schizophrenic families requesting such guidance" (p. 145).

The second project [162], that on the deaf, had many clinical ramifications and led to outpatient and inpatient services for totally deaf psychiatric patients—as Kallmann phrased it, "a psychiatric genetic department in action." From the point of view of research, the study made some significant observations about the prevalence of schizophrenia among this severely deprived population. The best determination of the schizophrenia rate for the deaf in New York turned out to be 2.5 per cent, a figure higher than the one per cent figure for the general population, but not as high as that for any category of relative of a schizophrenic patient. A group of 138 deaf schizophrenic patients were taken as index cases and the schizophrenia rate among their sibs determined (Table 8). The age-corrected schizophrenia rate for all the sibs of deaf schizophrenic patients was 11.6 per cent. Most interesting was the absence of a statistically significant difference between the schizophrenia risks for hearing and deaf sibs; among the hearing sibs the corrected expectancy rate was 11.2 per cent, among the deaf sibs 15.8 per cent. All groups of sibs, the hearing as well as the deaf, and their combined total, had significantly greater risks for schizophrenia than the general population, risks which were the same order of magnitude as those computed in other family studies. In particular, they closely approximated the rate of 14.3 per cent reported by Kallmann in 1946 for the sibs of schizophrenic index cases without a hearing loss. In this deafness study which was

TABLE 8. Schizophrenic Risk Data in the Siblings of Deaf and Hearing
Schizophrenics

	Number of siblings		Corrected frame of reference	Crude risk (%)	Corrected risk (%)
	Surviving age 15	Definitely schizophrenic			
Of deaf index cases:					
Hearing siblings	303	25	223	8.3	11.2
Deaf siblings[a]	28	3	19	10.7	15.8
Total	331	28	242	8.5	11.6[b]
Of hearing index cases (Kallmann 1946):	2014	184	1288	9.14	14.3

Adapted from Rainer, Altshuler, and Kallmann (Eds.) *Family and Mental Health Problems
in a Deaf Population*, New York: New York State Psychiatric Institute, 1963, p. 211.
[a]Includes seven cases with marked hearing loss.
[b]14.1% if probable cases are included.

carried out by Doctors Altshuler and Sarlin, it therefore appeared from the data
that the severe and various stresses associated with early total deafness do little to
increase the chance of developing clinical symptoms of schizophrenia.

Aside from these important statistical studies, Kallmann's vision in the
last years turned toward the future goals and tasks of psychiatric-genetic research,
and in so doing he dwelled again on the elucidation of the biology of schizophrenia.
Having seen his work serve as the stimulus for much current pharmacological and
biochemical research, he said in an unpublished progress report to the Scottish Rite
Committee on Research in Schizophrenia: "There is a French proverb to the effect
that people, as they grow older, tend to return to their first loves. If this saying is
correct, you may be inclined to evaluate my report in terms of my advancing age."
He went on to say that "over 30 years ago, while still working in pathological
laboratories, I became interested in the function of the reticuloendothelial system
as the system which is man's principle organ of defense. At that time we studied in
laboratories the regulatory and defensive activities of cell elements in connection
with artificially produced fever reactions, first in neurosyphilitics and later in
schizophrenics. Ever since, I have had the nagging thought that some systematic
work should be done along these lines in order to make it possible to have a con-
necting link between the psychodynamic and physiodynamic approaches to a
pathological state that may be referred to in terms of lack of ego strength in coping
with stress as well as in those of the organism's inability to maintain a state of

adaptiveness in the presence of a noncontainable deficit in its biochemistry. Since I have always assumed that genes exert their effects through control of metabolism, although there is a long chain of reactions between the initial gene effects and its observable end products, I never thought of the genetic factor for schizophrenia as something that is static or fixed at birth. Instead we defined it as a specific, complex, and quite dynamic vulnerability factor that leads to adaptive incompetence only when the defense system collapses. Unfortunately, it took many years to obtain whatever statistical data were needed to substantiate the theory that this collapse is more likely to occur in some people than in others. With this task accomplished, many fine clinicians had an incentive to search for the biological substrate of this presumably gene-specific metabolic deficiency while we finally got a chance to look again into the mysterious activity of reticuloendothelial cells." Kallmann went on to describe the function of monocytes, histiocytes, and phagocytes, the importance of studying such cell activity in potential schizophrenics prior to the manifestation of clinical symptoms, and the need to study therefore the offspring of two schizophrenic patients and the identical twins of persons with schizophrenic symptoms. Counting cellular elements in blister fluids, he observed tentatively a marked reduction in the total cell count and a rise in the percentage of lymphocytes. Kallmann soon established a cytogenetic laboratory in his department which began and continues now to investigate the phenomena, using modern tissue culture techniques.

If Kallmann were here, he would probably end his report to you by presenting his conception of the future of psychiatry in the perspective of genetics. He addressed himself to this theme in 1962 [146] at the meeting of the American Psychopathological Association where he outlined the new discoveries in human cytogenetics and molecular genetics and voiced his belief that genetic research at the molecular and chromosomal levels had the capacity of greatly enhancing our understanding of human behavior disorders. The length of time needed to achieve these ends, he said, could not be predicted because it depended upon our willingness to make heavier investments in terms of brain and manpower, laboratories, and above all, moral commitments. "If this stipulation can be met," he said, "no one among our junior colleagues should be even mildly astonished if much of the decoding of genetic elements in the etiology of puzzling psychiatric disorders is accomplished within his own lifetime" (p. 197). In another 1962 paper [151] he said, "the availability of precise experimental methods justifies the general conclusion that exploration of quantitative, as well as qualitative, changes in the genetic components of personality organization is now well within our reach" (p. 89).

I can only add that I think that if Franz Kallmann had had the time, he would have tried to systematize his ideas. I know he planned, if he ever retired, to put on paper some of his broader philosophical and theoretical views. Meanwhile, always curious, always searching, always critical, he must have exasperated those of his friends of every psychiatric persuasion, who wanted him to endorse dogmatically their theories or positions; yet they remained his friends. His absorption in research did not prevent him from developing a wide variety of interests, ranging from a deep appreciation of music and art to travel and sports—and above all, human relations.

A particularly poignant glimpse into his character and style of mind came to light last week, when the night nurse who tended him during his last illness gave Mrs. Kallmann two notes he had written her a year ago. Mrs. Kallmann has given me permission to read them to you on this occasion. The first says, "Mrs. No Flashlight (what a pleasure). Must one make an early reservation if one wants to join a few old friends on Mt. Olympus, and how does one get there?" and the second, a few days later. "Since you did not give me the proper directions to Mt. Olympus, I unfortunately missed the boat today. Wann geht der Nächste Schwan? from the Lohengrin Clan."

REFERENCES

1. Kallmann, F. J. Accidental stab wounds as cause of death. *Ärztl. Sachverständingen Ztschr.* (German), *22*:126, 1921.
2. Kallmann, F. J. On the psychopathology of the superstitious criminal. *Mschr. Psychiat. Neurol.* (German), *72*:37, 1929.
3. Kallmann, F. J. On the symptomatology of cerebral cysticercosis. *Mschr. Psychiat. Neurol.* (German), *72*:324, 1929.
4. Kallmann, F. J. and Marcuse, H. Sulfosin therapy in general paresis and schizophrenia. *Nervenarzt* (German), *2*:149, 1929.
5. Kallmann, F. J. and Salinger, F. Accident-conditioned metastasis in malignant tumors. *Ärztl. Sachverständigen Ztschr.* (German), *9*:1, 1929.
6. Kallmann, F. J. and Salinger, F. Diagnostic and forensic aspects of cerebral cysticercosis. *Mschr. Psychiat. Neurol.* (German), *76*:38, 1930.
7. Kallmann, F. J. The methods of fever treatment. *Hospitalsitende* (German), *75*:1, 1932.
8. Kallmann, F. J. The methods of fever treatment in neurosyphilis. *Mediz. Welt* (German), *43*:1, 1932.
9. Kallmann, F. J. The results of fever treatment in the Herzberg Hospital. In: *Ergebnisse der Reiztherapie bei progressiver Paralyse*, (German), edited by K. Bonhoeffer and P. Jossmann. Berlin: S. Karger, 1932.
10. Kallmann, F. J. The fertility of schizophrenics. *Bevölkerungsfragen*, (German), edited by H. Harmsen and F. Lohse. Munich: J. F. Lehmanns, 1936.
11. Kallmann, F. J. Eugenic birth control in schizophrenic families. *J. Contracept. 3*:195, 1938.
12. Kallmann, F. J. *The genetics of schizophrenia*. New York: J. J. Augustin, 1938.
13. Kallmann, F. J. Heredity, reproduction and eugenic procedure in the field of schizophrenia. *Eugen News 23*:105, 1938.

14. Kallmann, F. J. In memorian: Dr. med. Wilhelm Weinberg, *J. Nerv. Ment. Dis.* Vol. 87, 1938.
15. Kallmann, F. J. Informal discussion of "Sources of mental disease. Their amelioration and prevention." In: *Mental health*, edited by F. R. Moulton and P. O. Komora. Publication No. 9, American Association for the Advancement of Science, 1939.
16. Kallmann, F. J. Collaborator in *Psychiatric dictionary* (L. E. Hinsie and J. Shatzky). New York: Oxford University Press, 1940.
17. Kallmann, F. J., Barrera, S. E., and Metzger, H. The association of hereditary microphthalmia with mental deficiency. *Amer. J. Ment. Defic. 45*:25, 1940.
18. Kallmann, F. J. Editorial. *Amer. J. Ment. Defic. 46*:165, 1941.
19. Kallmann, F. J. Knowledge about the significance of psychopathology in family relations. *Marr. Fam. Living 3*:81, 1941.
20. Kallmann, F. J. The operation of genetic factors in the pathogenesis of mental disorders. *N.Y.S. J. Med. 41*:1352, 1941.
21. Kallmann, F. J., Barrera, S. E., Hoch, P. H., and Kelley, D. M. The role of mental deficiency in the incidence of schizophrenia. *Amer. J. Ment. Defic. 45*:514, 1941.
22. Kallmann, F. J. and Barrera, S. E. The heredoconstitutional mechanisms of predisposition and resistance to schizophrenia. *Am. J. Psych. 98*:544, 1942.
23. Kallmann, F. J. Genetic mechanisms in resistance to tuberculosis. *Psych. Quart. Suppl. 17*:32, 1943.
24. Kallmann, F. J. Twin studies on genetic variations in resistance to tuberculosis. *J. Hered. 34*:269, 1943.
25. Kallmann, F. J. and Reisner, D. Twin studies on the significance of genetic factors in tuberculosis. *Amer. Rev. Tuberc. 47*:549, 1943.
26. Kallmann, F. J. and Schonfeld, W. A. Psychiatric problems in the treatment of eunuchoidism. *Amer. J. Ment. Defic. 47*:386, 1943.
27. Kallmann, F. J. Review of psychiatric progress. 1943: Heredity and eugenics. *Am. J. Psych. 100*:551, 1944.
28. Kallmann, F. J. Schonfeld, W. A., and Barrera, S. E. The genetic aspects of primary eunuchoidism. *Amer. J. Ment. Defic. 48*:203, 1944.
29. Kallmann, F. J. Review of psychiatric progress, 1944: Heredity and eugenics. *Am. J. Psych. 101*:536, 1945.
30. Kallmann, F. J. The genetic theory of schizophrenia. *Am. J. Psych. 103*:309, 1946.
31. Kallmann, F. J. Review of psychiatric progress, 1945; Heredity and eugenics. *Am. J. Psych. 102*:522, 1946.
32. Kallmann, F. J. and Anastasio, M. M. Twin studies on the psychopathology of suicide. *J. Hered. 37*:171, 1946.
33. Kallmann, F. J. and Mickey, J. S. The concept of induced insanity in family units. *J. Nerv. Ment. Dis. 104*:303, 1946.
34. Kallmann, F. J. and Mickey, J. S. Genetic concepts and folie à deux. *J. Hered. 37*:298, 1946.
35. Kallmann, F. J. Genetics in relation to mental disorders. *Eugen. News 32*:51, 1947.

36. Kallmann, F. J. Modern concepts of genetics in relation to mental health and abnormal personality development. *Psych. Quart. 21*:1, 1947.
37. Kallmann, F. J. Review of psychiatric progress, 1946: Heredity and eugenics. *Am. J. Psych. 103*:513, 1947.
38. Kallmann, F. J. and Anastasio, M. M. Twin studies on the psychopathology of suicide. *J. Nerv. Ment. Dis. 105*:40, 1947.
39. Kallmann, F. J. and Sander, G. The genetics of epilepsy. In: *Epilepsy*, edited by P. H. Hoch and R. P. Knight. New York: Grune & Stratton, 1947.
40. Kallmann, F. J. Applicability of modern genetic concepts in the management of schizophrenia. *J. Hered. 39*:339, 1948.
41. Kallmann, F. J. Current trends in psychiatry and social medicine as observed at the International Conference of Physicians. *Psych. Quart. Suppl. 22*:326, 1948.
42. Kallmann, F. J. The genetic aspects of senescence in the light of twin studies. *Mschr. Psych. Neurol.* (German), *116*:58, 1948.
43. Kallmann, F. J. Genetics in relation to mental disorders. *J. Ment. Sci. 94*: 250, 1948.
44. Kallmann, F. J. The genetic theory of schizophrenia. In: *Personality in nature, society, and culture*, edited by C. Kluckhohn and H. A. Murray. New York: Alfred A. Knopf, 1948.
45. Kallmann, F. J. Heredity and constitution in relation to the treatment of mental disorders. In: *Failures in psychiatric treatment*, edited by P. H. Hoch. New York: Grune & Stratton, 1948.
46. Kallmann, F. J. Review of psychiatric progress, 1947: Heredity and eugenics. *Am. J. Psych. 104*:448, 1948.
47. Kallmann, F. J. and Sander, G. Twin studies on aging and longevity. *J. Hered. 39*:349, 1948.
48. Kallmann, F. J. Constitutional relationship between schizophrenia and tuberculosis. *De Paul J.* Vol. 3, No. 8, 1949.
49. Kallmann, F. J. Medical genetics and eugenics in relation to mental health problems and senescence. *Eugen. News 33*:15, 1949.
50. Kallmann, F. J. On the frequency of suicide in twins and nontwins. *Mschr. Psych. Neurol.* (German), *117*:280, 1949.
51. Kallmann, F. J. Review of psychiatric progress, 1948: Heredity and eugenics. *Am. J. Psych. 105*:497, 1949.
52. Kallmann, F. J., De Porte, J., De Porte, E., and Feingold, L. Suicide in twins and only children. *Am. J. Hum. Genet. 1*:113, 1949.
53. Kallmann, F. J. and Feingold, L. Principles of human genetics in relation to insurance medicine and public health. *J. Insur. Med.* Vol. 4, No. 1, 1949.
54. Kallmann, F. J. and Planansky, K. Utilization of genetic data in insurance medicine. *J. Insur. Med.* Vol. 4, No. 4, 1949.
55. Kallmann, F. J. and Sander, G. Twin studies on senescence. *Am. J. Psych. 106*:29, 1949.
56. Kallmann, F. J. The genetics of psychoses. In: *Proceedings, First International Congress of Psychiatry, Sect. VI*. Paris: Hermann & Cie, 1950.
57. Kallmann, F. J. The genetics of psychoses. *Am. J. Hum. Genet. 2*:385, 1950.
58. Kallmann, F. J. Review of psychiatric progress, 1949: Heredity and eugenics. *Am. J. Psych. 106*:501, 1950.

59. Kallmann, F. J. and Svendsen, B. B. Progress of genetics in relation to oligo-phrenia, epilepsy and other neurological disorders, 1939–1946. *Ztschr. ges. Neurol. Psychiat.* (German), *110*:1, 1950.

60. Kallmann, F. J. Recent progress in relation to the genetic aspects of mental deficiency. *Am. J. Ment. Defic. 56*:375, 1951.

61. Kallmann, F. J. Relationship between schizophrenia and somatotype. *Mod. Med. 19*:140, 1951.

62. Kallmann, F. J. Review of psychiatric progress, 1950: Heredity and eugenics. *Am. J. Psych. 107*:503, 1951.

63. Kallmann, F. J. Twin studies in relation to adjustive problems in man. *Trans. N.Y. Acad. Sci. 13*:270, 1951.

64. Kallmann, F. J., Feingold, L., and Bondy, E. Comparative adaptational, social, and psychometric data on the life histories of senescent twin pairs. *Am. J. Hum. Genet. 3*:65, 1951.

65. Kallmann, F. J. Comparative twin study on the genetic aspects of male homosexuality. *J. Nerv. Ment. Dis. 115*:283, 1952.

66. Kallmann, F. J. The genetic aspects of mental disorders in the aging. *J. Hered. 43*:89, 1952.

67. Kallmann, F. J. Genetic aspects of psychoses. In: *The biology of mental health and disease*. New York: Paul B. Hoeber, 1952.

68. Kallmann, F. J. The genetics of psychoses. In: *Proceedings, First International Congress of Psychiatry, Sect. VI.* Paris: Hermann Cie, 1952.

69. Kallmann, F. J. Human genetics as a science, as a profession, and as a social-minded trend of orientation. *Am. J. Hum. Genet. 4*:237, 1952.

70. Kallmann, F. J. Introductory statement of moderator. Round table discussion on "Psychiatric guidance in problems of marriage and parenthood" (résumé). *Eugen. News. 37*:55, 1952.

71. Kallmann, F. J. Review of psychiatric progress, 1951: Heredity and eugenics. *Am. J. Psych. 108*:500, 1952.

72. Kallmann, F. J. Twin and sibship study of overt male homosexuality. *Amer. J. Hum. Genet. 4*:136, 1952.

73. Kallmann, F. J. and Bondy, E. Applicability of the twin study method in the analysis of variations in mate selection and marital adjustment. *Am. J. Hum. Genet. 4*:209, 1952.

74. Kallmann, F. J. and Sander, G. Twin studies of senescence. In: *Psychological studies of human development*, edited by R. G. Kuhlen and G. G. Thompson. New York: Appleton-Century-Crofts, 1952.

75. Aschner, B. M., Kallmann, F. J., and Roizin, L. Concurrence of Morgagni's syndrome, schizophrenia and adenomatous goiter in monozygotic twins. *Acta Genet. Med. Gemellolog 2*:431, 1953.

76. Kallmann, F. J. *Heredity in health and mental disorder*. New York: W. W. Norton, 1954.

77. Kallmann, F. J. Review of psychiatric progress, 1952: Heredity and eugenics. *Am. J. Psych. 109*:491, 1953.

78. Kallmann, F. J. Genetic principles in manic-depressive psychosis. In: *Depression*, edited by P. H. Hoch and J. Zubin. New York: Grune & Stratton, 1954.

79. Kallmann, F. J. The genetics of psychotic behavior patterns. In: *Genetics and the inheritance of integrated neurological and psychiatric patterns*, edited by D. Hooker and C. C. Hare. Baltimore: Williams & Wilkins, 1954.

80. Kallmann, F. J. Heredity in health and mental disorder. In: *Genetica Medica*, edited by L. Gedda. Rome: Gregor Mendel Institute, 1954.

81. Kallmann, F. J. Review of psychiatric progress, 1953: Heredity and eugenics. *Am. J. Psych. 110*:489, 1954.

82. Kallmann, F. J. Twin data in the analysis of mechanisms of inheritance. *Am. J. Hum. Genet. 6*:157, 1954.

83. Allen, G. and Kallmann, F. J. Frequency and types of mental retardation in twins. *Am. J. Hum. Genet. 7*:15, 1955.

84. Kallmann, F. J. Review of psychiatric progress, 1954: Heredity and eugenics. *Am. J. Psych. 111*:502, 1955.

85. Kallmann, F. J. and Baroff, G. S. Abnormalities of behavior (in the light of psychogenetic studies). In: *Annual review of psychology, Vol. VI*, edited by C. P. Stone. Stanford: Annual Reviews, 1955.

86. Kallmann, F. J. Genetic aspects of mental disorders in later life. In: *Mental disorders in later life*, 2nd Ed., edited by O. J. Kaplan. Stanford: Stanford University Press, 1956.

87. Kallmann, F. J. The genetics of aging. *J. Chronic Dis 4*:140, 1956.

88. Kallmann, F. J. The genetics of aging. In: *The neurologic and psychiatric aspects of the disorders of aging*, edited by J. E. Moore, H. H. Merritt, and R. J. Masselink. Baltimore: Williams & Wilkins, 1956.

89. Kallmann, F. J. The genetics of human behavior. *Am. J. Psych. 113*:496, 1956.

90. Kallmann, F. J. Genetic variations in adjustment to aging. In: *Psychological aspects of aging*, edited by J. E. Anderson. Washington, D.C.: American Psychological Association, 1956.

91. Kallmann, F. J. Heredity in disturbed mentality, eugenic aspects. In: *Enciclopedia Medica Italiana, Vol. VIII*, (Italian). Florence: Sansoni Edizioni Scientifiche, 1956.

92. Kallmann, F. J. Objectives of the mental health project for the deaf. In: *Proceedings, 37th Convention of American Instructors of the Deaf*. Senate Document No. 99. Washington, D.C.: U.S. Government Printing Office, 1956.

93. Kallmann, F. J. Psychiatric aspects of genetic counseling. *Amer. J. Hum. Genet. 8*:97, 1956.

94. Kallmann, F. J. Review of psychiatric progress, 1955: Heredity and eugenics. *Am. J. Psych. 112*:510, 1956.

95. Kallmann, F. J., Aschner, B. M., and Falek, A. Comparative data on longevity, adjustment to aging, and causes of death in a senescent twin population. In: *Novant'anni delle Leggi Mendeliane*, edited by L. Gedda. Rome: Gregor Mendel Institute, 1956.

96. Kallmann, F. J. and Roth, B. Genetic aspects of preadolescent schizophrenia. *Am. J. Psych. 112*:599, 1956.

97. Sank, D. and Kallmann, F. J. Genetic and eugenic aspects of early total deafness. *Eugen. Quart. 3*:69, 1956.

98. Allen, G. and Kallmann, F. J. Mongolism in twin sibships. *Acta Genet. Stat. Med.* 7:385, 1957.

99. Jarvik, L. F., Kallmann, F. J., Falek, A., and Klaber, M. M. Changing intellectual functions in senescent twins. *Acta Genet. Stat. Med.* 7:421, 1957.

100. Jungeblut, C. W., Kallmann, F. J., Roth, B., and Goodman, H. O. Preliminary twin data on the salivary excretion of a receptor-destroying enzyme. *Acta Genet. Stat. Med.* 7:191, 1957.

101. Kallmann, F. J. Heredity and aging. *Newsl. Geront. Soc.* 4:5, 1957.

102. Kallmann, F. J. Medical Arts Congress of Turin: With International Symposium of Medical Genetics. *Eugen. News.* 4:162, 1957.

103. Kallmann, F. J. Review of psychiatric progress, 1956: Heredity and eugenics. *Am. J. Psych.* 113:595, 1957.

104. Kallmann, F. J. The role of genetics in psychiatry. *Am. J. Psychother.* 11:885, 1957.

105. Kallmann, F. J. Twin data on the genetics of aging. In: *Methodology of the study of aging*, edited by G. E. W. Wolstenholme and C. M. O'Connor. Ciba Foundation Colloquia on Ageing, Vol. III. London: J. & A. Churchill, 1957.

106. Kallmann, F. J. and Baroff, G. S. Heredity and variations in human behavior patterns. *Acta Genet. Stat. Med.* 7:410, 1957.

107. Rainer, J. D. and Kallmann, F. J. Behavior disorder patterns in a deaf population. *U.S.P.H.S. Rep.* 72:585, 1957.

108. Kallmann, F. J. An appraisal of psychogenetic twin data. *Dis Nerv. Syst. Suppl.* 19:9, 1958.

109. Kallmann, F. J. Comments on eugenic abortion from the viewpoint of psychiatric genetics. *Med. Klin.* (Germany), 53:2064, 1958.

110. Kallmann, F. J. Genetic aspects of schizophrenia. *Med. Hyg.* (French), 393:173, 1958.

111. Kallmann, F. J. The genetic viewpoint of the etiology of mental illness. In: *Proceedings, Joint Commission on Mental Health.* Boston: JCMH, 1958.

112. Kallmann, F. J. In memoriam: Bruno Schulz, 1890–1958. *Arch. Psych. Ztschr. Ges. Neurol.* 197:121, 1958.

113. Kallmann, F. J. Review of psychiatric progress, 1957: Heredity and eugenics. *Am. J. Psych.* 114:586, 1958.

114. Kallmann, F. J. Types of advice given by heredity counselors. *Eugen. Quart.* 5:48, 1958.

115. Kallmann, F. J. The use of genetics in psychiatry. *J. Ment. Sci.* 104:542, 1958.

116. Kallmann, F. J. and Jarvik, L. F. Twin data on genetic variations in resistance to tuberculosis. In: *Genetica della Tubercolosi e dei Tumori*, edited by L. Gedda. Rome: Gregor Mendel Institute, 1958.

117. Kallmann, F. J. and Sank, D. Genetics, eugenics and psychohygiene. In: *Psychohygienische Vorlesungen*, (German), edited by H. Meng. Basel: Benno Schwabe, 1958.

118. Rainer, J. D. and Kallmann, F. J. Constructive psychiatric program for a deaf population. In: *Proceedings, 38th Convention of American Instructors of the Deaf*. Senate Document No. 66. Washington, D.C.: U.S. Government Printing Office, 1958.

119. Rainer, J. D. and Kallmann, F. J. The role of genetics in psychiatry. *J. Nerv. Ment. Dis. 126*:403, 1958.

120. Kallmann, F. J. Genetic aspects of schizophrenia. In: *Proceedings, Second International Congress for Psychiatry*, Vol. IV. Zurich: Orell Füssli Arts Graphiques, 1959.

121. Kallmann, F. J. The genetics of mental illness. In: *American handbook of psychiatry*, edited by S. Arieti. New York: Basic Books, 1959.

122. Kallmann, F. J. Psychogenetic studies of twins. In: *Psychology: A study of a science*, Vol. III, edited by S. Koch. New York: McGraw-Hill, 1959.

123. Kallmann, F. J. Review of psychiatric progress, 1958: Heredity and eugenics. *Am. J. Psych. 115*:586, 1959.

124. Kallmann, F. J. Types of advice given by heredity counselors. In: *Heredity counseling*, edited by H. G. Hammons. New York: Paul B. Hoeber, 1959.

125. Kallmann, F. J. and Jarvik, L. F. Individual differences in constitution and genetic background. In: *Handbook of aging and the individual: Psychological and biological aspects*, edited by J. E. Birren. Chicago: University of Chicago Press, 1959.

126. Kallmann, F. J. and Rainer, J. D. Genetics and demography. In: *The study of population*, edited by P. M. Huaser and O. D. Duncan. Chicago: University of Chicago Press, 1959.

127. Rainer, J. D. and Kallmann, F. J. Genetic and demographic aspects of disordered behavior patterns in a deaf population. In: *Epidemiology of mental disorder*, edited by B. Pasamanick. Publication No. 60. Washington, D.C.: American Association for the Advancement of Science, 1959.

128. Rainer, J. D. and Kallmann, F. J. Observations, facts and recommendations derived from a mental health project for the deaf. *Trans. Amer. Acad. Ophth. Otolar. 63*:179, 1959.

129. Falek, A., Kallmann, F. J., Lorge, I., and Jarvik, L. F. Longevity and intellectual variation in a senescent twin population. *J. Geront. 15*:305, 1960.

130. Jarvik, L. F., Falek, A., Kallmann, F. J., and Lorge, I. Survival trends in a senescent twin population. *Amer. J. Hum. Genet. 12*:170, 1960.

131. Kallmann, F. J. Important events in genetics. *A.M.A. Arch. Neurol. 2*:363, 1960.

132. Kallmann, F. J. Review of psychiatric progress, 1959: Heredity and eugenics. *Am. J. Psych. 116*:577, 1960.

133. Kallmann, F. J. Twin studies (human genetics). In: *Encyclopedia of science and technology*, Vol. VI, edited by W. H. Crouse. New York: McGraw-Hill, 1960.

134. Kallmann, F. J. Discussion of "Defining the unit of study in field investigations in the mental disorders." In: *Field studies in the mental disorders*, edited by J. Zubin. New York: Grune & Stratton, 1961.

135. Kallmann, F. J. Discussion of two psychiatric twin studies. *Am. J. Psych. 117*:804, 1961.

136. Kallmann, F. J. Genetic factors in aging: Comparative and longitudinal observations on a senescent twin population. In: *Psychopathology of aging*, edited by P. H. Hoch and J. Zubin. New York: Grune & Stratton, 1961.

137. Kallmann, F. J. Genetic factors in the etiology of mental disorders. *Am. J. Orthopsych. 31*:445, 1961.

138. Kallmann, F. J. Heredity in the etiology of disordered behavior. In: *Comparative epidemiology of the mental disorders*, edited by P. H. Hoch and J. Zubin. New York: Grune & Stratton, 1962.

139. Kallmann, F. J. New goals and perspectives in human genetics. *Acta Genet. Med. Gemellolog. 10*:377, 1961.

140. Kallmann, F. J. Review of psychiatric progress, 1960: Heredity and eugenics. *Am. J. Psych. 117*:577, 1961.

141. Jarvik, L. F., Kallmann, F. J., and Falek, A. Intellectual changes in aged twins. *J. Geront. 17*:289, 1962a.

142. Jarvik, L. F., Kallmann, F. J., and Falek, A. Psychiatric genetics and aging. *Gerontologist 2*:164, 1962b.

143. Jarvik, L. F., Kallmann, F. J., Lorge, I., and Falek, A. Longitudinal study of intellectual changes in senescent twins. In: *Social and psychological aspects of aging*, edited by C. Tibbits and W. Donahue. New York: Columbia University Press, 1962.

144. Kallmann, F. J. Discussion of "Somatic chromosomes in mongolism". In: *Mental Retardation*, edited by L. C. Kolb, R. L. Masland, and R. E. Cooke. Baltimore: Williams & Wilkins, 1962.

145. Kallmann, F. J. (Ed.). *Expanding goals of genetics in psychiatry*. New York: Grune & Stratton, 1962a.

146. Kallmann, F. J. The future of psychiatry in the perspective of genetics. In: *The future of psychiatry*, edited by P. H. Hoch and J. Zubin. New York: Grune & Stratton, 1962b.

147. Kallmann, J. F. Genetic factors in relation to psychiatric diagnosis. *Dis. Nerv. Syst. 23*:594, 1962.

148. Kallmann, F. J. Genetic research and counseling in the mental health field, present and future. In: *Expanding goals of genetics in psychiatry*, edited by F. J. Kallmann. New York: Grune & Stratton, 1962.

149. Kallmann, F. J. The hybrid speciality of psychiatric genetics. *Acta Genet. Med. Gemellolog. 11*:317, 1962.

150. Kallmann, F. J. Introduction, research in genetics. In: *Psychiatric research in public service*, edited by R. M. Steinhilber and G. A. Ulett. Psychiatric Research Report No. 15. Washington, D.C.: American Psychiatric Association, 1962.

151. Kallmann, F. J. New genetic approaches to psychiatric disorders. In: *Research approaches to psychiatric problems*, T. T. Tourlentes, S. L. Pollack, and H. E. Himwich. New York: Grune & Stratton, 1962.

152. Kallmann, F. J. Recent cytogenetic advances in psychiatry. In: *Proceedings, Third World Congress of Psychiatry, Vol. 1*. Toronto: University of Toronto Press, 1962.

153. Kallmann, F. J. The William Allan Memorial Award for outstanding work in human genetics. *Am. J. Hum. Genet. 14*:95, 1962.

154. Kallmann, F. J., Baroff, G. S., and Sank, D. Etiology of mental subnormality in twins. In: *Expanding goals of genetics in psychiatry*, edited by F. J. Kallmann. New York: Grune & Stratton, 1962.

155. Kallmann, F. J. and Glanville, E. V. Review of psychiatric progress, 1961: Heredity and eugenics. *Am. J. Psych. 118*:577, 1962.

156. Rainer, J. D. and Kallmann, F. J. The role of genetics in thyroid disease. In: *The thyroid*, 2nd ed., edited by S. C. Werner. New York: Hoeber, Harper & Row, 1962.

157. Kallmann, F. J. Genetic aspects of sex determination and sexual maturation portentials in man. In: *Determinants of human sexual behavior*, edited by G. Winokur. Springfield, Il: Charles C: Thomas, 1963.

158. Kallmann, F. J. Specialized psychiatric services for the deaf. In: *The psychiatric problems of deaf children and adolescents*, edited by G. M. L. Smith. London: The National Deaf Children's Society, 1963.

159. Kallmann, F. J. and Glanville, E. V. Review of psychiatric progress, 1962: Heredity and eugenics. *Am. J. Psych. 119*:601, 1963.

160. Kallmann, F. J. and Rainer, J. D. Psychotherapeutically oriented counseling techniques in the setting of a medical genetics department. In: *Proceedings, Fifth International Congress of Psychotherapy, Vol. IV. Topical problems in psychotherapy*, edited by B. Stokvis. Basel: S. Karger, 1963.

161. Rainer, J. D., Altshuler, K. Z. and Kallmann, F. J. Psychotherapy for the deaf. In: *Proceedings, Fifth International Congress of Psychotherapy, Vol. III. Advances in Psychosomatic medicine*, edited by B. Stokvis. Basel: S. Karger, 1963.

162. Rainer, J. D., Altshuler, K. Z., and Kallmann, F. J. (Eds.). *Family and mental health problems in a deaf population*. New York: New York State Psychiatric Institute, 1963.

163. Kallmann, F. J. Main findings and some projections. In: *Family and mental health problems in a deaf population*, edited by J. D. Rainer, K. Z. Altshuler, and F. J. Kallmann. New York: New York State Psychiatric Institute, 1963.

164. Rainer, J. D., Altshuler, K. Z., and Kallmann, F. J. Psychotherapy for the deaf. In: *Family and mental health problems in a deaf population*, edited by J. D. Rainer, K. Z. Altshuler, and F. J. Kallmann. New York: New York State Psychiatric Institute, 1963.

165. Rainer, J. D. and Kallmann, F. J. Preventive mental health planning, In: *Family and mental health problems in a deaf population*, edited by J. D. Rainer, K. Z. Altshuler, and F. J. Kallmann. New York: New York State Psychiatric Institute, 1963.

166. Sank, D. and Kallmann, F. J. The role of heredity in early total deafness, *Volta Rev. 65*:461, 1963.

167. Kallmann, F. J. Some genetic aspects of deafness and their implications for family counseling. In: *Proceedings, International Congress on Education of the Deaf*. Senate Document No. 106. Washington, D. C.: U.S. Government Printing Office, 1964.

168. Kallmann, F. J., Falek, A., Hurzeler, M., and Erlenmeyer-Kimling, L. The developmental aspects of children with two schizophrenic parents. In:

Recent research on schizophrenia, edited by P. Solomon and B. C. Glueck, Jr. Psychiatric Research Report No. 19, Washington, D.C.: American Psychiatric Association, 1964.

169. Kallmann, F. J. and Goldfarb, C. Review of psychiatric progress, 1963: Heredity and eugenics. *Am. J. Psych. 120*:625, 1964.

170. Kallmann, F. J. and Rainer, J. D. The genetic approach to schizophrenia: Clinical, demographic and family guidance problems. In: *Schizophrenia*, edited by L. C. Kolb, F. J. Kallmann, and P. Polatin. International Psychiatry Clinics, Vol. I, No. 4. Boston: Little, Brown, 1964.

171. Kolb, L. C., Kallmann, F. J., and Polatin, P. *Schizophrenia*. International Psychiatry Clinics, Vol. I, No. 4. Boston: Little, Brown, 1964.

172. Kallmann, F. J. Contributor to *Aging and levels of biological organization*, edited by A. M. Brues and G. A. Sacher. Chicago: University of Chicago Press, 1965.

173. Kallmann, F. J. The genetic theory of schizophrenia. In: *Readings for introductory psychology*, edited by R. C. Teevan and R. C. Birney. New York: Harcourt, Brace & World, 1965.

174. Kallmann, F. J. Review of psychiatric progress, 1964: Heredity and eugenics. *Am. J. Psych. 121*:628, 1965.

175. Kallmann, F. J. Some aspects of genetic counseling. In: *Genetics and the epidemiology of chronic diseases*, edited by J. V. Neel, M. W. Shaw, and W. J. Schull. Public Health Service Publication No. 1163. Washington, D.C.: U.S. Government Printing Office, 1965.

176. Erlenmeyer-Kimling, L., Rainer, J. D., and Kallman, F. J. Current reproductive trends in schizophrenia: A psychiatric-genetic survey. In: *Psychopathology of Schizophrenia*, edited by P. H. Hoch and J. Zubin. New York: Grune & Stratton, 1966.

AUTHOR'S COMMENTS

Since Franz Kallmann's career was cut short a decade and a half ago, genetics has become firmly established as one of the foundation stones of psychiatric research. Many important large scale investigations in schizophrenia genetics have been undertaken—some still ongoing or not yet completely reported—and a number of critical review articles have appeared [17, 24, 35, 37, 45]. Rather than offer here another general overview, I thought it would be more appropriate to have another look at certain aspects of Kallmann's work in schizophrenia, consider their subsequent influence or modification, and present a review based on some of the leads he provided and some of the research subsequently conducted in the department he founded at the New York State Psychiatric Institute.

At the 25th anniversary of that department, in 1961, Sandor Rado had said "we are now standing at the threshold of an era in which the entire proud edifice of medicine, including psychiatry as a whole as well as psychoanalysis, will rest on genetics" [30]. Ten years later, at the 75th anniversary of the New York State Psychiatric Institute, I had occasion to speak on Kallmann's contributions, and pointed to "the modern flavor of Kallmann's thought and writings. . . . In these writings he anticipated and foreshadowed many of the current approaches and formulations in the study of the genetics of schizophrenia. . . . It is still possible to go back to the wealth of material he presented and to find there hypotheses and leads and even some answers to questions that are still troubling researchers today" [31].

At the time of Kallmann's death in 1965, Eliot Slater wrote that it had been very largely as a result of his efforts that the biological aspects of psychiatry had been kept alive in the United States [39]. Fifteen years later, his contributions to the genetic study of schizophrenia still provide a context for the subsequent upsurge of scientific attention to the field in which he pioneered. If the methods have been refined and the technology enlarged, the goals remain those which he set forth, their extensions those which he foresaw or even initiated.

DIAGNOSIS

Diagnosis was always considered by Kallmann, both in his family study and his twin study, as a source of potential "noise" in the interpretation of research data, but genetic studies were also seen as opportunities to learn more about nosology and the interrelationship among various clinical syndromes, including those subsumed under the schizophrenic disorders. Kallmann tended to use the Kraepelinian categories in the subtyping of the schizophrenias, colored by his conception of a "check in the development of personality" as contrasted with the cyclic course of the mood disorders. In his 1938 volume [20], he provided succinct thumbnail descriptions of some of his index cases and all of his secondary cases. In the interest of conservatism, he tended to omit "doubtful psychoses and other borderline cases" from his primary analysis, but recognized "schizoidia" as a trait with genetic significance in the schizophrenia "heredity-circle." (In his graphic description, "schizoid types . . . comprise the borderline cases, the cranks and eccentrics suggesting schizophrenia, and all psychopathic types with schizoid personality—that is, stubborn and perverse recalcitrants, malicious and cold-hearted despots, superstitious and pietistic religio-maniacs, secretive recluses, sectarian dreamers out of touch with reality, and the over-pedantic, avaricious and literal-minded people.")

In the years following Kallmann's work, problems of diagnosis loomed large in genetic studies, at times to the point of almost discrediting their validity. During his life and for a time after he died, there were criticisms that his diagnostic standards were unreliable and even that they might have been contaminated by his knowledge of zygosity in the twin studies. In 1967 these doubts were largely laid to rest by Shields and co-workers [38] who were able to review Kallmann's raw data and found that all but a very few co-twins (monozygotic and dizygotic) were themselves in a mental hospital by the end of the study.

Kallmann's concepts of schizoid personality, of nuclear and peripheral, and of deteriorating or mild schizophrenia parallelled and foreshadowed more recent concepts of the schizophrenic spectrum, as well as the typical-atypical, process-reactive, chronic-acute distinctions that are being made more of in current research. Rado [29] advanced the concept of "schizotype" on the basis of Kallmann's work, which he knew well having been both his teacher (and analyst) in Berlin and a keen disciple and admirer of his career in genetics. Defined as an abbreviation of the term "schizophrenic phenotype," schizotypal organization covered the psychodynamic expression of the genotype "from birth to death" and included an "integrative pleasure deficiency"—defect in the organizing role of pleasure—and a "predisposition to proprioceptive disorder." While this formulation (and the allied concept of pseudoneurotic schizophrenia) have not been widely used in genetic family studies, "schizotype" as schizophrenic equivalent has been discussed by Meehl [25] and by Gottesman and Shields [16], and "schizotypal personality disorder" with "some evidence that chronic schizophrenia is more common among family members . . . than in the general population" has become part of the DSM-III classifications [2].

The spectrum concept advanced in the major US-Danish adoption studies [36] is based on the strategy of starting with a wide net (acute, chronic, and latent or borderline schizophrenia, definite or uncertain) and allowing the genetic findings to say something about heterogeneity. Acute schizophrenia, for example, as defined by this group, seems to be genetically unrelated to the others. Winokur [44] on the basis of a family study has indeed suggested that acute schizophrenia is a variant of affective disorder. A similar clinically derived classification was proposed by Mitsuda and his colleagues [27]. In their studies, "typical" schizophrenia is marked by gradual onset, emotional blunting, personality deterioration, and a chronic progressive course, with thought disorder but no confusion or disturbed consciousness. In "atypical" forms, the course is episodic and periodic, with disturbances of orientation, confusion or change of consciousness, and some insight during recovery. The typical form was believed to be genetically distinct from the possibly heterogeneous atypical forms.

Currently there has been a return to narrower criteria so that the DSM-III requires very specific quantitative and qualitative standards. Some old and not-so-old genetic studies have had to be re-evaluated with attention to the exclusion or special treatment of manic and schizo-affective cases. The mutual feedback between genetics and nosology continues.

GENETIC EXPRESSION

While family and twin studies from Kallmann's to the present and studies of adoptees here and abroad have established with little doubt the role of genetic expression and transmission in schizophrenic illnesses, pathways of expression, the mode (or modes) of transmission, and the details of gene-environment interaction are still obscure.

Various strategies have been used in searching for the basic nature of what is inherited and how the responsible gene or genes exert their action during growth and development, leading to clinically apparent schizophrenic illness. In Kallmann's thought, the study of twins which he undertook in New York was designed mainly to uncover a group of genetically disposed but clinically well subjects, namely the group of healthy monozygotic co-twins. More recently, studies of subjects at empirically high genetic risk, mainly children of two schizophrenic parents (or to a lesser degree, of one), seek also to determine premorbid or protective factors which might foreshadow illness on the one hand or prevent or modify it on the other.

In all twin studies from Kallmann's on, more severely affected index cases were more likely to have affected co-twins than less severely affected index cases. In Kallmann's study, the rates of concordance in MZ twins varied from 26 per cent to 100 per cent, and of DZ twins from two per cent to 17 per cent, depending on the degree of deterioration in the index twin [21]; in the Gottesman and Shields study at the Maudsley, concordance rates varied with diagnostic criteria used for the co-twins, and with severity or premorbid personality of the index cases [16]. Since monozygotic twins presumably had the same genome at conception, subsequent differences would have to be attributed to the nature of the inherited gene-encoded aberration and its interaction, prenatally and postnatally, with the genetic and nongenetic setting through which it finds expression. More severe index cases might have a more noxious gene-encoded defect, a larger number of pathology-disposing genes (in a polygenic system), or a less effective set of modifying or regulating genes ("constitution"); hence their co-twin would more likely become ill regardless of environment.

Investigation of nonschizophrenic deviations in co-twins or in high-risk children may give other clues to the nature of the basic defect (or one of the basic defects) in the illness as found in the given family or across a set of families with comparable findings. In the Maudsley data, for example, MZ nonschizophrenic co-twins included those with personality disorders (psychopathic, inadequate, hypochronidrical, hysterical, depressive), neurotic depression, and anxiety neurosis, though they did not as a group have a "schizophrenia or schizoid looking profile" on the MMPI [16]. In the study by Heston [18], the offspring of schizophrenic mothers raised in adoptive or foster homes included not only a significant excess of diagnosed schizophrenics as compared to a control group, but also of individuals with sociopathic personality (Kallmann's "schizoid psychopath") and of persons with emotionally unstable personality (including a number with panic attacks). The "spectrum" diagnoses found in the Danish adoption studies have already been described. In some early results of a high-risk study currently underway at the New York State Psychiatric Institute [8] children of a parent with a schizophrenia diagnosis, as a group, did poorly in tests of neuromotor functioning, and sustained attention. They also showed positive findings in cortical event-related potentials related to attentional process [14]. The children are being followed to see if extreme deviance on a composite of such tests is related to the development of psychopathology in adolescence.

Since there is no reason to assume etiological homogeneity in schizophrenia [5], two strategies have been used, singly or in combination, which attempt to avoid combining dissimilar groups: one, the study of individual, large pedigrees with two or more affected persons, and two, ascertainment by other than clinical criteria, that is, biochemical, neurological. It would go beyond the scope of this review to outline the current biochemical theories, but the suggestion that such distinction be considered the independent variable has been made by a number of researchers [23, 26] and recalls Kallmann's hope for a "Wassermann test" for schizophrenia. Actually Kallmann's suggestion of immunological defect still remains under consideration [42] alongside such "genetic markers" as measured by enzymatic abnormalities or disorders of receptors associated with the dopamine theory.

MODE OF TRANSMISSION

While the search for the underlying phenotype or set of phenotypes goes on, there have been a range of theories regarding modes of inheritance. Although generally identified with the recessive theory, Kallmann always postulated the role of polygenic modifiers conferring susceptibility or resistance: he also considered the possibility of schizoid symptoms in the heterozygote, which is essentially a partial dominance theory. As I noted in the Dean lecture, he came to believe that "the modern concept of human heredity does not hinge upon the counting of single Mendelian ratios, and it seems that most pathological conditions in man are neither fully dominant nor completely recessive but only relatively so, depending on the facility with which the trait is detectable in the heterozygote. The question of how to designate a mode of transmission that seems distinguished by a specific unit factor for the entire syndrome, plus a number of modifying genes responsible for the variable clinical expressions for the main genotype, is more or less a matter of semantics [22]." With the risk figures in family studies inconsistent with simple Mendelian ratios and lower than would be expected under single-gene inheritance, the old concept of "penetrance" has been invoked, but this concept has also been expanded by considering more sophisticated models of transmission incorporating genetic and environmental components. Recent formulations include a dominant mode of inheritance with incomplete penetrance in heterozygotes [40], a polygenic form of inheritance [15], and a dominant gene usually inherited from the mother with potentiating polygenes usually from the father [25]. Kallmann's twin data have been analyzed by Elston and co-workers [7, 41] using families of two generations or more and maximum likelihood models for the individual pedigrees. So far, single-gene inheritance has been ruled out in their analysis. Liability-threshold models [12] were introduced into schizophrenia research [15] and multiple threshold models, single-gene and polygenic [34], are currently being tested [3] to see if one or the other can be ruled out as incompatible with the data under the assumptions of the model. Such an approach makes it possible to test the hypothesis that various clinical types represent wide and narrow forms of an illness with different cut-off points on an assumed normal distribution of an underlying genetically and environmentally determined liability. This model can be adapted as well to account for differential biological findings, quantitative or qualitative.

Data on changing marriage and fertility rates of hospitalized schizophrenic patients in New York [9] and on rates of schizophrenia in and out of hospital [6] have represented population-genetic and epidemiological extensions of Kallmann's concern for the social implications of his genetic theories. On the evolutionary scale there has been interest in understanding the survival of schizophrenia in spite of diminished marriage and fertility rates, at least until very recent times [10, 19]. The concept of heterogeneity is pertinent here; if a number of genes were variously responsible, it would take only a small heterozygote advantage of each to maintain an equilibrium level [10].

ENVIRONMENTAL INTERACTION

While it has been acknowledged by Kallmann and many others that the environment plays a necessary part in the development of a schizophrenic illness, a detailed description of environmental interaction awaits not only further genetic knowledge as described above but also more precise specification of noxious environmental factors.

Monozygotic twins indeed show differences at birth, particularly in weight; such differences and their effect on family role and self-image were implicated in the National Institute of Mental Health studies of Pollin and co-workers [4, 28]. Differential parental treatment of MZ twins sometimes occurs in the desire to distinguish them, but no series has been studied with this in mind. It is hard to reconcile simple environmental theories of schizogenesis with the observation of Fischer [13] on the children of monozygotic twins from pairs with schizophrenia in one or both members. In such children, the schizophrenia risk did not differ from the accepted risk for children of one schizophrenic parent, and did not vary according to whether the child's parent was schizophrenic or not, or belonged to a concordant or discordant pair. It is also hard to explain from a purely environmental point of view why the risk in children of two schizophrenic parents is as *low* as it is (about 40 per cent).

Adoption studies also seem to rule out the environmental effect of rearing by a schizophrenic parent, especially the cross—fostering study of Wender et al. [43]. Early profound deafness does not increase the risk of schizophrenia [33] and among the siblings of deaf schizophrenic patients, the risk is comparable to the accepted risk for siblings and does not differ significantly for deaf siblings or hearing siblings [1]. The roles of family communication and of stressful life-events deserve further elucidation, both in themselves and in the context of genetic-environmental interaction.

SOCIAL AND PSYCHOTHERAPEUTIC ISSUES

The memorial plaque for Kallmann installed at the New York State Psychiatric Institute reads "Creative scientist, wise teacher, devoted counselor, he pioneered in the new field of psychiatric genetics, searching untiringly for clarity and truth, always applying his scientific data to people." One of his last endeavors was a study of the broken homes and neglect suffered by children whose parents were hospitalized for schizophrenia. This concern, as much as the potential for genetic research, motivated the ensuing marriage and fertility study and the current study of high-risk children [11]. To find a way to protect families, children, and society as a whole from the tragedies, the suicides, and other violent or disintegrative consequences of schizophrenic illness has been one of Kallmann's most poignant charges to his successors in schizophrenia research. Genetic counseling in a psychotherapeutic setting was one of Kallmann's most persistent preoccupations and with the scientific explosion that the last 15 years has seen, it has become more than ever the measure of effective research and the humanitarian application thereof [32].

REFERENCES

1. Altshuler, K. Z. and Sarlin, M. B. Deafness and schizophrenia, interaction of communication stress, maturation lag and schizophrenia risk. In: *Expanding Goals of Genetics in Psychiatry*, edited by F. J. Kallmann. New York: Grune & Stratton, 1962, pp. 52–62.
2. American Psychiatric Association. *Diagnostic and Statistical Manual, 3rd Edition*, Washington, American Psychiatric Association, 1980, p. 312.
3. Baron, M. Genetic models of schizophrenia. *Arch. Gen. Psych.*, (in press).
4. Belmaker, R., Pollin, W., Wyatt, R. J., and Cohen, S. A follow-up of monozygotic twins discordant for schizophrenia. *Arch. Gen. Psych. 30*:219–222, 1974.
5. Cancro, R. Genetic Evidence for the existence of subgroups of the schizophrenia syndrome. *Schizo. Bull. 5*:453–459, 1979.
6. Deming, W. E. A recursion formula for the proportion of persons having a first admission as schizophrenic. *Behav. Sci. 13*:467–476, 1968.
7. Elston, R. C., Namboodiri, K. K., Spence, M. A., and Rainer, J. D. A genetic study of schizophrenia pedigrees II One-locus hypothesis. *Neuro-psychobiology 4*:193–206, 1978.
8. Erlenmeyer-Kimling, L., Cornblatt, B., and Fleiss, J. High-risk research in schizophrenia. *Psych. Ann. 9*:79–102, 1979.

9. Erlenmeyer-Kimling, L., Nicol, S., Rainer, J. D., and Deming, W. E. Changes in fertility rates of schizophrenia in New York State. *Am. J. Psych. 125*: 916–927, 1969.

10. Erlenmeyer-Kimling, L. and Paradowski, W. Selection and schizophrenia. *Am. Naturalist 100*:651–665, 1966.

11. Erlenmeyer-Kimling, L., Wunsch-Hitzig, R. A., and Deutsch, S. Family formation by schizophrenics. In: *The Social Consequences of Psychiatric Illness*, edited by L. Robins, P. Clayton, and J. Wing. New York: Brunner/Mazel, 1980.

12. Falconer, D. S. The inheritance of liability to certain diseases, estimated from the incidence among relatives. *Ann. Hum. Genet. 29*:51–76, 1965.

13. Fischer, M. Psychoses in the offspring of schizophrenic monozygotic twins and their normal co-twins. *Br. J. Psych. 118*:43–52, 1971.

14. Friedman, D., Frosch, A., and Erlenmeyer-Kimling, L. Auditory evoked potentials in children at high risk for schizophrenia. In: *Evoked Brain Potentials and Behavior*, edited by H. Begleiter. New York: Plenum Press, 1979, pp. 385–400.

15. Gottesman, I. I. and Shields, J. A polygenic theory of schizophrenia. *Proc. Nat. Acad. Sci. 58*:199–205, 1967.

16. Gottesman, I. I. and Shields, J. *Schizophrenia and Genetics*, New York: Academic Press, 1972.

17. Gottesman, I. I. and Shields, J. A critical review of recent adoption, twin, and family studies of schizophrenia: behavioral genetics perspectives. *Schiz. Bull. 2*:360–398, 1976.

18. Heston, L. L. Psychiatric disorders in foster home reared children of schizophrenic mothers. *Br. J. Psych. 112*:819–825, 1966.

19. Jarvik, L. and Chadwick, S. B. Schizophrenia and survival. In: *Psychopathology*, edited by M. Hammer, K. Salzinger, and S. Sutton. New York: John Wiley & Sons, 1972, pp. 57–73.

20. Kallmann, F. J. *The Genetics of Schizophrenia*. New York: JJ Augustin, 1938.

21. Kallmann, F. J. The genetic theory of schizophrenia. *Am. J. Psych. 103*: 309–322, 1946.

22. Kallmann, F. J. The genetics of mental illness. In: *American Handbook of Psychiatry*, edited by S. Arieti. New York, Basic Books, 1959, p. 191.

23. Kidd, K. K. and Matthysse, S. Research designs for the study of gene-environment interactions in psychiatric disorders. *Arch. Gen. Psych. 35*:925–940, 1978.

24. Kessler, S. The genetics of schizophrenia: a review. *Schizo. Bull. 6*:404–416, 1980.

25. Meehl, P. E. Schizotaxia, schizotypy, schizophrenia. *Am. Psychol. 17*:827–838, 1962.

26. Meltzer, H. Biology of schizophrenia subtypes. *Schizo. Bull. 5*:460–479, 1979.

27. Mitsuda, H. Clinico-genetic study of schizophrenia. In: *Clinical Genetics in Psychiatry*, edited by H. Mitsuda. Tokyo: Igaku Shoin, 1967, pp. 49–90.

28. Pollin, W. and Stabenau, J. R. Biological, psychological and historical differences in a series of monozygotic twins discordant for schizophrenia. In: *The Transmission of Schizophrenia*, edited by D. Rosenthal and S. S. Kety. Oxford: Pergamon, 1969, pp. 317–332.

29. Rado, S. Dynamics and classification of disordered behavior. *Am. J. Psych.* *110*:406–416, 1953.

30. Rado, S. Remarks at anniversary banquet. In: *Expanding Goals of Genetics in Psychiatry*, edited by F. H. Kallmann. New York: Grune & Stratton, 1962, p. 259.

31. Rainer, J. D. Perspectives on the genetics of schizophrenia: a re-evaluation of Kallmann's contribution, its influence and current relevance. *Psych. Quarterly 46*:356–362, 1972.

32. Rainer, J. D. Genetic knowledge and heredity counseling: New responsibilities for psychiatry. In: *Genetic Research in Psychiatry*, edited by R. R. Fieve, D. Rosenthal, and H. Brill. Baltimore: Johns Hopkins Press, 1975, pp. 289–295.

33. Rainer, J. D. and Kallmann, F. J. Genetic and demographic aspects of disordered behavior patterns in a deaf population. In: *Epidemiology of Mental Disorders*, edited by B. Pasamanick. Washington: Amer. Assoc. for the Advancement of Science, 1959, pp. 229–239.

34. Reich, T., James, J. W., and Morris, C. A. The use of multiple thresholds in determining the mode of transmission of semi-continuous traits. *Ann. Hum. Genet. 36*:163–184, 1972.

35. Rosenthal, D. Genetic research in the schizophrenia syndrome. In: *The Schizophrenic Reactions*, edited by R. Cancro. New York: Brunner/Mazel, 1970, pp. 245–258.

36. Rosenthal, D., Wender, P. H., Kety, S. S., Schulsinger, F., Welner, J., and Ostergaard, L. Schizophrenics' offspring reared in adoptive homes. In: *The Transmission of Schizophrenia*, edited by D. Rosenthal and S. S. Kety. Oxford: Pergamon Press, 1968, pp. 377–391.

37. Shields, J. Genetics in Schizophrenia. In: *Schizophrenia: Towards a New Synthesis*, edited by J. K. Wing. New York: Academic Press, 1978, pp. 57–70.

38. Shields, J., Gottesman, I. I., and Slater, E. Kallmann's 1946 schizophrenic twin study in the light of new information. *Acta. Psych. Scand. 43*:385–396, 1967.

39. Slater, E. Obituary Notices, F. J., Kallmann, M.D. *Br. Med. J. 1*:1440, 1965.

40. Slater, E. and Cowie, V. *The Genetics of Mental Disorders*, London: Oxford University Press, 1971.

41. Spence, M. A., Elston, R. C., Namboodiri, K. K., and Rainer, J. D. A genetic study of schizophrenia pedigrees I Demographic studies: sample description, age of onset, ascertainment and classification, *Neuropsychobiology 2*:328–340, 1976.

42. Vartanian, M. A. and Gindilis, V. M. Genetic models and biological research in schizophrenia. In: *Genetic Factors in "Schizophrenia,"* edited by A. Kaplan. Springfield IL: Charles C. Thomas, 1972, pp. 327–338.

43. Wender, P. H., Rosenthal, D., Kety, S. S., Schulsinger, F., and Welner, J. Crossfostering: a research strategy for clarifying the role of genetic and experiential factors in the etiology of schizophrenia. *Arch. Gen. Psych. 30*:121–128, 1974.
44. Winokur, G. The use of genetic studies in clarifying clinical issues in schizophrenia. In: *Biological Mechanisms of Schizophrenia and Schizophrenia-Like Psychoses*, edited by H. Mitsuda, and T. Fukuda. Tokyo: Igaku Shoin, 1974, pp. 241–247.
45. Zerbin-Rüdin, E. Genetic research and theory of schizophrenia, *Int. J. Ment. Health 1*:42–62, 1972.

FRANZ J. KALLMANN was born in Neumarkt, Germany in 1897, received his medical degree at the University of Breslau, and began psychiatric training and practice in that city. He later joined the clinic of Professor Karl Bonhoeffer in Berlin, studied with Ernst Rüdin in Munich, and started his investigations into the inheritance of schizophrenia. By the time of the publication of his major Berlin family study in 1938, he had migrated to the United States and for the remainder of his career was associated with the New York State Psychiatric Institute and Columbia University. He became noted for studies of twins, and in addition to his continued work in schizophrenia, he provided evidence for the role of genetic factors in manic-depressive psychosis, involutional psychosis, homosexuality, and the aging process. He was one of the first to be concerned with problems of genetic counseling, and was a pioneer in the establishment of psychiatric services for the deaf. Dr. Kallmann died in 1965 at the age of 67.

A Program of Research on Heredity and Schizophrenia

David Rosenthal, Ph.D.

INTRODUCTION AND BACKGROUND

For good and various reasons, though not compelling ones, American behavioral scientists have long shown a remarkable indifference to the possible role of heredity in the etiology of behavioral disorders. The reasons included: a healthy skepticism regarding the validity and reliability of assessing traditionally defined diagnostic categories, such as schizophrenia, manic-depressive psychosis, psychoneurosis, psychopathy, and others; the association of fallacious, hereditary theories with the political ideology of the Nazis; the fact that genetic research has sometimes been linked to the suppression of black people; the repugnance to Americans of any theory that implied a genetic determination of behavior, even in part, in that it threatened to delimit our concept of personal freedom as well as our subjective or collective consciousness of such freedom; the fact that so-called genetic research has often been cavalier in its disregard of basic, accepted methodological practices, such as the use of a control group, or making assessments while blind with respect to the relationship between a subject and the index case in a given study; the popular but mistaken belief that if a disorder had a genetic basis, it was *ipso facto* untreatable; the absorption of psychologist in psychodynamic explanations of psychopathology and in principles of learning left little room for an ego-alien notion such as genetics in their conceptualization

Research in the Schizophrenic Disorders: The Stanley R. Dean Award Lectures, vol. 2, edited by R. Cancro and S. R. Dean. Copyright © 1985 by Spectrum Publications.

of behavioral disorder; and the fact that none of the behavioral disorders followed any clear Mendelian distribution.

Nevertheless, during the past 50 years, evidence for an hereditary contribution to the psychopathologies had been gathering steadily in Europe, and to a lesser extent in the United States. The evidence might have been fallible because of methodological insufficiencies, but its accumulating weight began to demand attention here. The evidence was based essentially on two kinds of studies:

Consanguinity Studies

Here the assumption was that if a disorder occurred more frequently in the relatives of an affected individual than in the population at large, this finding provided evidence for an hereditary contribution to that disorder. Moreover, if the frequency of the disorder was greater in first degree relatives as compared to second or third degree relatives, this finding reinforced the evidence for an hereditary contribution. However, investigators who made these assumptions were ignoring the possibility that nongenetic factors could also have accounted for such distributions of the disorder. Such nongenetic factors could be psychological, such as parental behavior that has been described as attention fragmenting, chaotic, or double-binding, to name a few of the terms used by psychodynamic environmentalists, or these factors could be sociocultural, so that a trait such as poverty might show the same patterns of correlation between degree of consanguinity and degree of poverty that one might find with various forms of psychopathology.

Twin Studies

The classical twin study design is based on the fact that monozygotic twins have exactly the same heredity whereas dizygotic twins have only about half their genes in common. Therefore, it has been assumed that if pairs of monozygotic twins are concordant—for example, have the same psychopathology—more often than pairs of dizygotic twins, then such a finding constitutes evidence of a genetic contribution to the disorder. This inference is based on the assumption that intrapair environmental factors are the same for both monozygotic and dizygotic twins. Usually the environmental factors that have been considered most relevant involved a common rearing in the same home. Since both members of the monozygotic and the dizygotic pairs were reared together, then the assumption of equal intrapair environmental variance across groups was considered to have been met.

However, psychological factors unique to monozygotic twins, especially that of shared identity, have been described vividly by several investigators who maintain therefore that the equal environment assumption is ill-founded, that solely on psychological grounds one would predict a higher concordance rate for mono-zygotic twins, and that the inference of a genetic contribution to the disorder is not warranted based on such findings alone. We should note too that the classical twin study design invokes a unidirectional hypothesis; the prediction is always that there will be greater intrapair similarity for monozygotic twins. But there is almost never any reason to predict greater intrapair similarity for dizygotic twins, whether for genetic or for environmental reasons. Therefore, the traditional twin studies of psychopathology have been suggestive but not conclusive. Studies of twins reared apart could be helpful in that the problems associated with shared identity cannot arise in separated twins. However, it is difficult to obtain representative samples of separated twins, and the happenstance case by case reporting of such twins might involve selective bias.

For these reasons, Dr. Seymour Kety and I began a series of conversations about ten years ago in which we decided to embark on a different research strategy in attempting to resolve the old controversies. We planned to use naturally occur-ring adoptions to tease apart the hereditary and environmental factors that were thought to be implicated in most forms of psychopathology. Not long afterward we were joined by Dr. Paul Wender who had had the same idea. The psychopathology that we chose for our investigations was the one called schizophrenia. Most of the previous genetic research by far had been devoted to this disorder, and it was the one of greatest concern to the mental health professions and to the population at large. The environmental variable we chose to work on involved type of rearing, which many psychiatrists, psychologists, and laymen felt was the primary etiologic agent in the schizophrenic disorders. Since rearing involves a hugh subset of variables, we chose to focus more specifically on rearing by or with a schizophrenic relative.

Of course, the idea of using adoption to separate the genetic and rearing vari-ables is not new. Psychologists have employed this research strategy liberally in the study of intelligence [3,6,12-14]. One adoption study has been carried out with respect to alcoholism [9] and one with respect to antisocial behavior [16]. How-ever, considering the potential value of such research, the adoption strategy has been used very sparsely. There were good reasons for this apparent neglect. Adop-tion agencies and the courts have been zealous in their desire to protect all parties to the adoption, the biological parents, the adopting parents, and the child, and the agencies have usually been unwilling to divulge any information about them to out-siders. Without the agencies' cooperation, it becomes extremely difficult to mount any adoption study at all, although we have generated one research strategy that circumvents this problem. Nevertheless, to carry out our studies in the way we wanted, we eventually felt obliged to go abroad, where cooperation was possible. Perhaps in the future there will be some liberalization of American agencies' rules with respect to information released to researchers. The researchers, in turn, will have to commit themselves to prescribed practices and constraints that must be acceptable to the agencies.

Although the adoption strategy is not new, Dr. Kety, Dr. Wender and I have developed research designs that build upon and amplify the potential usefulness of this strategy. We did not have all these designs in mind when we started, but as happens often in research, once we were enmeshed in the work itself, new findings and problems that arose suggested new methodological approaches. I focus on these research designs rather than on the details of our research findings. However, I touch briefly upon the findings, where they are available, using them primarily to indicate the power of the designs and the nature of the information they yield. Through all these designs, we treat heredity as though it were an independent variable, just like any other independent variable in a well carried out laboratory experiment.

EXPERIMENTAL DESIGNS

The first group of studies was carried out in Denmark under the excellent supervision of Dr. Fini Schulsinger. In the first of these studies [7], we were interested in the incidence of schizophrenic disorders in the biological and adoptive relatives of schizophrenics and nonschizophrenics, respectively. The design of the study is shown in Figure 1. Because the focus of the study is the relatives of the adoptees rather than the adoptees themselves, we may call it the Adoptees' Families Design. In this study we are testing two opposed hypothesis. (1) If schizophrenic disorders are heritable, we should find a higher prevalence of such disorders among the biological relatives of our schizophrenic index cases than among the biological relatives of the matched controls. (2) If schizophrenic disorders are transmitted behaviorally, and at least in good part by rearing parents whose own behavior is confused, disorganized, erratic or chaotic, to mention some of the terms cited in the literature, we should expect that the index cases would have a greater number of adoptive relatives with schizophrenic disorders than would be found in the adoptive relatives of the controls.

To carry out this design, we began by collecting identifying information on all persons who had formally been given up for nonfamily adoption at an early age in the greater Copenhagen area between 1923 and 1947. There were almost 5,500 such adoptions. From the records we learned the name and birthdate of the adoptee, and the names and other identifying information of the adopting and biological parents. From the Psychiatric Register, we found out which of the approximately 5,500 adoptees had been admitted to a psychiatric facility. The hospital records for each admitted adoptee were examined by two Danish psychiatrists, and the main information provided by one psychiatrist was sent to the American investigators. All five made their own independent diagnoses. By this procedure, we were able to select 33 index cases. Of these, 16 were chronic or process schizophrenics; seven were acute schizophrenic reactions of the schizophreniform, schizo-affective or paranoid type, and ten were cases of borderline schizophrenia.

	Probands	RELATIVES Biological	Adoptive
Schizophrenic			
Control (nonschizophrenic)			

FIGURE 1. Adoptees' families design.

We selected from among the remaining adoptees in the total pool a control group who did not have a file in the Psychiatric Register and who were matched individually to the index cases with respect to sex, age, pretransfer history, and socioeconomic status of the rearing family.

In determining the rates of schizophrenic disorders among the relatives of our 66 probands, we did not examine the relatives personally. Instead, we first identified each biological or adoptive relative who was either a parent, sib or half-sib of a proband, and we then identified each one of these relatives who had a known psychiatric history. These histories were abstracted from records by a Danish psychiatrist who did not know if the individual case he was abstracting was the biological or adoptive relative of an index case or a control. The psychiatric abstract was then sent to the United States investigators who independently made their own diagnoses while they were similarly blind regarding the relationship of the relative to the proband. Diagnostic differences among the four major investigators were settled by discussions based on more complete data from the records before we broke the relationship code.

At this point, I would like to call to your attention two important features of our research. The first is that, wherever possible, we keep all examiners blind with respect to the index or control status of the subject under examination. We are almost always successful in this respect. This procedure insures against the possibility of bias either for or against any preferred hypothesis that the examiner may hold. The second feature has to do with the fact that we have included a broad spectrum of disorders in the ones I am calling schizophrenic. These include not only the classical chronic, process types of cases, but patients called doubtful schizophrenic, reactive, schizo-affective, borderline or pseudoneurotic schizophrenic, or schizoid or paranoid. If we dealt only with hardcore schizophrenia, our Ns would be too small to make any of these studies meaningful. However, a more positive reason for including the spectrum of disorders is that in the process, we hope to be able to determine whether any or all of these disorders, which phenotypically have strong resemblances to hardcore schizophrenia, are genetically related to it as well [10].

With respect to Figure 1, the major finding was that we obtained the highest concentration of schizophrenic spectrum disorders among the biological relatives of the schizophrenic index cases. The rates for such disorders did not differ appreciably in the other three cells. Thus, this finding provides strong evidence for an

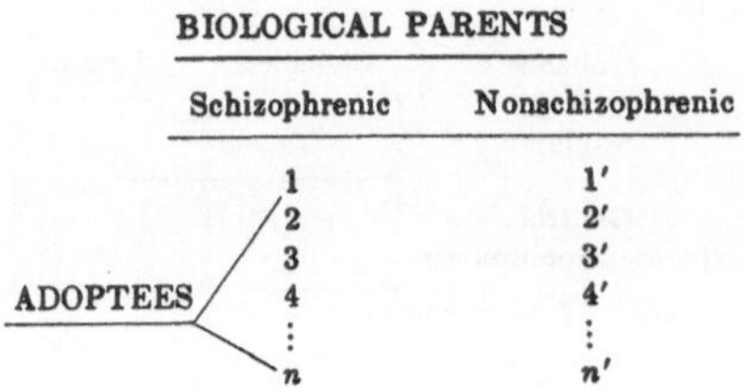

BIOLOGICAL PARENTS

Schizophrenic	Nonschizophrenic
1	1'
2	2'
3	3'
4	4'
$\vdots$	$\vdots$
n	n'

ADOPTEES

FIGURE 2. Adoptees study design.

hereditary contribution to such disorders. However, I want to point out that Figure 1 does not comprise a true fourfold table. That is why a double line is drawn to separate the biological and adoptive halves. The reasons for this are practical rather than theoretical. I will mention only the major reason. It is important to understand that both the adopting and biological parents of our adoptees represent screened populations. The screening with respect to adopting parents is well known, since adoption agencies have long taken the view that mentally ill people do not make the kinds of parents that serve the best interests of the child. But biological parents are also screened in that if they are known to be schizophrenic, adoption agencies may be reluctant to place their children for formal adoption. Instead, the children may be reared in foster homes or in institutions. Moreover, at least in Denmark, schizophrenic women, or women with schizophrenia in their families, may request and have legal abortions. Thus, fewer such children are born and cannot come into the pool of probands in Figure 1. We do not known the extent of screening in the biological and adopting families, but the screening may be unequal. This fact limits the possible range of differences that might otherwise be found in this type of study, but it does not invalidate the procedure. It also means that we can compare the two groups of biological relatives, and the two groups of adoptive relatives, but we cannot now make valid comparisons between biological and adoptive relatives.

The second model [11] is shown in Figure 2. We call it the Adoptees Study Design because the focus of study is the adoptees themselves rather than their relatives. The design asks the question: What is the fate of offspring of schizophrenic parents when these offspring are reared adoptively? In this study, the starting point is the approximately 10,000 biological parents of our pool of adoptees. A search was conducted to see who among these parents had a file in the Psychiatric Register. The hospital records of each such parent were reviewed in detail by a psychiatrist who completed a prescribed form which was reviewed independently by the American investigators. If we agreed that the parent's diagnosis belonged in our spectrum of schizophrenic disorders, or was a clearcut or possible case of manic-depressive psychosis, the adopted-away child of that parent was chosen as an index case. From among the remaining adoptees, we chose as controls those whose both biological parents had no known psychiatric history, that is, neither parent had a file in the Psychiatric Register. Controls were matched to the index cases for sex, age, age at transfer to the adopting family, and the socioeconomic status of the adopting family.

The index and control subjects were invited to participate in a study of the relationship between environment and health. We were able to achieve almost 80 per cent cooperation, an acceptable figure, and the two groups did not differ in this respect. At this time, we are able to report on 76 index cases and 67 controls. The subjects were given a semi-structured psychiatric interview by Dr. Joseph Welner that lasted from three to five hours. Each subject also had one and a half days of psychological testing, but we will not be able to present the test findings now. The examinations of all subjects spanned a period of four years.

The main finding of this study is that there is a significantly greater number of schizophrenic spectrum disorders among the index cases than among the controls. Three cases were called clear-cut schizophrenia by Dr. Welner. All three were index cases. However, only one of these had been hospitalized for the disorder. As a matter of fact, the rate for hospitalized schizophrenia and for diagnosed schizoprenia tends to be appreciably lower than the rates usually found in Scandinavia for the nonadopted offspring of schizophrenics. Therefore, this study leads us to the twofold conclusion that heredity contributes significantly to the development of schizophrenic spectrum disorder, and that adoptive rearing may contribute to the reduced expressivity of such disorder. In both studies presented, evidence is accumulating that the disorders in our spectrum are genetically related, with the probable exception of reactive schizophrenia, which may have to be excluded from the spectrum.

Now I would like to show you a research model that is based on an experimental design that has been used in the past by behavioral geneticists [2,4,5,8]. It has generally been referred to as a cross-fostering or reciprocal fostering model. To review briefly the essentials of this model, let us assume that the experimenter is interested in learning whether he can breed in a trait such as social dominance. He would first decide on a test or criterion for the trait. He would then run his starting pool of animals through this test and separate those who test high (called dominant) and those who test low (called submissive). He would then inbreed the dominant animals and inbreed the submissive animals, and repeat the test with the next generation. This procedure is continued as long as the respective inbreedings continue to increase the test discrimination between the dominant and submissive groups. Let us say that at the nth generation, the experimenter decides that he can no longer increase the discrimination. At this point, he must ask himself whether he has successfully bred in the trait or whether each generation had become more dominant or more submissive because it had, in turn, been reared with successively more dominant or more submissive parent populations. Therefore, he checks this possibility with the $n + 1$ generation. He does this by transposing the $n + 1$ generation. He does this by transposing the $n + 1$ dominant dams. Then he runs the $n + 1$ adult generation through his test to see what effect the transposed rearing may have with respect to the test performances.

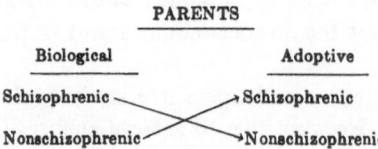

FIGURE 3.   Cross-fostering design.

We cannot control human breeding, but we can follow the model somewhat by thinking of our pool of adoptees as an $n + 1$ generation. The design is shown in Figure 3. We begin with the biological parent generation. From among them we select those who are schizophrenic and who are presumably breeding the trait. Among their adult offspring, we select those who were reared by adoptive parents who had had no schizophrenic disorder, as far as we can tell. These offspring constitute one testing group. Then we select from among those biological parents who had had no schizophrenic disorder, as far as we can tell, those whose offspring had been reared by an adopting parent who did have some schizophrenic disorder. These offspring comprise our second testing group. The two groups of offspring are then compared with respect to the trait in question. Although we have not yet analyzed the data in this study, a preliminary look at the data suggests that the incidence of schizophrenic spectrum disorders tends to be about equal for the two cross-fostered groups. Should this tentative observation prove true, it would not mean that heredity is irrelevant, but rather that rearing by a schizophrenically disordered parent may also be influential in the development of spectrum disorders.

Although the cross-fostering design has its own built-in elegance, what it does in effect is to pit two competing hypotheses against one another. However, we also want to know in more detail the effect of each independent variable considered separately. Now that some statistical evidence is accumulating to the effect that rearing by a schizophrenic parent may itself produce spectrum disorders in offspring, it is important that we have a research model that provides a clean test of this hypothesis. This model is shown in Figure 4.

In this model, we begin with biological parents who do not have any schizophrenic spectrum disorder, as far as we can tell. This is done to insure to the maximal extent possible that the offspring under study are as free of genetic contamination as we can make them. Preferably, all biological parents should be examined personally and in depth to make the determination of no spectrum disorder, but we have not as yet been able to do this. Now we ask the question: When there is minimal or no genetic predisposition in the child, will rearing by a schizoprenic parent induce spectrum disorders in the child? Thus, we have two groups of adoptees. The first or index group are reared by a schizophrenic spectrum parent, the second or control group by rearing parents who are free of spectrum disorder. The second group, which is matched to the first group for various relevant variables constitutes as ideal a control group as we can find in that both their

	BIOLOGICAL PARENTS	
	Nonschizophrenic	
	REARING PARENTS	
	Schizophrenic	Nonschizophrenic
ADOPTEES	1	1'
	2	2'
	3	3'
	4	4'
	⋮	⋮
	n	n'

FIGURE 4. Design to test the "pure" environmentalist hypothesis.

biological and rearing parents are free of spectrum disorders. Any psychopathology that we find in these offspring should arise from other factors. Any psychopathology in the index group in excess of that occurring in this idealized control group represents the contribution of rearing by a schizophrenic parent. We cannot at this time report any findings on this study, but will do so in the future.

We must be alert to another alternative. It may be that rearing by a schizoprenic parent is insufficient per se to induce spectrum disorders in offspring, but that such rearing could raise havoc with genetically predisposed individuals. To test this possibility, we require a research model such as that shown in Figure 5.

This design is exactly like that of the previous design, with one important exception: this time all subjects must have a biological parent who has schizophrenic spectrum disorder. Thus, from a genetic standpoint, the amount of hereditary predisposition for such disorder should be the same for our two groups, and it should be considerable. Again the difference between the groups occurs in the rearing variable. Actually, it is not possible to carry out such a design in pure form, at least not in Denmark, since this would require that both groups of subjects should be adoptees. However, the likelihood of generating a sample in which the subjects have a biological parent who is schizophrenic and are then given up for adoption to a rearing parent who is also schizophrenic is, fortunately, very small. Thus, to carry out the intent of the design, we have had to substitute for adoptees a group of subjects who had a schizophrenic parent and who were reared in the parental home at least during their first fifteen years of life. This represents the group in which the hypothesized genetic and rearing factors would be truly coacting to produce the schizophrenic phenotype. The comparison group of adoptees provides a baseline that represents only the genetic contribution, without the superimposition of rearing by a schizophrenic parent. Any difference between the two groups should represent the coaction or true interaction effect. We have collected a matched sample of nonadoptees to carry out this design, but the research material has not yet been subjected to analysis.

We are now in a position to bring together several of the samples of subjects

| BIOLOGICAL PARENTS | |
| Schizophrenic | |

REARING PARENTS	
Schizophrenic	Nonschizophrenic
1	1'
2	2'
3	3'
ADOPTEES 4	4'
⋮	⋮
n	n'

FIGURE 5. Design to test the effects
of the hypothesized environmental
variable coacting with the
genetic variable.

we have collected and arrange them in a fourfold table that represents the various combinations of genetic and rearing variables, as shown in Figure 6.

Thus, we have two types of rearing variables, schizophrenic and nonschizophrenic, and two types of genetic variables, schizophrenic and nonschizophrenic. Three of the four cells contain adoptees. Two diagonal cells represent subjects in the cross-fostering design. The adoptees in the lower right cell are obtained from the control group in the adoptees study design. The upper left cell, unfortunately, has to be represented by the nonadoptees obtained in the previous design, and that is why it is represented by a double line. Thus, in one cell the factor of adoption does not hold and we do not know to what extent this fact invalidates the findings of this otherwise neat design. Nevertheless, we may carry out such an analysis if we have reason to think it will be worthwhile.

The next study I want to present was carried out in Bethesda [15]. It represents the kind of study that can be done without requiring the cooperation of adoption agencies. The design is based on the following rationale. Many investigators have maintained that a child develops a schizophrenic disorder because his parents have subjected him to various kinds of noxious rearing. In accounting for the elevated incidence of schizophrenia among the parents of schizophrenics, they point out that such parents are more likely than normal parents to emit these noxious behaviors in regard to their children and that, therefore, the elevated incidence of schizophrenia among parents of schizophrenics is to be expected on rearing grounds alone. Alanen, Rekola, Stewen, Takala, and Touvinen [1] reported that parents of schizophrenics had a higher rate of severe psychopathology than did the parents of neurotics, and inferred that the correlation between parents and children regarding severity of psychopathology represented evidence for behavioral transmission. However, such findings could equally well imply that the elevated rates of schizophrenia and severe psychopathology in the parents represent genetic factors that are transmitted to offspring who in turn manifest schizophrenic disorder. To test these alternative hypotheses, we invoke the design shown in Figure 7.

In this design, we are concerned with the parents of schizophrenics. Since in the type of study done by Alanen and other investigators the genetic and rearing variables are confounded in the same parents, we again resort to adoption to

Genetic Background	TYPE OF REARING	
	SCHIZOPHRENIC	NONSCHIZOPHRENIC
Schizophrenic	Nonadoptees	Cross Fostering
Nonschizophrenic	Cross Fostering	Controls

FIGURE 6. Modified design to test for statistical interaction between the hypothesized genetic and environmental variables.

separate the two variables. We begin by finding young adult schizophrenics who had been given up for adoption early in life. This can best be done by interviewing all new admissions to mental hospitals, and their parents. It is a tedious job, but it is feasible. From among other schizophrenic admissions who were home reared, we find a group that is matched to the adopted schizophrenics with respect to the variables deemed most relevant. The third group in the design shown is used to control for the factor of adoption. However, this group of adoptees is free of schizophrenic disorder. The subjects studied are not the offspring, but the parents. The particular focus of the study is the adoptive parents of the schizophrenics. Our reasoning goes like this: If the schizophrenia in the children represents primarily genetic influences, the degree of psychopathology in their adoptive parents should not be severe, as Alanen had reported, and should be less than that of the biological parents of schiophrenics. If the schizophrenia in the children represents the effects of noxious behavioral influences, the degree of psychopathology among the adoptive parents should be the same as that found in the biological parents.

Our findings indicated that the degree of psychopathology in the adoptive parents of schizophrenics was significantly less than that of the biological parents of schizophrenics but significantly greater than that of the adoptive parents of normal subjects. Thus, it is possible to have schizophrenia in offspring who are not subjected to the noxious influences associated with severe psychopathology in the rearing parents. The finding of a difference in degree of psychopathology between the adoptive parents of schizophrenics and the adoptive parents of normals could have any of several explanations which I will not take the time to discuss here. It is interesting that on a word association test, the biological parents of schizoprenics produced more unusual responses than did the adoptive parents of schizophrenics.

To carry out the last design that I shall present, I went to Israel. This study was done without the collaboration of my two brilliant colleaques, Dr. Kety and Dr. Wender. As noted earlier, our attempts to generate a clean fourfold—table design by using adoption fell short of our goals. However, we can forego the advantages conferred by adoption if we can specify two different environments that bear

	DIAGNOSIS OF PROBANDS		
	Schizophrenic	Schizophrenic	Nonschizophrenic
REARING PARENTS	Adoptive	Biological	Adoptive

FIGURE 7. The adoptive parents study design.

on the kinds of rearing we have been talking about. Israel provided two such environments, the kibbutz and the nuclear family types of rearing. The reasoning underlying the study is: In the typical nuclear family, if a parent—let us say the mother—is schizophrenic, the child is likely to endure the following psychological hazards: the mother may be too autistic to attend or to be responsive to the child's needs and she may program reinforcements haphazardly and unpredictably, thus impairing the child's cognitive training and his affective and motivational integration; during the times she is not hospitalized, she is likely to be the only person in the child's environment during most of each day, so that during the greater part of the time that he is awake, the child has no other model with whom to identify during his formative years; sometimes the parent undergoes successive hospitalizations, so that the child may suffer increased insecurity each time he loses her, and he may develop a deep sense of mistrust of the world around him; sometimes the home will be broken, the child may be reared by relatives or friends, in institutions, or he may be shuttled back and forth in various combinations of such rearing; he may be isolated from other children; if he has siblings, they are likely to be similarly influenced and they may tend to influence each other noxiously in turn.

Although kibbutzim vary among themselves in a number of ways, in the main they may provide greater protection for the child who has a schizophrenic parent. For example, the child grows up in children's houses under the guidance of trained caretakers. During the greater part of each day, he receives the same tutelage and training as do other children. During evenings and holidays when the children and parents visit together, the child will visit with both the well parent and the sick parent, and the well parent may help to neutralize any noxious impact of the sick parent. Usually, the child is well known to other adults in the kibbutz and they may serve as parent surrogates. If the sick parent requires hospitalization, the child suffers minimal disruption of his life. He remains in the same children's house with the same caretakers, teachers, and friends. He lives in the same community, and he can still visit with the well parent during the evenings and holidays.

The design for this study is shown in Figure 8.

This study was carried out under the supervision of Dr. Shmuel Nagler and Dr. Sol Kugelmass. The key cell is in the upper lefthand corner. We had to find children who were born and reared in a kibbutz and who had a schizophrenic parent. We were able to find 25 such cases. We then found 25 matched cases who

PARENTAGE	TYPE OF REARING	
	Kibbutz	Nuclear Family
Schizophrenic	25	25
Nonschizophrenic	25	25

Source	df
Parentage	1
Rearing	1
Parentage × Rearing	1
Error	96
Total	99

FIGURE 8. Generalized design for estimating the relative contributions of heredity, environment, and heredity-environment interactions.

lived in the usual nuclear family situation and who also had a schizophrenic parent. For kibbutz controls, we selected a group of children who were reared in the same children's houses as the index cases, but whose parents had no spectrum disorder, and for nuclear family controls we selected children from the same neighborhood and classroom, but without a schizophrenic parent. The children had two days of examination. They were brought in pairs, each index case and his control, but all examiners were kept blind as to which child was which. Thus, we had the rare opportunity to observe and test both the index and control subjects in the same situation. The children ranged in age from eight to 14. Our major dependent variables involve the degree and type of psychophathology found in the four groups of subjects. With respect to each variable, we can apportion the amount of variance contributed by genetic background or parentage, the amount contributed by type of rearing environment, and the amount contributed by the genetic-rearing interaction. At this time, data are being analyzed. We hope to begin reporting our findings in the next year. It is worth noting that this is a generalized design that avoids the problems confronting us in adoption studies, and that can be applied whenever the investigator can specify two contrasting types of environment, whether they have to do with rearing or with other kinds of environmental or experiential phenomena. The latter can be conceptualized narrowly or broadly, depending on the investigator's theoretical predilection.

OUTLOOK

Only a decade ago there existed a widespread air of pessimism about the possibility of ever unraveling the hereditary and environmental factors involved in the etiology of the behavioral disorders. Today, the outlook is completely opposite. During the seventies we should see a marked acceleration in the accumulation of knowledge in this important field.

REFERENCES

1. Alanen, Y. O., Rekola, J. K., Stewen, A., Takala, K., and Tuovinen, M. The family in the pathogenesis of schizophrenic and neurotic disorders, *Acta Psych. Scand., Copenhagen*, Suppl. 189, *42*, 1966.
2. Broadhurst, P. L. Analysis of maternal effects in the inheritance of behavior. *Anim. Behav. 9*:129–141, 1961.
3. Burks, B. S. The relative influence of nature and nurture upon mental developments. A comparative study of foster parent-foster child resemblance and true parent-true child resemblance. *27th yearbook of the national society for the study of education*, 1928, pp. 219–316.
4. Fredericson, E. Reciprocal fostering of two inbred mouse strains and its effect on the modification of aggressive behavior. *Am. Psychol. 7*:241–242, 1952.
5. Ginsburg, B. E. and Allee, W. C. Some effects of conditioning on social dominance and subordination in inbred strains of mice. *Physiol. Zoology. 15*:485–506, 1942.
6. Honzik, M. P. Developmental studies of parent–child resemblance in intelligence. *Child Devel. 28*:215–228, 1957.
7. Kety, S. S., Rosenthal, D., Wender, P. H., and Schulsinger, F. The types and prevalence of mental illness in the biological and adoptive families of adopted schizophrenics. In: *The transmission of schizophrenia*, edited by D. Rosenthal and S. S. Kety. London: Pergamon Press, 1968, pp. 345–362.
8. Ressler, R. H. Genotype-correlated parental influences in two strains of mice. *J. Comp. Physiol. Psychol. 56*:882–886, 1963.
9. Roe, A. Children of alcoholic parentage raised in foster homes. In: *Alcoholism, Science, and society*, published by Quarterly Journal of Studies on Alcohol, 1945, pp. 115–127.
10. Rosenthal, D. *Genetic theory and abnormal behavior*. New York: McGraw-Hill, 1970.
11. Rosenthal, D., Wender, P. H., Kety, S. S., Schulsinger, F., Welner, J., and Ostergaard, L. Schizophrenics' offspring reared in adoptive homes. In: *The transmission of schizophrenia*, edited by D. Rosenthal and S. S. Kety. London: Pergamon Press, 1968, pp. 377–391.
12. Skeels, H. M. Mental development of children in foster homes. *J. Genet. Psychol. 49*:91–106, 1936.
13. Skodak, M. Children in foster homes: A study of mental development. *University of Iowa Studies of Child Welfare 16*, No. 1., 1939.
14. Skodak, M. and Skeels, H. M. A final follow-up study of one hundred adopted children. *J. Genet. Psychol. 75*:85–125, 1949.
15. Wender, P. H., Rosenthal, D., and Kety, S. S. A psychiatric assessment of the adoptive parents of schizophrenics. In: *The transmission of schizophrenia*, edited by D. Rosenthal and S. S. Kety. London: Pergamon Press, 1968, pp. 235–250.
16. zur Nieden, M. The influence of constitution and environment upon the development of adopted children. *J. Psychol. 31*:91–95, 1951.

Toward a Unified View of Schizophrenic Disorders

Robert Cancro, M.D.

The theme of this chapter is an effort toward a unified view of the concept of the schizophrenic disorders. This task obviously is more easily entertained than achieved. I present a historical view of what the term "schizophrenia" has meant and how the illness has been diagnosed, some caveats about what is known concerning its etiology and pathogenesis and, finally, the reasons why such a synthesis is premature if it is not to be trivial.

Any history of a particular form of madness must be arbitrary in its choice of a starting point. Madness, undoubtedly, has been with us as long as man. Peculiar and maladaptive behaviors are not restricted to homo sapiens. Yet, because of the sapient nature of the species, madness has been traditionally conceived as including cognitive disturbances. Even severe disturbances of mood, such as are seen in melancholia, are not usually considered madness until the cognitive apparatus breaks down in the form of impaired reality testing, or delusions, or in some similar fashion. It is for this reason that much of the early 19th century psychiatric literature used the term "dementia" so generously. Morel [22] in 1852 used the word dementia to describe the condition of a patient with onset of illness at age 14. Because of the peculiarity of seeing a process of dementia in one so young, he coined the phrase "démence précoce." Morel, although Belgian, was very much in the tradition of French psychiatry and used the words as a descriptive phrase which

Research in the Schizophrenic Disorders: The Stanley R. Dean Award Lectures, vol. 2, edited by R. Cancro and S. R. Dean. Copyright © 1985 by Spectrum Publications.

expressed the salient clinical characteristics of his young patient and not as a nosologic term. He did not presume to be creating a category of disease, but rather to be
describing symptoms in an individual.

German psychiatry was much more reflective of a search for disease entities.
For example, in 1871, Hecker [11] described the category of hebephrenia and
Kahlbaum [13], in 1874, described the category of catatonia. Both authors used
nouns to label disease entities which they presumed were being identified and
captured in their descriptions. The presentation by Koch [14] of his brilliant
experiments on the bacterial transmission of tuberculosis in 1882 was an occasion
for the disease entity approach to become even more powerful in continental
psychiatry. Within the year after Koch's work received its appropriate scientific
recognition. Kraepelin [16] published a small volume in which his lifelong quest
for disease entities in psychiatry was already apparent. In the fifth edition [16] of
his famous textbook, he placed dementia praecox in a category of disorders characterized by a process of confused and illogical thinking. In 1899, in the sixth edition
[17] of this textbook, known to all students of the schizophrenic disorders, he
gave up the fourfold classification of the fifth edition and separated dementia
praecox from the manic-depressive psychoses. The manic-depressive psychoses
shared with dementia praecox cognitive impairment including depressions but
tended to have a better prognosis. This division of major mental disorders has
remained with us virtually unchanged for more than eight decades. While Kraepelin
abandoned his early position about age of onset and the inevitability of dementia,
he never abandoned the search for entities.

Bleuler's [2] monograph on dementia praecox appeared in print in 1911,
three years after its completion. He intended his book to be, in part, an attempt to
apply the ideas of Freud to dementia praecox. Despite this effort to bring psychodynamic and biologic thinking together, the major contribution of this book was
in the utilization of a syndrome concept. He rejected dementia praecox as a single
disease and saw it as a group of disorders which included several diseases. In a
remarkable insight, he argued that they were a group "in the same sense as the
organic psychoses." He recognized that the etiologies of the group members could
differ as could their courses. They shared in common, according to his formulation, the feature of splitting. This essential feature consisted of a loss of harmony
amongst the various groups of mental functions. The unity of the mind was lost.
And feelings, ideas, facial expressions, posture, body movement, and emotional
tone could be dissociated from each other.

Bleuler made the disorder of association or thought disorder of central
importance as a diagnostic criterion. The formal signs of thought disorder are
difficult, but not impossible, to quantify in an operational manner [4]. Their
recognition does require clinical judgment. The presence and the severity of formal signs of thought disorder do not lend themselves readily to the patient's self-
assessment but rather require measurement by an observer. The real limitations in
applying Bleuler's criteria, coupled with recent trends in scientific fashion, have led
to a diminished utilization of these criteria.

MIND VERSUS BODY

Bleuler occupied a position midway between Kraepelin and Meyer. He believed the primary cause of these disorders to be organic but included psychogenic factors as determinants of symptom content and perhaps even symptom form. Meyer [10] placed the schizophrenias exclusively into a reaction mode. They were the reactions of an individual dealing with the vicissitudes of living. Meyer's thinking influenced DSM-I heavily and the official nosology spoke of schizophrenic reactions rather than the schizophrenia of DSM-II or the schizophrenic disorders of DSM-III.

The apparent but almost certainly illusory distinction between organic and psychogenic or endogenous and exogenous etiology has been a recurrent one in nosologic thinking. In the eighth edition of his textbook Kraepelin [18] referred to the endogenous vs. exogenous etiology of mental disorders but credited the distinction to Möbius, Jaspers [12] in 1913, stated that this division was not helpful in the schizophrenias. Bleuler also shared in this general lack of enthusiasm for the approach. In the fourth edition of his textbook [2], published in 1923, he commented on the difference between reactive and process schizophrenia but indicated that "no division can be based on these classes because the two symptomatologies intermingle."

In 1932, Frank [9] returned to this strategy when he spoke of the nuclear group of schizophrenics. He assumed that the population of schizophrenics included a group of true or essential cases which could be distinguished from others who share a similar symptomatology but differ in origin. It is the same model of classification which has been used in epilepsy, hypertension, and alcoholism. It is the same approach taken by Bleuler in classifying the symptoms of the disorder into a fundamental and an accessory group. It is the same strategy utilized by Langfeldt [20] who prefers the terms "true schizophrenzia" and "schizophreniform psychoses." All of these observers argue that, although patients may present with similar symptoms, there is a group which tends to have a different premorbid history, a different course, a poorer outcome, and therefore a different illness. This is the kind of thinking that led Sullivan to make a distinction between dementia praecox and schizophrenia. Clearly, the autistic use of language is not restricted to schizophrenics.

DIAGNOSIS

The goal of improving diagnosis by achieving greater reliability has at times achieved near mystical proportions. The most reliable ratings are obtained by asking the patient directly as to the presence or absence of a symptom and immediately recording the response. It is important to understand that this so-called reliability is merely interrater agreement and not a measure of the reliability of the presence of the symptom. The patient who denies hallucinations may not be telling the truth, but the interrater reliability will be superb. This is a frequent source of confusion. Interrater reliability does not equal clinical certainty.

As recently as 1972, an NIMH report [23] indicated that improved diagnosis would help us to understand what schizophrenia is. Improved diagnosis will almost certainly not increase our understanding of the nature of schizophrenia. Our misconceptions can be standardized in such a way that the same patient will be identically mislabeled by an even larger number of colleagues than in the past. A fundamental improvement in diagnostic practice would not be merely improved reliability but a deeper knowledge and understanding of the patient. Improved diagnosis would involve longitudinal rather than cross-sectional study. The very meaning of the word diagnosis expresses the necessity of a thorough knowledge of the patient. This can only occur over a period of time and with repeated observations in a number of different situations and circumstances. The making of a snap judgment based on a cross-sectional assessment of signs and symptoms is in no way compatible with the establishment of a diagnosis. What ever schizophrenia may be, it is not just a collection of momentary clinical findings. It is both a process which occurs over time and the clinical picture produced by that process. It is important to remember that even though the process and the clinical picture may be given the same name they are not the same thing. It is also wise to remember Kubie's [19] warning of 1971, in which he stated that a "nosological system which is based on symptom clusters leads to inappropriate diagnostic 'fashions.'"

The development of symptom lists, in which each item on a list has equal diagnostic value, leads to an approach which is not unlike the menu of a Chinese restaurant—one from column A and two from column B. Unfortunately, this tactic leads to increased sample heterogeneity, because the diagnostically equivalent symptoms in fact derive from different genetic and biochemical mechanisms. It may well be wisest to select one clinically identifiable mental system which is believed to reflect biologic differences and use signs of its altered function as the single diagnostic criterion. An initial effort at doing this was published in 1968 [4] and a more elaborate one was published in 1970 [5]. It was assumed that a disturbance of thinking paralleled the hypothesized information processing differences between schizophrenics and nonschizophrenics. Thought disorder was, therefore, selected as a good candidate to serve as the diagnostic criterion. Using the classical formal signs of schizophrenic thought disorder, it was shown that sample heterogeneity could be reduced in terms of premorbid social adjustment, rapidity of onset of illness, severity of presenting symptomatology, impairment of abstraction on proverbs, and duration of hospital stay. Obviously, homogeneous groups were not established but homogeneity was significantly increased.

The effort to create meaningful classifications of schizophrenia has always suffered from a persistent tendency toward reification. One cannot help but be reminded of Wordsworth's [26] warning:

That false secondary power
By which we multiply distinctions, then
Deem that our puny boundaries are things
That we perceive, and not that we have made.

This quotation should be printed on every copy of DSM-III, if not on every page.

Various diagnostic schemata have been developed reflecting the thinking of many of the people cited earlier and others who have not been mentioned. The problem of the extent of overlap amongst different classifications was addressed in a study done by Arthur Sugerman and me and reported initially in 1962 [3] and in more detail in 1968 [7]. A consecutive series of patients admitted to a psychiatric hospital with a diagnosis of schizophrenia were rated using the Phillips scale of premorbid adjustment, Langfeldt criteria, and Bleulerian criteria. The intercorrelations amongst these three diagnostic schemata were all moderate to high. The premorbid score correlated with the Langfeldt classification at 0.6 and with Bleuler's classification at 0.7. The Langfeldt and Bleulerian classifications correlated at 0.82. Nevertheless, the amount of variance in one classification accounted for by the second classification ranged from a low of 36 per cent to a high of 67 per cent. In other words, while the intercorrelations between the three classifications were all highly significant, there was a failure of overlap ranging from one-third to two-thirds of the cases. These data remind the observer painfully of Wordsworth's acumen in warning us of the danger of deceiving ourselves into believing "that our puny boundaries are things that we perceive, and not that we have made."

ETIOLOGY

Obviously, when we cannot identify the population suffering from a disorder with reliability let alone with validity, it becomes difficult to speak of etiology with any real confidence. It is in many ways a testimony to the scientific courage of a number of investigators who explored etiologic issues in the schizophrenias when there was little reason to believe that the signal to noise ratio was sufficiently favorable to yield success. The earliest studies on etiology looked at genetic factors while more recent ones have explored both biologic parameters and social factors as well. The literature tends to describe factors as genetic, biochemical, physiologic, psychologic, and social. There is a tendency to think of these as real and separate entities which are independent sources of variance. These so-called factors are nothing more than different levels of abstraction to explain the same observed phenomenon. It is sometimes more convenient, because of the way our cognitive apparatus functions, to think in terms of biochemistry and at other times in terms of physiology. This does not mean that the organism has real biochemical and physiologic levels. The organism functions as a unitary whole, and the observer imposes disciplinary conceptualizations and abstractions on it. It is helpful and productive to think of organismic functions in terms of these disciplines, but as has been observed, there is no evidence to believe that God created the world to correspond to the departmental pattern of a university. It is easier to think in this fashion, but it is necessary to recognize the dangers of both reification and dualism so that they can be reduced if not avoided.

Genetic factors are involved in the transmission of the schizophrenic disorders. This has been statistically demonstrated to the satisfaction of most observers. The studies on which this conclusion is based are clinical and, therefore, not as rigorous as laboratory studies. Nevertheless, the totality of the consanguinity, twin, and adoptive studies which have been published since 1916 compel one to the conclusion that genetic factors are operative in transmission.

Even in the case of the most simple trait, the genotype does not unilaterally nor immutably determine the phenotype. In complex traits there are many more developmental steps between the genotype and the phenotype with still more room for variation. The genotype is only one of the factors which accounts for individual differences. Every gene, be it single or operating in combination, has inherent within it a range of possible phenotypic outcomes. This range is determined by the characteristics of the gene, the environmental factors which activate it, and the timing of the activation. The gene is not an homunculus. It is encoded information. In many ways it is best thought of as a potential instruction. Until that instruction is activated by the environment, the gene has no function. Most genes are never activated during the lifetime of the organism. Different environments will selectively activate different genes. In a very real sense, then, the environment determines which genes are activated and thereby determines the operating or functional genotypic configuration of the organism.

The separation of gene from environment is a semantic convenience. The gene is a real structure, and everything outside that structure is considered the environment. More accurately, the environment is the biochemical "bath" in which the gene sits and which will serve to activate it. That "bath" is influenced by a variety of factors, including psychosocial events. The precise nature of the pathways by which these psychosocial events are translated into biochemical differences in the "bath" are not known.

Every gene inherently contains a range of possible outcomes. If a given gene is activated by a particular environment, it will produce a particular phenotype. A second environment activating the same gene will produce a different phenotype. The number of possibilities within the genotype is finite and the difference in the final characteristic can be small. Conversely, the difference can be of significant proportions, including the presence or absence of a pathologic trait such as audiogenic seizure susceptibility [10]. As can be seen from the twin studies, the very genes which contribute to a schizophrenic illness in a given person can contribute to a nonschizophrenic and even a highly adaptive outcome in another. Having the appropriate genes for a schizophrenic illness is not sufficient to produce it.

Just as a knowledge of the genotype does not permit the prediction of the phenotype, a knowledge of the phenotype does not permit the inference of the genotype. Any phenotype can be arrived at from different genotypes. There is no immutable relationship between a phenotype and genotype. Different environments acting upon different genotypes can produce the identical or different phenotypes. The identical characteristic can be arrived at through different mechanisms in the same individual as well as in different individuals. There is enormous plasticity in every biologic system, and this plasticity results in the rich diversity that characterizes complex organisms. Each member of a species is a unique biologic-environmental experiment never to be perfectly reproduced again. Even identical twins will show differences in phenotypes based on differences in the conditions of the evoking environment and the timing of the activations.

The phenotypes which are necessary but not sufficient for a schizoprenic illness need not be inherently abnormal. They can represent Gaussian variants of the expression of the trait in question. More importantly, they do not have to be immutably pathogenic. There are many steps between a trait and an illness. The expression of the traits which are transmitted genetically does not have to be a disturbance but can be nothing more than a statistically unusual variant in expression. The critical fact is that the phenotypes were present long before the illness emerged. The presence of the phenotype does not explain why only selected individuals decompensate, let alone why a given individual becomes psychotic at age 20 while another waits until age 30. Finally, it does not explain why one individual has long periods of remission and another does not. The identical twin studies demonstrate clearly the insufficiency of the genotype as the single cause of schizophrenia. Similarly, clinical reality demonstrates the insufficiency of the phenotype as the single cause. Behavior genetics can give insights into why the decompensation takes a particular form but does not yet explain why only certain people decompensate rather than others who have similar genetic makeups. Stated more simply, behavior genetics contributes to an understanding of predisposing factors but not of precipitating and/or sustaining factors.

The clinical syndrome of schizophrenia is not invariably associated with any known phenotype. There has been no true genetic marker demonstrated nor any specific chromosome identified which may be involved. While the evidence for the existence of a genetically transmitted phenotypic predisposition is clear, the nature of that trait, the characteristics of the genotype, and the mode of transmission remain obscure. The phenotype can be characterized in a number of ways. This was done traditionally on the basis of a visible characteristic such as color or size. Ideally, the phenotype should be identified at a precise biochemical level. In the behavioral sciences this ideal represents much more of an aspiration than a reality. For the present time, it may be best to conceptualize phenotypes as psychologic traits which can be defined and measured in some operational fashion.

The search for a psychologic phenotype is a reasonable activity but has been conducted at an excessively molar level. The schizophrenic disorders are not phenotypes and should not be so called. The choice of a personality organization such as schizoid to represent the phenotype is only modestly less molar. It is necessary to seek psychologic processes which are more like atoms rather than like complex mixtures of different compounds. The selected process should be one which can be studied both genetically and psychologically. The process must relate to the clinical signs and symptoms of the disorder in a logical fashion. The use of different aspects of attention as the psychologic phenotype in schizophrenia has received increasing interest [25]. Attention can be studied genetically, psychologically, and psychophysiologically. It can be related to the clinical picture of the illness, and its careful measurement may serve to reduce the heterogeneity in populations diagnosed as schizophrenic.

Some of my own work may be relevant here. Certain individuals are more oriented toward inner sources of sensory stimulation, while others preferentially attend to outer sources. It was assumed that individuals who differ in preferential attention to the source of stimuli would differ in their form of mental illness. It was also assumed that those who are more inner oriented would develop—during a mental illness—signs more consistent with those used for the diagnosis of schizophrenia, that is, withdrawal, autism. Through the use of measures of visual fixation and other measures of information processing, it was shown that hospitalized schizophrenics take in fewer bits of information about the visual environment than hospitalized nonschizophrenics and normal controls [24]. Schizophrenics behave at their resting level in the way that nonschizophrenics and normals behave when performing a mental task. Once again the differences are *not* qualitative, and schizophrenics behave like nonschizophrenics under different conditions.

While there is only very modest evidence to support the idea of the preschizophrenic having an abnormal nervous system, there is good reason to assume that the nervous system will behave differently during illness. The recent development of positron emission tomography (PET) offers a relatively nonintrusive technique to study the metabolic activity of the nervous system. PET is a technique that combines computerized axial tomography (CT) and radioactive isotope tracing. The result of the combination of these two methods is that the technique produces the precise anatomical localization achieved by computerized tomographic imagery construction, with the biochemical assessment made possible through the use of radioactive isotopes. The administration of radioactively labeled substances allows the investigator to follow the fate of that compound through the body using instruments which detect the decay of the isotope. The major limitation of radioactive isotope tracers lies in the fact that the information is available to the investigator only in two dimensions rather than in the three dimensions which exist in the body.

CT allows the investigator to reconstruct an image at a specific level of interest and, thereby, adds the third dimension. CT scans alone yield information about tissue density, for example, structure. Obviously, by the time a pathologic condition advances to the stage of structural change, it is usually quite well advanced and does not permit the investigator the opportunity to study the biochemical processes which have contributed to its development.

In PET, a chemical compound with the desired biologic activity is labeled with a radioactive isotope. This particular isotope emits a positron when it decays. The positron combines almost immediately with an available electron. This combination results in the mutual annihilation of the two particles with the emission of two gamma rays. Because the gamma rays tend to fly off in directions very nearly opposite to each other, the measurement of the gamma ray activity allows the reconstruction of the approximate location of where the positron–electron annihilation took place. One inherent limitation of the technique is that the positron travels a variable distance before it encounters an electron. This is an unavoidable deficiency within the method and limits the localization of the region of interest to a theoretical minimum of several millimeters. Because of a variety of other factors, it is not realistic to believe that the chemical activity can be localized more accurately than approximately eight to ten millimeters at this time.

Utilizing a computer, a spatial reconstruction is made of the radioactivity within the subject at a selected plane. The results are displayed either as numerical printouts of activity at regions of interest or as a visual image on a display device. PET, therefore, gives a relatively noninvasive technique for studying different regions of a given organ in terms of the biochemical processes which the investigator wishes to examine. Most of the compounds that have been used are isotopes of oxygen, nitrogen, and carbon. There is no positron-emitting isotope of hydrogen, but water can be labeled with O^{15}, and thereby, information on the transport of water becomes available.

There are many problems involved in this technology, not the least of which is the useful half-life of a positron-emitting isotope. It is necessary to be able to produce and handle very short-lived isotopes in a manner that does not do injury to the subject. The PET unit requires not only a cyclotron to accelerate the particles, but also a data acquisition system, a computer, and display devices. The cyclotron must be encased in several feet of concrete. Finally, it is necessary to have a large team of radiochemists and other specialized personnel in order to make such a unit functional. It is not a technology which lends itself to the bedside but rather must be restricted to research centers for the foreseeable future.

The principles of imagery construction require that a number of projections be taken at different angles. The accuracy of the reconstruction is proportional to the number of projections. Most PET systems utilize between 100 and 300 projections, which yield a spatial resolution of a few millimeters. The more up-to-date systems can also record up to seven tomographic images of the particular organ simultaneously.

The measurement of regional metabolism is more complex than the measurement of regional blood volume. Nevertheless, it is this quantitative metabolic appraoch which is particularly exciting to psychiatrists. Quantitative studies of brain metabolism have been conducted with tagged oxygen, with fluorine-tagged glucose, and other compounds as well. These measurements are powerful methods of quantifying and comparing brain activity both within and between subjects.

In 1978, I asked Jonathan Brodie to undertake the development of a program of collaboration between New York University Medical Center and Brookhaven National Laboratories. Under the overall leadership of Alfred Wolf of Brookhaven National Laboratories this collaboration has resulted in a variety of studies. The principal investigator for the schizophrenia section of the collaboration has been Tibor Farkas. Thirteen schizophrenic patients have been studied in the NYU Medical Center-Brookhaven National Laboratory project [8]. These patients all met RDC criteria for the diagnosis of schizophrenia and seven were drug-free. Eleven normal volunteers were studied as controls. All subjects were studied with their eyes closed. One tomographic slice was taken at the level of the basal ganglia including the thalamus and lateral ventricles. A second slice was taken at the centrum semiovale level which included the cingulate gyrus. Because 11 of the schizophrenic subjects met Crow Type II criteria, they have been treated as a single group and compared to the 11 normal controls. It was found that there were no significant differences between the normals and schizophrenics in terms of the metabolic activity of the posterior regions of the brain at these two levels. On the other hand, the frontal region showed a significant diminution in both slices at beyond the .05 and the .01 levels. These preliminary data suggest that there is diminished frontal lobe activity relative to the posterior portion of the brain in schizophrenics characterized by negative symptoms.

It is important to stress the preliminary nature of the results of the NYU-Brookhaven collaboration. In many ways, these data are comparable to Rüdin's [24] initial report in 1916 of increased rates of schizophrenia in the relatives of index cases. The finding needs to be replicated and extended. It will be many years before the meaning is clear. Even if the finding of diminished frontal lobe metabolic activity were to be replicable, it is not known at this time whether this antedates or postdates the illness. It is certainly possible that this diminished level of metabolic activity is still within the normal range and that schizophenia is a disorder of people with low-normal frontal lobe metabolic activity. It is also possible that these people have, for compensatory and/or defensive reasons, withdrawn from contact with the exterior world and, therefore, show lowered activity in the frontal lobes. There is a multiplicity of explanations for the findings, and until they have been replicated and extended in a scientifically vigorous manner, it would be premature to draw any conclusions. It is fair to say, however, that the findings are intriguing and suggestive of altered metabolic activity in the brain of at least certain individuals who meet RDC criteria for the diagnosis of schizophrenia.

THE CONCEPT OF SCHIZOPHRENIA

The repeated intrusions of the healer's ambivalence into the treatment of the mentally ill is a sobering insight. Tragic harm has been done in the name of treatment. If the negatively charged affect elicited by madness is to be reduced, there must be a better understanding of the disorder being treated. To understand schizophrenia fully would yield an understanding of all mental functions—normal and abnormal. The search for a total understanding is futile and destined for frustration. Nevertheless, there is a need for the development of theoretical models or conceptual formulations of the disorder. Yet, the question will arise as to the necessity for such theoretical models. The realities of trying to help people living through the human disaster of the schizophrenic psychosis make many clinicians suspect of what often sounds like useless academic theorizing. This tension between the practice and academic communities is real. Researchers frequently do not treat patients, particularly over prolonged periods of time, and clinicians often do not understand the implications and limitations of research studies. Splitting, too, is not restricted to the schizophrenias.

Clinicians must diagnose and prognosticate and, in order to do so, must have conceptions of the disorder. Restricting our attention exclusively to the so-called practical concerns, such as reliable diagnostic criteria, would be an error. The effort must be made to develop more homogeneous subgroups of schizophrenics, and this requires a better conceptualization of the category. The development of these subgroups is essential if the field is to have any success in identifying the biologic and environmental realities which unite at least some patients who are presently called schizophrenic. There can be little doubt that many patients so labeled in fact do not have the disorder. Yet, there will never be the equivalent of a glucose tolerance test for schizophrenia, until those cases which are clinically similar but biologically different are excluded. The need for an adequate theory is, therefore, obvious.

THEORIES OF SCHIZOPHRENIA

There are almost as many theories of schizophrenia as there are individuals who have thought about the subject. It is very natural to wonder which if any of these formulations is true. Despite the naturalness of this curiosity, it is scientifically unsound. To ask for a platonic or absolutely true theory of schizophrenia is not a scientific request. The theory need not and probably cannot be a perfect representation of the disorder. If we cannot rely on truth as the barometer, then what criteria shall we follow? The theory must be useful. It must be useful in the sense that it leads to testable hypotheses which in turn further our knowledge. Perhaps even more importantly, it must help the student conceptually to organize the data, and thus the theory serves as a cognitive crutch. Every theory will have its own unique array of advantages and disadvantages which must be matched to the requirements of particular situations. It can be anticipated that different theories will be better "fits" for different data. It may even be of value to utilize different theories under different conditions. Not all the criteria, however, relate to the data. An important consideration is personal preference. It is necessary to be comfortable with a particular formulation before it can be utilized effectively. If the theory is to serve as a cognitive crutch, it must fit comfortably with the cognitive style, personality, and values of the person utilizing that crutch. It would be difficult to improve on the logic of Einstein's rejection of the theory of indeterminism when he stated that he chose not to believe it.

Unfortunately, theories, like all other products of human activity, can become hypercathected. Theorists often confuse their speculations with their vital organs. Sadly, they then defend the former with a vigor that would be more appropriate for the latter. These highly partisan positions can be presented as the truth rather than one of many alternative explanations that fit the data equally well. While the only limitations to the number of theoretical formulations that correspond to the data at any given time is the imaginativeness of the available theorists, alternative explanations are often seen as rivals and treated accordingly. The so-called medical model has in recent years suffered the most. Many of the attacks have utilized a hypothetical medical model which is a variant of Koch's postulates and is pre-Kraepelinian in its simplicity. No serious biologic theorist of schizophrenia has advanced such a naive windmill recently, but this has not prevented much needless tilting.

The view being offered here is that the schizophrenic psychoses form a group of disorders which are only moderately homogeneous clinically and share to some degree certain signs and symptoms, but only intermittently. This group of end states has derived from a variety of initial conditions through a variety of pathways. There is no evidence to support the contention that under certain conditions anyone can develop a schizophrenic psychosis and therefore, the number of initial conditions which can lead to this disorder are finite. The initial conditions are conceived of as necessary but not sufficient for the illness. Some of the paths lead from these initial conditions to end states called schizophrenia, while others do not. The initial states need not be pathologic, and even the end states that are labeled "schizophrenia" have important adaptive features. The individual with a capacity for a schizophrenic psychosis may well have a Gaussian variation rather than an abnormal trait as the predisposing factor. The factors that precipitate and/or sustain the illness do not have to be the same as those that predispose. My bias is that biologic differences play a relatively greater role in accounting for individual variance in capacity for the form of the illness, and environmental differences play a relatively greater role in accounting for individual variance in precipitating and sustaining the illness.

It is obvious that any symptom-based nosology cannot be a scientific classification, because it does not identify the consistent differences in etiopathogenesis necessary to establish valid categories. Research efforts should be to move away from a symptom-based nosology toward classification based on stable findings which are more reflective of underlying mechanisms. Increasingly, measures of information processing and metabolic activity in the nervous system offer promise for a numerical taxonomy of the schizophrenias which will be based on the statistical analysis of physiologic differences. These taxonomic efforts will produce a very different nosology than our present ones. At that time a unified view of the schizophrenias may well be possible. In the interim we must use and improve upon symptom-based nosology so as to make it more useful for research and/or clinical purposes. This is a debt we owe to our patients which cannot be postponed.

PERSONAL NOTE

I have treated people labeled "schizophrenic" for the past 26 years and have done research on them for all but the first of those years. Certain impressions emerge from that experience, which may not appear as scientific as an experiment, but which in a very real sense may reflect far greater stability and reproducibility than is the case in much published research. Mental illness as we define it today consists of arbitrary categories. Our categories are hypothetical; the human suffering is real. The schizophrenic disorders range from tragic to catastrophic in their impact on human life. It is true that some patients who have recovered have benefited from the experience. Nevertheless, the cost is excessive, and we must be cautious not to romanticize madness in others, while praying for its absence in those whom we love.

I have been deeply impressed by the integrity of many of these people called schizophrenic who struggle with quiet courage and dignity against an oppressive alien force, which often appears to them neither alien nor oppressive. I have also been impressed by how difficult it often is for the families to live with the consequences of the illness. Professionals frequently are quick to criticize families for excluding such a member when they themselves are more quick to divorce a family member of their own for far less. Why we always expect greater nobility in the damned than in ourselves is an intriguing question deserving more careful study.

The tendency to make schizophrenia the noun remains with us. It is the person who is the noun and not the illness. The shared human experience of the therapist and patient is not significantly diminished by the peculiar form of the patient's anguish. Anguish remains a feeling state with which the psychiatrist can empathize. No matter how devastating the illness, the essential humanity of the patient remains.

Science and medical care suffer from fads and fashions as much as the automobile or fashion industries. Unfortunately, there is more denial operating in mental health and because of our ahistoricality we fail to recognize how arbitrary, if not silly, our current activities may appear in the future. There can be no substitute for the doctor-patient relationship. Egalitarian words about team approaches lend themselves well to after-dinner speeches, particularly if wine has been served with the meal and critical faculties are appropriately dulled. Patients do not want to relate to teams. They want to relate to other people because they—like all of us—are social creatures and need their own kind. The schizophrenic person may need other people more ambivalently but not less intensely. Patients need a stable physician who will assume responsibility for their total care and be available to them as long as they deem necessary.

The schizophrenic patient brings an impaired mind to the therapeutic relationship. This impairment manifests itself in altered modes of communication. In a curious way the schizophrenic is bilingual and the doctor is not. It is useful to try to learn the language of the schizophrenic in the same sense that it is useful to strive to achieve our ego ideal. The process of striving is important, but the goal is never attained. The therapist cannot bridge the communicative gap by learning to speak a version of primary process. The schizophrenic must come to the therapist and will only do so when he feels able and willing. There is no substitute for trust, patience, and hope in this relationship, and these qualities can only emerge in the presence of the doctor's honesty, candor, and willingness to make the patient's needs primary. Trust, particularly, cannot emerge in a relationship which is unilateral. The doctor must learn to trust the patient first.

In an effort to describe the curious relationship that must develop between the doctor and the schizophrenic patient, I used an analogy some years ago which is still helpful. We must let the schizophrenic play Virgil to our Dante, even though we understand no Latin. It is only with this attitude of openness and trusting uncertainty, initially on our part, that we may bridge the gap in communication and find our way through the inferno tegether.

REFERENCES

1. Bleuler, E. Dementia Praecox oder die Gruppe der Schizophrenien. In: *Handbuch der psychiatrie*, edited by G. Aschaffenburg. Leipzig: Deuticke, 1911.
2. Bleuler, E. *Lehrbuch der Psychiatrie*, 4th ed. Berlin: Springer, 1923.
3. Cancro, R. *A comparison of process and reactive schizophrenia.* Ann Arbor: University Microfilms, 1962, pp. 69–5166.
4. Cancro, R. Thought disorder and schizophrenia. *Dis. Nerv. Syst. 29*:846–849, 1968.
5. Cancro, R. A classificatory principle in schizophrenia. *Am. J. Psych. 26*: 1655–1659, 1970.
6. Cancro, R., Glazer, W., and Van Gelder, P. Patterns of visual attention in schizophrenia. In: *Cognitive defects in the development of mental illness*, edited by G. Serban. New York: Brunner/Mazel, 1978, pp. 304–313.
7. Cancro, R. and Sugerman, A. A. Classification and outcome in process-reactive schizophrenia. *Comprehen. Psychiat. 9*:227–232, 1968.
8. Farkas, T., Wolf, A. P., Jaeger, J., Brodie, J. D., DeLeon, M., DeFina, P., Christman, D. R., Fowler, J. S., MacGregor, R. R., Goldman, A., Yonekuka, Y.,

Bull, A. B., Schwartz, M., Logan, J., and Cancro, R. Regional brain glucose metabolism in the study of chronic schizophrenia: I. The frontal lobes. In press.

9. Frank, J. Psychoanalyse und Psychiatrie. *Sammlung Psychoanalytischer Aufsatze*, 99–102, 1932.

10. Ginsburg, B. E., Cowen, J. S., Maxson, S. C., and Sze, P. Y. Neurochemical effects of gene mutations associated with audiogenic seizures. In: *Progress in Neuro-Genetics*, edited by A. Barbeau and J. R. Brunette. Amsterdam: Excerpta Medica, 1969, pp. 695–701.

11. Hecker, E. Die Hebephrenie. *Arch. Path. Anat. Physiol. Klin. Med. 52*:394–429, 1871.

12. Jaspers, K. *Allgemeine Psychopathologie*. Berlin: Springer, 1913.

13. Kahlbaum, K. L. *Die Katonie oder das Spannungsirresein*. Berlin: Hirschwald, 1874.

14. Koch, R. Presentation to the Physiological Society in Berlin, 1882.

15. Kraepelin, E. *Compendium der Psychiatrie*. Leipzig: Abel, 1883.

16. Kraepelin, E. *Psychiatrie, Ein Lehrbuch für Studierende und Ärzte*, 5th ed. Leipzig: Barth, 1896.

17. Kraepelin, E. *Psychiatrie, Ein Lehrbuch für Studierende und Ärzte*, 6th ed. Leipzig: Barth, 1899.

18. Kraepelin, E. *Psychiatrie, Ein Lehrbuch für Studierende und Ärzte*, 8th ed. Leipzig: Barth, 1909–1915.

19. Kubie, L. S. Multiple fallacies in the concept of schizophrenia. *J. Nerv. Ment. Dis. 153*:331–342, 1971.

20. Langfeldt, G. The prognosis in schizophrenia. *Acta Psychiat. Scand.*, Suppl. 110, 1956.

21. Mayer, A. The life chart and the obligation of specifying positive data in psychopathological diagnosis. In: *The Collected Papers of Adolf Meyer, Vol. III, Medical Teaching*, edited by E. E. Winters. Baltimore: Johns Hopkins Press, 1951, pp. 50–54.

22. Morel, B. *A. Etudes Cliniques: Traité Théorique et Pratique des Maladies Mentales*. Paris: Masson, 1852–1853.

23. Mosher, L. and Feinsilver, D. *Special Report on Schizophrenia*. Washington: HEW, 1972.

24. Rüdin, E. *Zur Vererbung und Neuentstehung der Dementia Praecox*. Berlin: Springer, 1916.

25. Spring, B. J. and Zubin, J. Attention and information processing as indicators of vulnerability to schizophrenic episodes. In: *The Nature of Schizophrenia: New Approaches to Research and Treatment*, edited by L. C. Wynne, R. L. Cromwell, and S. Matthysse. New York: Wiley, 1978, pp. 366–375.

26. Wordsworth, W. *The Preludes*. Book II. Boston: D.C. Heath and Co., 1899.

Index

Academic competence, 91–93
Acetylcholine, 213–214
Adolescent schizophrenics
  areas of conflict for, 72
  attributes of, 53–62
  family environment of, 50
  parents of, 53–62
  research on, design of, 51–53
Adoptive studies of schizophrenia,
  296–304
Adrenaline
  biosynthesis of, 219
  historical perspectives on, 213
  metabolism of
    route of, 215
    in schizophrenia, 162–164
  stress affecting, 221
Adrenergic receptors, 204
Adrenochrome, 162, 163, 214
Adrenolutin, 162, 163
Alcoholism, 119
Amino acid metabolism, 161–162
Amphetamines, 119, 171, 185, 201
  in dopamine release, 230–231
  metabolism of, 214
  mode of action of, 217
  paranoid psychosis from, 230–231
Anticholinergic agents, 177
Antidepressants
  historical perspectives on, 214
  mode of action of, 217–218
Antipsychotic drugs
  catecholamine effects of, 219
  in dopa formation, 179
  mode of action of, 175–177

[Antipsychotic drugs]
  nature of effect of, 183–185
  phenothiazines. *See* Phenothiazines
  receptor-blocking, 175–177
  for relapse prevention, 108–109
  research on, 173
  tyrosine hydroxylase inhibitor,
    181–183
Anxiety
  basic vs secondary, 46
  phenothiazine effect on, 80
Apomorphine, 200, 207, 238, 239
Asperger's syndrome, 119
Aspiration behavior
  competence criteria and, 89
  in psychosocial deficit analysis in
    schizophrenia, 33–34
Assay for neuroleptics, 208
Association tests
  for children at risk for schizo-
    phrenia, 94–95
  for chronic syndromes of schizo-
    phrenia, 120
  Kent-Rosanoff, 31–33
  for psychosocial deficit analysis,
    31–33
Ataractic drugs, 159
Atropine, 178
Autism, 119
Autonomic reactivity
  in expressed emotionality analysis
    within family, 108
  in psychosocial deficit analysis, 17–
    20
Avoidance behavior, 79–81

323